ew Edition

advanced expert CAE

TEACHER'S RESOURCE BOOK

Drew Hyde • Jan Bell • Roger Gower • Nick Kenn
Photocopiable activities by Rawdon Wyatt

Pearson Education Limited
Edinburgh Gate
Harlow
Essex CM20 2JE
England
and Associated Companies throughout the world.

www.pearsonlongman.com

First published 2005

Second edition published 2008

ISBN 978-4058-4838-1

Set in 10/12pt Times New Roman

Printed in Malaysia (CTP-VVP)

Illustrated by Chris Pavely

Designed by Jennifer Coles
Second edition layout by 320 Design Ltd

Edited and project managed by
Catriona Watson-Brown

New edition edited by Alice Kasasian

Acknowledgements

We are grateful to the following for permission to reproduce copyright material:

Bloomsbury Publishing Plc for an extract adapted from *They F*** you up: How to Survive Family Life* by Oliver James published by Bloomsbury; Classic FM Magazine for an extract adapted from 'Henry meets Evelyn Glennie' by Henry Kelly published in *Classic FM Magazine* April 2003 © www.classicfm.com; Paul Daniel at English National Opera and Jane Gilchrist for the text of an interview about *Operatunity*; Amanda Holloway at BBC Worldwide for an extract adapted from 'Operatunity' published in *BBC Music Magazine*; The Independent for an extract adapted from 'The last menageries' by Michael McCarthy published in *The Independent* 3rd November 2001 © The Independent; Solo Syndication Limited for extracts adapted from 'Giving some direction to some artistic career paths' by Tina Bexson published in *The Evening Standard* 26th November 2001 and 'Is little Jack really on the road to a life of crime?' by David Cohen published in *The Evening Standard* 11th December 2002 © The Evening Standard; Telegraph Group Ltd for extracts adapted from 'A tale of two Everests' by Charlotte Cory published in *The Sunday Telegraph* 13th May 2001 © Telegraph Group Limited 2001; 'Now, they ask for my autograph' by Elizabeth Grice published in *The Daily Telegraph* 10th October 2002 and 'Turning the page' by Sally Williams published in *The Telegraph Magazine* 12th October 2002 © Telegraph Group Limited 2002; and 'Be lucky – it's an easy skill to learn' by Richard Wiseman published in *The Daily Telegraph* 8th January 2003 © Telegraph Group Limited 2003; Times Newspapers Limited for extracts from the articles 'We love T-Shirts' by Sheryl Garratt published in *The Times* 22nd June 2002 © The Times 2002; 'The Cash Lady Sings' by Richard Morrison published in *The Times* 26th February 2003 and 'Lifeline across the sea' by Rosalind Renshaw published in *The Times* 30th July 2003 © The Times 2003; Meagan Tudge, Ethically Me Ltd and National Magazine Company for an extract adapted from 'What counts: job satisfaction not salary' by Emily Partridge published in *Prima Magazine* May 2003; A P Watt Ltd on behalf of Sadie Plant for an extract from *How the mobile changed the world* by Sadie Plant; *BBC Wildlife Magazine* for extracts adapted from 'Digital Noah' by Pam Beddard published in BBC Wildlife Magazine May 2003; and Richard Wiseman, Psychologist, University of Hertfordshire for extracts adapted from The 2003 UK Superstition Survey published on www.luckfactor.co.uk.

Sample answer sheets are reproduced by kind permission of Cambridge ESOL.

Every effort has been made to trace the copyright holders and we apologise in advance for any unintentional omissions. We would be pleased to insert the appropriate acknowledgement in any subsequent edition of this publication.

Contents

Introduction to the course

Components

Advanced Expert CAE consists of:

a Coursebook for classroom use with CD ROM

a Student's Resource Book for homework, private study or classroom use (with or without Answer key) with audio CD

Teacher's audio CD pack, containing the Listening material for the Coursebook, the Student's Resource Book and the Teacher's Resource Book

this Teacher's Resource Book.

Interactive exam practice at **www.iTests.com** allows students to test themselves, monitor their progress and improve exam performance.

Six key features

1 *Advanced Expert CAE* is flexible. It is designed in a modular way so that teachers can either follow the order of the material in the book or choose their own route through the course to meet the needs of specific classes. Each page or double-page spread is free-standing, and almost always follows the same order in each module, making it easy to access and isolate separate elements of the course and integrate them in different ways.

   So, a teacher might follow a linear route through each module, and through the book. Alternatively, you might decide to follow different, tailored routes through each module, for example starting with Speaking or Listening rather than Reading. And you might choose to do the modules in a different sequence, depending on your students' interests.

2 While each section can be taught independently, there are usually links between the sections to provide a coherent progression when the linear route is chosen. For example, the Language development in the 'A' units is contextualised in the previous Reading, as well as the Use of English text. Writing 1 usually provides useful skills related to Writing 2 in the 'B' unit. The Speaking usually has a topic which relates to the Listening in the same unit. The Language development in the 'B' units often provides language which will be useful for students in the following Writing section.

3 Most of the Use of English/Language development pages follow a test-teach approach, in which the language is first tested by means of a Use of English task, then focused on in the Language development section, using the examples from the Use of English task to clarify form and meaning. Students are referred to the Grammar reference for a detailed summary of the grammar and to the Student's Resource Book for extra practice.

4 The sub-skills needed for the different parts of Paper 4 Listening and Paper 2 Writing are systematically practised in Listening 1 and Writing 1 sections in the 'A' units. These include:
   Listening: understanding text structure, identifying attitudes and opinions, etc.;
   Writing: using appropriate register, planning your writing, coherence and cohesion, etc.

5 The Writing and Speaking strands in the 'B' units provide practice for each part of Paper 2 Writing and Paper 5 Speaking. However, the focus is more on process than end product. In other words, students are trained to build up good habits, develop the skill of self-monitoring and so become more independent learners.

6 Most sections contain a Help feature, with clues which help students complete the task at hand. These often focus attention on how the task is constructed, and thus help students to help themselves in the exam.

Coursebook

The Coursebook consists of ten modules, each divided into two units (A and B). Each module practises all the papers of the exam, and includes grammar and vocabulary consolidation and development.

Each module is designed around a theme. There is a lead-in page, with an Overview listing what the module contains, which facilitates planning. Photos and questions prompt discussion aimed at getting students interested in the theme.

Then each of the two units in the module is based on a topic linked to the overall theme of the module.

At the end of each module, there is a Module review with revision and further practice of the language covered in the module.

After Modules 2, 4, 6, 8 and 10, there is an Exam practice section in the Teacher's Resource Book, each containing a selection of tasks from Paper 1 Reading, Paper 2 Writing, Paper 3 Use of English and Paper 4 Listening.

Other elements of the Coursebook are:

- Exam overview at the front of the book, giving an at-a-glance outline of what is included in each paper and what skills are tested in each section

- Exam reference section at the end of the book, giving more detailed information about what to expect in each part of the paper, plus a list of recommended strategies for each task type
- Grammar reference section, giving more detailed information about the main grammar points practised in each module
- Writing reference, which provides:
  - a mark scheme, showing what the examiners are looking for when determining the three pass grades (Bands 3, 4 and 5)
  - a checklist to help students monitor and edit their own writing
  - a sample question for each type of writing task in the exam, with model answer, specific guidance, and another question for further student practice
  - sections to give useful support on the following areas practised in the Writing sections: linking devices; punctuation; spelling; attitude clauses and phrases; sentence structure
- Speaking material/Keys section with extra material (e.g. photographs and audio script extracts) needed for the modules

Module and unit structure

Each module contains the following sections. For ease of use and flexibility, the spreads are in the same order in each unit. The teaching notes indicate

- when the photocopiable activities at the back of this Teacher's Resource Book may be used to expand or supplement the lessons
- when there is additional follow-up material in the Student's Resource Book.

Overview and Lead-in questions

Use the Overview to introduce the module contents. You could discuss with the class in what order they would like to cover the module.

Use the photos and Lead-in questions to generate interest in the overall theme of the module.

'A' units

Reading

The texts have been chosen for their interest value as well as their potential to provide a 'window on the world' and generate discussion. There is a three-stage approach.

Stage 1 A *Before you read* exercise establishes the topic and gives a purpose for reading the text through a first time. For example, in Module 7A (*Against the odds*), students discuss questions about the headline and subheading of a newspaper article. They then skim the article to compare how the writer answers the questions and discuss whether they agree with him.

Stage 2 Students do an exam-style reading task. They should be referred to the relevant Task strategy points before attempting the task for the first time. These can be found at the back of the book in the Exam reference. Each task is followed by a Task analysis exercise, in which students are encouraged to discuss and compare how they performed and which strategies they found useful.

Stage 3 A discussion activity, based on the text, which may incorporate some vocabulary. Questions for discussion relate to the students' own lives and encourage them to give their opinions.

Vocabulary

This section practises and sometimes extends language from the reading text in the previous section. Areas focused on include collocation, idiomatic expressions, phrasal verbs, prepositions and easily confused or similar words. Students are given opportunities to use the vocabulary in a different context and personalise it.

This section is always followed up in the Student's Resource Book by a Paper 3 Use of English-style multiple-choice cloze or word formation task, topically linked to the Reading text.

Listening 1

In Modules 1–6, this section develops the listening skills needed for different parts of Listening Paper 4. The last four modules (7–10) provide further exam-style practice tasks.

Stage 1 The *Before you listen* exercise encourages students to think about the topic, and introduces, or gets the students to generate, vocabulary.

Stage 2 Exercises that focus on and develop the sub-skills needed for Paper 4 Listening. For example, Module 1A (*Learning experiences*) practises distinguishing main points from details (important skills for Parts 2 and 3) and identifying attitudes and opinions (important for Part 3).

Stage 3 A discussion activity based on the text.

Use of English 1

This section practises one of the tasks found in Paper 3 Use of English, using a text which relates to the topic of the unit. There is further practice of all parts of Paper 3 in the Student's Resource Book and in the Teacher's Resource Book (Exam practice).

In terms of language development, the aim is to follow a test-teach procedure, as some of the language tested in the Use of English task is focused on and practised in Language development 1.

Stage 1 The *Lead-in* exercise aims to build up motivation in relation to the topic of the text and generate some of the vocabulary needed. This is usually done through a short discussion.

Stage 2 Students are referred to the Task strategy at the back of the book and asked to complete the task. Graded guidance is supplied, e.g. students are usually asked to read the text quickly first for general comprehension. They are further supported by Help clues, which give specific guidance for individual answers. This support is reduced throughout the book.

Students are then asked to analyse the language tested in the task. The questions often relate to language focused on in the following Language development section.

Stage 3 A discussion based on the content of the text.

Language development 1

This section generally focuses on an aspect of the language tested in the Use of English section in the same module. For example, in Module 4A (*Life's rich tapestry*), the focus is on word formation which is tested in Use of English 1.

These sections contain a range of controlled and less controlled practice activities, linked to the topic of the unit. There are opportunities for personalisation.

There is a cross reference to the Grammar reference section at the end the book, which provides a detailed summary of the language point being practised. Students should be encouraged to use this resource to check their answers.

Further communicative practice of the language area is often provided in the photocopiable activities at the back of the Teacher's Resource Book.

This section is always followed up in the Student's Resource Book by further language practice, plus another exam-style Use of English task that provides a further test of students' understanding of the language, as well as exam practice.

Writing 1

Each Writing 1 section practises a sub-skill required for the Writing tasks that students may be required to do in the exam. For example, Module 5A (*In the slow lane*) focuses on selecting and ordering information, an important skill for Paper 2 Part 1, the compulsory task. Module 9A (*Something to say*) focuses on editing your work for accuracy.

Stage 1 A 'Writing strategy' note explaining to students how practice of the sub-skill helps prepare them for Paper 2, plus an activity aimed at raising awareness of the issue.

Stage 2 A controlled practice task. The task might consist of sequencing notes in the most appropriate order, rewriting a paragraph, choosing the most appropriate language, identifying the key information in a text or matching information in a student answer with language in the input.

Stage 3 A freer practice activity. This could consist of a film review, a letter or a newspaper article. In many cases, students are encouraged to exchange their work with a partner.

This section is sometimes followed up in the Student's Resource Book, with practice of relevant vocabulary for the task, or further sub-skills practice.

'B' units

Listening 2

This introduces the topic of the 'B' unit. It may be covered before or after the Speaking section, which usually has a linked topic.

Stage 1 A *Before you listen* activity. This aims to establish the context, to get students to predict the content and to generate the vocabulary needed for the task.

Stage 2 An exam task, with relevant strategies provided in the Exam reference section. Students are usually provided with Help clues, which give them guidance as to how to answer some of the questions. A Task analysis exercise encourages students to reflect on the task they have completed and share the strategies they have used.

Stage 3 Discussion based on the topic of the text.

This section is sometimes followed up in the Student's Resource Book, e.g. with practice of relevant vocabulary for the task.

Speaking

Each section provides relevant vocabulary for the exam-style task students have to do, covers the strategies needed for the task and provides useful functional exponents. In Modules 1–8, there are recorded sample answers for students to evaluate from the point of view of appropriate language and effective strategies.

Stage 1 Vocabulary that students might find useful to the Speaking task is introduced and practised. For example, in Module 1, students practise adjectives which describe personal qualities. The exam task is to discuss the qualities needed for certain jobs illustrated in the photographs.

Stage 2 Sample answer. Students are referred to appropriate Task strategies at the back of the book. They then listen to the examiner's instructions and an example of a student or students doing the task. They evaluate the performance of the speaker or speakers. Then they listen again to focus on useful language exponents.

Stage 3 Students perform the exam task themselves, using the same photos, or different ones at the end of the book. A Task analysis exercise encourages them to reflect on how well they performed.

This section is sometimes followed up in the Student's Resource Book, e.g. with practice of relevant vocabulary for the task.

In Module 9B (*Making a statement*) there is a complete Speaking paper.

Use of English 2

The task in this section focuses on another part of Paper 3. The texts are related to the topic of the sesction, and the exercises follow a similar structure to Use of English 1 (see p.5).

Language development 2

As in Language development 1, this section usually practises an aspect of the language tested in the preceding Use of English section.

This section is always followed up in the Student's Resource Book by further language practice, plus another exam-style Use of English task that provides a further test of students' understanding of the language, as well as exam practice.

Writing 2

The Writing 2 sections cover all the types of writing that students may be required to do in the exam. There is particular emphasis on the compulsory Part 1, which requires analysis of input material such as letters and adverts.

The principle behind the section is to establish 'good practice' through a clear set of procedures consistently applied, which can be used when answering any exam Writing task.

The approach focuses on process more than end product. Each spread is graded, and the aim is to give carefully guided preparation, so that students build up to complete the main task at the end of the section. In each section, there is considerable language support; in particular, a range of functional exponents is given and linked to the task.

In the Writing reference, there is a model answer for each type of writing in the exam, plus notes for guidance and a second task for further practice if desired.

The procedure in the Writing sections is as follows:

1 Lead-in
2 Understanding the task
3 Planning the task
4 Thinking about the language and content
5 Writing
6 Checking and improving the writing output

This section is always followed up in the Student's Resource Book with further practice of the sub-skills required by the task.

Module reviews

These revise the grammar and vocabulary of the previous module in non-exam formats. The exercises can be used as practice in the classroom, given as a test or set as homework.

Student's Resource Book

The Student's Resource Book is an integral part of the *Advanced Expert CAE* course. It contains ten modules that mirror the themes and contents of the Coursebook units. It can be used:

- by teachers to supplement and extend the Coursebook lessons
- by students on their own to consolidate and enrich their language and practise exam skills
- as an intensive course, for example, in the last term before the exam.

It provides: extensive grammar practice, following up the Language development sections of the Coursebook; vocabulary consolidation and extension; and additional topic-related exam practice for Papers 1–4. There is a complete Practice Exam at the end of the book.

The Student's Resource Book comes with an audio CD. The Listening material is also available on the Teacher's CD pack.

Each module contains the following sections.

'A' units

Use of English

There are two pages related to this part of the exam. The first contains a Paper 3 Use of English-style multiple-choice cloze or word-formation task, topically linked to the Reading text in the Coursebook. It is intended as a follow-up to the Vocabulary section in the Coursebook, providing an opportunity to practise related language in an exam format.

The second contains another exam-style task which provides further exam practice and also tests the language area in the Language development section. In early modules, the task type is the same as in the Coursebook, but in later modules, there are different task types.

Language development

This section provides extensive practice of the language area focused on in the Language development section of the Coursebook.

Listening

This provides further practice of a Paper 4 exam task type already practised in the Coursebook. The content is always linked to the topic of the 'A' unit in the Coursebook.

Modules 2A, 3A, 6A and 7A also contain a section following up Writing 1 in the Coursebook. For example:

- Module 2A: Vocabulary of film reviews
- Module 3A: Coherence

'B' units

Language development

Further practice of the language area focused on in the Language development section of the Coursebook.

Use of English

Another exam-style task which provides further exam practice and also tests the language area in the Language development section. In early modules, the task type is the same as in the Coursebook, but in later modules, there are different task types.

Reading

Further practice of a Paper 1 exam task type already practised in the Coursebook. The content is always linked to the topic of the 'B' unit in the Coursebook.

Writing

Further practice of the sub-skills required by the exam task type covered in the Coursebook module.

Modules 2B, 4B, 5B, 7B, 8B, 9B and 10B also contain a section following up Listening or Speaking in the Coursebook.

Complete practice exam

A complete exam provides the opportunity for timed exam practice. It can be used at the end of the course, or at any stage of the course.

Teacher's Resource Book

As well as this introduction, the Teacher's Resource Book contains:

Unit-by-unit teacher notes

Guidance on how to use the Coursebook material; 'books closed' activities to get things going at the beginning of modules and sections; background information on the texts; ideas for additional activities; and answers to all exercises with explanations where helpful.

OMR answer sheets (photocopiable)

Replicas of the answer sheets students have to use in the exam for the Reading, Listening and Use of English papers. They can be photocopied and given to students when they do the Exam practice sections (see below) or the Practice exam in the Student's Resource Book.

Photocopiable activities

A pre-course exam quiz to see how much students know about the CAE exam; three photocopiable activities to supplement each Coursebook module, providing communicative classroom practice for grammar, vocabulary and skills; full teacher's notes and answer keys for each activity.

Exam practice (photocopiable)

Five exam practice sections for use after Modules 2, 4, 6, 8 and 10. Each section provides practice tasks for Paper 1 Reading, Paper 2 Writing, Paper 3 Use of English and Paper 4 Listening. The recordings for the Paper 4 Listening tests can be found on the Coursebook audio CDs after the appropriate module, and answer keys are at the appropriate points in the unit-by-unit teacher's notes.

Audio scripts (photocopiable)

These are all at the back of the book for ease of reference and photocopying.

Teacher's CD pack

The Teacher's CD pack contains all the listening material from the Coursebook, Teacher's Resource Book and Student's Resource Book. There are three CDs for the Coursebook, which include the listening tests in the Teacher's Resource Book, and one CD for the Student's Resource Book, which includes the complete Practice exam.

Abbreviations used in the Teacher's Resource Book

CB = Coursebook
SRB = Student's Resource Book
TRB = Teacher's Resource Book
OMR = Optical Mark Reader
cf. = compare
l./ll. = line/lines
p./pp. = page/pages
para. = paragraph

Module 1 New directions

This module includes topics such as learning experiences (including travel), education, university life, and work (getting and changing jobs).

Photocopiable activity

The pre-course photocopiable activity on page 137 provides an introduction to the CAE exam. Students answer questions about the exam referring to the Exam overview on page 6 or the Exam reference on page 168 of the Coursebook where necessary.

After the quiz, show students other features of the book. This could be done as a quick quiz with questions such as *Where can you find the Grammar reference? How is it organised? By module or by topic? What can you find on page 188?* (the Writing reference) Use the contents pages to ask questions such as *What type of writing is practised in Module 5?* (an article)

Lead-in p.7

The purpose of the lead-ins is to introduce the general theme of the module. Try to avoid giving too much away at this stage by keeping the discussion brief and by not focusing specifically on points that are covered later.

Get students to discuss the questions together. The main themes of the photos are travel, marriage and work. Encourage students to identify and discuss less obvious aspects such as growing up, freedom and independence, taking on responsibility, etc.

The photos show a couple on their wedding day (top right), a woman in a job interview (bottom right) and three young people backpacking in Machu Picchu, Peru.

1A Learning experiences

One way to begin would be, with books closed, to get students to discuss what they understand by the expression 'Lifelong learning' and brainstorm the people and places they are likely to learn from in their lives.

Reading p.8

1 Extend the initial discussion asking who they have learnt most from already and who they think they will continue to learn most from in the future.

2 If students are not familiar with the concept of skimming, explain that the purpose is to quickly get a general understanding of the text as well as to establish the type and style of the text and its general organisation and layout. One way to skim is to read selected sentences, such as the first and the last in each paragraph. Another way is to move the eye quickly over the page, picking out selected words to get an idea of the topic of each paragraph. Suggest a suitable time limit to skim each text (e.g. 30 seconds) to discover the text type and topic. Point out that Paper 1 Part 1 consists of three texts, of different types but linked thematically.

3 Most students will be familiar with multiple-choice questions, but, having read the rubric, start by looking at the task strategy on page 168. Use the strategy to do the first text together, before letting the class do the other two texts on their own.

4 In the Task Analysis ask the students how long they think they would have to do the task in the exam (about 12–15 minutes) and what the task tests (detail, tone, opinion, purpose, main idea, attitude and text organisation).

5 Extend the discussion by getting students to share some of their learning experiences.

Key

2 1 B
2 A (It is from Australian writer Clive James' autobiography, *Unreliable Memoirs.*)
3 C

3 **1** C – *It was then revealed I had Dropped Behind the Class*
2 D – *... why I grew up feeling I needed to cause laughter was perpetual fear of being its unintentional target.*
3 B – all the second paragraph
4 A – *... film is the other way round. The writing feels more detached in an environment that favours realism ...*
5 D – *... they taught him to gather information from a wide range of disciplines.*
6 A – *... film making is now a second career that converges and subsidises what he really loves doing best.*

5 **Text 1:** primary school, dealing with authority, using humour as a defence.
Text 2: secondary school, relationships between teachers and pupils.
Text 3: university, travel and exploration, organisation.

Vocabulary p.10

The vocabulary sections that follow the reading texts pick up on and extend the vocabulary in the text. They focus on topical areas as well as concepts such as collocation, connotation, phrasal verbs and other lexical forms. You could use this opportunity to give students some advice on keeping/learning vocabulary through the course.

1 Students search the text for phrases containing *on*. If necessary point out to students that such phrases are an important aspect of advanced vocabulary and are often tested at CAE. As a follow up ask students to search the texts for expressions with *in*. (Text A: *in those days, in the window*; Text B: *in translation, be set in (a period of time)*, in a medium; Text C: *in good stead, follow in someone's footsteps)*.

2 Highlight or put the following sentence from Text A on the whiteboard, 'I did everything to get out of facing up to Miss Turnbull.' Use it to teach or remind students what phrasal verbs are. Ask students to identify the phrasal verbs and to think how it could be rewritten without using them (e.g. *'I did everything to avoid confronting Miss Turnbull'*). Draw students' attention to the fact that *of* in *get out of* is a preposition and that the second phrasal verb is therefore a gerund. Exercise 2 could be followed by a quick review of the grammar of phrasal verbs; e.g. *bounce back* is intransitive, *start something off / turn down* are separable, *go on to / miss out on* are inseparable. Ask students to find other phrasal verbs in the text and to identify what type they are. (e.g. Text A: *end up, come up* both intransitive)

Students could be asked to make other sentences with the verbs but point out that *pluck up* is almost always used with *courage*. This would be a neat link to the next section on collocation.

3 If necessary, explain the concept of collocation (the way in which some words are often used together and others are not). Awareness of collocation will help students in various parts of the exam, as well as improving their writing and speaking. Follow up the question by getting students to think of words that collocate with the odd one out (e.g. *win a prize, there's a danger that*).

4 Point out that collocation also exists between adjectives and nouns as well as other word forms (e.g. verbs + adverbs).

5 Check that students know what a *potter* is. (Students will probably know the famous scene from the film *Ghost* where Demi Moore works on a potter's wheel with the ghost of her boyfriend (Patrick Swayze) behind her.)

6 Other questions students could discuss include: *Did you miss out on anything when you were younger? Do you find it easy to bounce back after a setback?*

Extra!
Now would be a good time to talk to students about keeping systematic vocabulary records that they can expand as the course progresses. Encourage them to group words thematically or by stem. They could start here with a page for prepositional phrases with *on*, adding to it as they come across more examples. Other pages could have expressions with *in, for,* etc. or types of collocation.
Remember that the module reviews and the photocopiable activities help to recycle vocabulary from the module, and that each Vocabulary section is followed up in the Student's Resource Book with a Use of English task on a related topic. This will either be multiple-choice cloze (to test collocations, fixed phrases, etc.) or a word-formation task.

▶ **Student's Resource Book, page 6**

Key

1 **1** heaps praise on **2** getting on (a bit) **3** reflect on **4** set (his) heart on **5** on a whim **6** doing research on **7** (some years) on

2a **1** start off **2** go on to **3** turned down **4** bounce back **5** make up for **6** pluck up **7** missed out on

3 **1** a prize **2** a danger **3** realise **4** finish **5** backache **6** notice

4 **1** ideal way **2** future generations **3** positive outlook **4** real world **5** personal goal **6** survival skills **7** sheltered upbringing **8** open mind
Note: *in* **the** *real world* (real and unique) compared with *in* **an** *ideal world* (hypothetical)

5 **1** gone on to **2** missed out on **3** on a whim **4** gathering/collecting **5** turned down **6** set my heart on **7** plucked up **8** make sacrifices **9** on **10** took the plunge

Listening 1 p.11

The purpose of the Listening 1 sections in Modules 1–6 is to develop various sub-skills that are needed to complete the exam tasks. For Parts 2 and 3, these include: distinguishing main points from details (gist) and, for Part 3, understanding attitude/opinion.

1 This could be done in groups or as a brainstorm with the whole class. Point out that this sort of prediction exercise helps focus on possible content and therefore understanding.

2 Repeat the introduction if necessary. Students could compare answers, identifying aspects that helped them decide. They need to be aware that they are looking for a summary, not a specific point.

3a Explain to students that this exercise helps them analyse how a good talk is organised. Understanding the structure of a talk will help them follow and understand the content more easily. Give students a minute to read the main points before playing part 2.

3b If necessary, play it again so that they can complete the examples and tips.

3c Students listen for useful language. These discourse markers help to signal aspects of her talk. Get students to identify which expressions are used for giving examples and which are used for tips.

4a Here, students are listening to identify the speakers' attitudes – this is tested in Part 3 of the Listening paper.

4b Here, they are listening for language, so play it two or more times.

5 Students could discuss the points in groups, followed by whole-class feedback.

Extra!
This would be a good opportunity to raise some expectations for this course, such as the amount of work students will be expected to do, how much homework they will have and how they can most effectively use their time out of class!

Note: There is further listening practice in the Student's Resource Book on a topic related to the first part of the module. You can use this for extra listening in class, or students can listen at home to give themselves extra practice.

Key

2a A
2b Rita is going to talk about different ways of approaching university studies and offer advice.
3a/b 3 Set your own learning objectives and deadlines
e.g. *decide how many hours a week to spend studying*
Tip: *build in a safety margin*
1 Make sure you know what you have to do.
e.g. *number of assignments and deadlines for them*
Tip: *use a wall planner or diary*
2 Check what standard of work is expected.
e.g. *how your work should be presented*
Tip: *get hold of some examples of good work*

3c Examples: *You know … and all that, for instance, things like*
Tips: *One way of doing this, It's a good idea to, I always find it useful*
4a Ann: B Nick: D
4b Ann: *I think she had a point when …*
Nick: *for me, the most relevant part was when … that was really what Rita helped them to grasp*

Use of English 1 p.12

1 Students could discuss the questions in groups. Then get them to expand their answers with examples.

Background

Jamie Oliver was born in 1975. Having worked in the family restaurant, he went to catering college and then trained in a number of top London restaurants. His lucky break came when his enthusiasm and personality was 'spotted' when he appeared in a TV documentary about the famous restaurant the River Café where he was working. This led to his own TV shows and books, called *The Naked Chef* because of his attempts to strip cooking down to its basics. He is especially keen on breads and salads. Lucrative advertising deals and regular newspaper columns keep him busy.

2a Begin by getting the students to read the title then just the first sentence before looking at the photo and answering the first question. Then get them to skim the rest and answer the other two questions.

2b When students have read the instructions, get them to look at the task strategy on page 169. Emphasise that they can use only one word.

2c Look at the example together and do question 1 together as a whole class before giving them ten minutes to complete the exercise.

3 These questions provide a good introduction to the following Language development section and could be left until then.

4 Discuss this in groups.

Key

2a 1 Jamie Oliver
2 When he was eight (in his parents' pub)
3 He set up a charity restaurant to train disadvantaged young people as chefs.

2b Write one word in each space to complete the text.

2c **1** until **2** for **3** the **4** before **5** had **6** if **7** be **8** such **9** where **10** into **11** around **12** for **13** out **14** has **15** despite

3 1 Questions 5, 7, 14
2 perfect verb forms: *have discovered, he had worked, Oliver has done, more Fifteens have opened*
continuous verb forms: *chefs are cooking*
simple verb forms: *Oliver went on, learnt, he started, he trains*
passive verb forms: *be given*

Language development 1 p.13

This is designed to be a rapid review of the major tenses. Students that have particular difficulties should be given suitable remedial exercises.

1 Check that students are familiar with perfect tenses by eliciting the form (*have* + past participle), use (linking two time frames) and examples of present/past/future perfect forms. One way to approach the exercise would be to get students to answer the questions individually before referring to the grammar reference to check their answers. Finish with whole-class feedback.

2a Repeat the process above for continuous forms.

2b Do this as pairwork.

3 The exercise includes simple, continuous and perfect forms of the same verb (*sleep*) to focus attention on the use of each particular form. The exercise could be extended by giving students some of the same or similar stems to complete themselves with other verbs (e.g. *This time last week … / I feel terrible/great today because …*).

4a The assumption is that students will be familiar with basic future forms, and this exercise is designed to increase awareness of the many other, often lexical, ways speakers use to refer to future events. When students have matched the expressions to their functions, ask them to spot the odd one out to highlight form (all are followed by infinitive except *be on the point of -ing*). Elicit examples with a phrase such as *when … happened.*

4b Tell students that there may be more than one way to answer these questions.

4c Most students are likely to disagree, so encourage them to think of both sides of the argument.

5a This section uses many of the same structures as in Exercise 4 but in the past form. Use concept questions to establish that they refer to events that were planned or predicted but then didn't happen. For example, in sentence 1 *Where did we meet?* (at my house) *Was I expecting you?* (no) *What had I planned to do?* (go out) *What would have happened if you had arrived much later?* (you would have missed me)

5b If students are having difficulty coming up with ideas of changes of plans, give them more specific prompts.

Note: There is further practice of tense forms in the Student's Resource Book, followed by another chance to test themselves through a Part 3 Use of English task.

▶ **Student's Resource Book, pages 7–8**

Key

1 1 *is* (now); *have ever been* (unfinished time)
2 *has lived* (unfinished action linking past and present); *moved* (finished action in specified point in past)
3 *has been* (unfinished time – then to now); *changed* (finished past action)
4 *got* (past action); *had left* (previous past event)
5 *have practised* (unfinished action); *did* (finished action)
6 *will have broken* (action completed by a point in the future); *has ended* (unfinished event)

2a 1 ✓(The meaning of *loving* here is similar to *enjoy* and can be used in the continuous; there are other meanings that cannot.)
2 At the moment, she**'s staying** (*temporary action*) at her sister's flat until she **finds** a place of her own.
3 Vanessa **enjoys** (*general*) entertaining, so she's always inviting (*habit with* always) people round.
4 Last week I **visited** (*single complete action*) her for dinner.
5 I **hadn't seen** (*previous to past action*) Vanessa for over a month and I was looking forward to it.
6 Vanessa **was cooking** (*action in progress*) when I **arrived** (*short complete action*) at the flat, so I offered to help.
7 ✓
8 Tonight **I'm cooking** (*personal arrangement*) for her. **I'm making / will be making** (*arrangement*) my speciality.

3 1 f (habit/routine) 2 c (action in progress) 3 e (action in progress at a point in past) 4 h (recent repeated past with present consequence) 5 b (action previous to past state) 6 d (single past action specified when) 7 a (past activity previous to other past action) 8 g (unfinished time)

4a 1 immediate future: *be about to, be on the point/verge of -ing.*
2 expected to happen at a particular time: *be due to*
3 official arrangements, etc.: *be to (not)*
4 probability/certainty: *be bound/sure to, be (un)likely to, expect sb. to*

4b 1 is to 2 are likely to / are expected to 3 are bound/sure to 4 expect fewer people to apply 5 is unlikely to 6 is on the point/verge of announcing

5a 1 was (just) about to 2 were going to 3 would have 4 was due to 5 would be 6 were to have / would have

Photocopiable activity

Activity 1A could be used here. It is a pairwork activity in which students identify and correct common structural mistakes.

Writing 1 p.14

Writing 1 sections are designed to develop writing sub-skills that will help students to produce better writing in the exam format Writing 2 sections. These include planning, organising and linking ideas in a logical sequence, selecting and ordering information (important for Part 1 tasks) and sentence skills. This section focuses on using an appropriate style for the type of writing.

1a One way in would be to ask students what they think training with Jamie Oliver would be like. Then get students to read the comments and identify which of the options are less formal before choosing the most likely expressions used. Read the Writing strategy note.

1b As well as checking the answers to Exercise 1a, students listen for the three stages in the training, which they need to complete Exercise 1d.

1c Students identify the features of formal writing and informal speech with examples.

1d Look at the example (0) and demonstrate how students need to both combine information from Exercises 1a, 1b and 1c and also change it to a more formal style. Some variations are possible here (e.g. *a 14-week course / a course which lasted 14 weeks*). Get students to compare answers in groups before whole-class feedback.

2 Do sentence 1 together, pointing out that for most sentences, they require two words from the box.

3a Give students a couple of minutes to decide on a suitable course and to think about its good or bad points; give more prompts if needed (e.g. first-aid course, sports/music coaching).

3b This can be done in class or at home as a contrast to the informal speaking exercise of Exercise 3a.

Extra!
At this point, you could spend a few minutes on an aspect of learner training, highlighting the importance of keeping a record of whether a new phrase/vocabulary is more or less formal.

Key

1a 1 There was no messing about. 2 go through; demanding 3 He got us into; top-class 4 round off; cooked 5 We were shattered half the time.

1b 1 college course 2 work placements in top London restaurants 3 cooking in Jamie's restaurant

1c In interview: phrasal verbs (*go through / round off*), colloquial expressions (*shattered*), question tags (*the training was really hard, wasn't it?*), contractions (*Jamie'd let*)
In formal writing: passive structures, clear sentence structure, linking words

1d **Suggested answers**
1 First, the trainees were required to attend a rigorous 14-week basic training course at a London college.
2 Then they were given two-month work placements in reputable London restaurants.
3 Finally, they completed their training by working as chefs in 'Fifteen', Oliver's London restaurant.
4 According to the trainees, it was an exhausting but rewarding experience.

2 1 Everyone thought the tuition was outstanding.
2 It wasn't suitable for beginners.
3 On completion of the course, everyone was presented with a certificate.
4 The practical parts of the course were very disorganised – they weren't well prepared.
5 The course was not well publicised, so not many people attended.
6 It was a (big) advantage having such an experienced teacher.
7 Unfortunately, he was sometimes a bit irritated.
8 I'd like to congratulate everyone involved.

Photocopiable activity

Activity 1B follows this section. Done in pairs or groups, students complete a crossword by thinking of alternative ways of expressing the same idea.

1B A job for life

The module theme continues with a focus on 'work', with the following topics: changing careers, speaking about jobs and the qualities needed for them, interview techniques and written references.

One way to begin would be, with books closed, to put the following words on the board – *job, work, occupation, career, livelihood* – and get students to discuss what they perceive are the differences in meaning/use.

Listening 2 p.15

1 For the first question, students could come up with factors as they discuss the question or they could be brainstormed first and then ranked in some way.

For the second one, ask students if they know of anyone who has retrained later in life.

2 Start by getting students to read the task rubric and rubric for each part. Check understanding of the task with suitable concept questions. Point out that both tasks refer to the same recording (so questions 6–10 still refer to speakers 1–5). Three of the answers in each case are not used. Then refer students to the Task strategy notes on pages 170 and 171. Emphasise that although it is best to answer Task 1 the first time they listen and Task 2 the second time, it will help them if they skim both parts before they listen. It might also be worth pointing out that although the people are speaking about career change, students do not have to identify what job they changed to, only why they changed and the negative aspects of the new job.

As this is the first time students have done this type of task, you could pause the recording before it is repeated and look at the Help notes together.

3 Students compare their answers in groups. If they can identify and remember the phrase that helped them, the comparison can be discussed in class. However, emphasise that it is not necessary in the exam task – they only need to have heard and recognised the similar points. If you have time, play the recording again now that students know what they are listening for and give examples of how unused answers act as distracters (e.g. Task 2 answer A: Speaker 1 talks about colleagues, but says they are terrific; Speaker 2 talks about colleagues in a negative way, but says it doesn't bother him).

4 Another discussion question could be: *Which person do you think is happiest?*

Key

2 TASK ONE
1 H (*I could make a useful contribution to society, superficial atmosphere*)
2 A (*the actual work didn't stretch me*)
3 B (*not … commercially viable, without a regular income*)
4 E (*I just happened to be present when …*)
5 G (*out of the blue, I was offered a part … It was too good to be true*)
TASK TWO
6 F (*how much is expected of you, working all hours*)
7 G (*out of my depth … under pressure … almost gave up*)
8 H (*the commuting came as a bit of a shock*)
9 B (*I can't afford to do half the things I used to, which is a pain*)
10 C (*having to give up my regular job for a six-month contract*)

Speaking p.16

The speaking sections aim to extend vocabulary in the topic area as well as providing exam practice.

1a Stronger students could do this by covering up the box and trying to guess the word from the definition. If students have problems with the pronunciation of any of the words, correct them but don't emphasise the point, as it is covered in Exercise 1d.

1b Encourage students to write their own definitions where possible and only use a dictionary for words that are unknown to them.

1c Students could read out definitions to a partner or to the whole class, who can shout out the answer as soon as they know it.

1d Check students understand the terms *syllable* and *stress* by using other words from the page (e.g. *personal, vocabulary*) as examples. Students could put the words into columns. Give examples of different ways students could mark syllables (e.g. circles, striking through silent syllables) and the word stress (e.g. underlining, highlight, mark abõve) in their notebooks.

Patterns they may notice are that two- and three-syllable adjectives ending in *-ible* are often stressed on the first syllable; other three-syllable words often have the stress on the middle syllable.

2a As in the examples, get students to think about both the necessary qualities and why other qualities would be no good for a particular job. Also encourage students to use the negative or opposite forms of the words where possible (e.g. *tactful – tactless*) or to change the word form (e.g. *you'd need a lot of tact …*).

2b/c Emphasise that here they should give their opinions, and reasons to justify them, something they need to get into the habit of doing for the speaking paper, and that there are no right or wrong answers.

3a Exercises 3–6 focus on exam technique for Paper 5 Part 2, the individual long turn. You could ask students what they know about the speaking exam before they look at the Task strategy on pages 171 and 172.

3b Students listen to the instructions and identify the two aspects of the task.

3c/d Play the sample answer twice. After the first time, the students can discuss the content and after the second time, how successfully the task was achieved. Refer them to the task strategies again and go through them, using them to help evaluate the sample answer.

4a Here the stems are the useful language that students will be able to use as they attempt this type of task in the future. See how much of each sentences the student can complete before you repeat the recording again.

4b Ask the students what effect it would have on the listener if the speaker had used *I think …* four times.

5 If necessary, remind students of the points in the Task strategy before they start. Remind the partner to stop the speaker after about a minute.

6 Task analysis could be done in the pairs or as a whole class, with students giving good (or bad!) examples from their partner's answers.

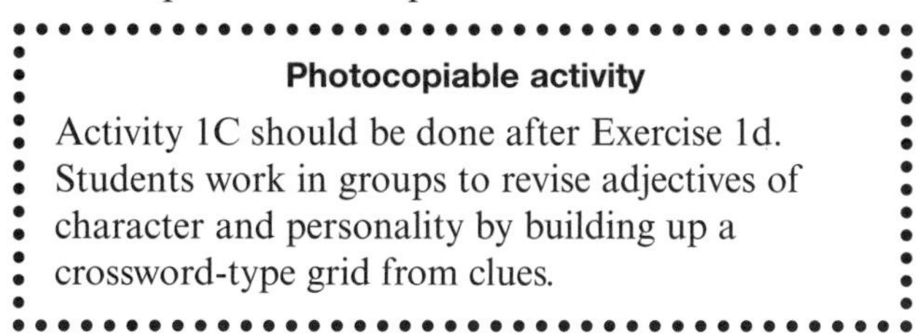

Photocopiable activity

Activity 1C should be done after Exercise 1d. Students work in groups to revise adjectives of character and personality by building up a crossword-type grid from clues.

Key

1a **1** tactful **2** sensitive **3** fair-minded **4** resilient **5** gregarious **6** tolerant **7** sensible **8** persistent

1d two syllables: **frien***dly,* **pa***tient,* **tact***ful*
three syllables: *as***ser***tive, cre***a***tive,* *de***ci***sive, ef***fi***cient, fair-***mind***ed,* **flex***ible, per***sis***tent,* **sen***sible,* **sen***sitive,* **tol***erant*
four syllables: *ener***get***ic, gre***gar***ious, re***sil***ient*

3b **1** Say what qualities the jobs would require and say why
2 One minute

3c He spoke about the doctor, the hairdresser and the teacher.

3d He completed the task well. He stuck to the relevant jobs, he mentioned the personal qualities they would need, comparing those that all would need and contrasting those that would be different. He had a couple of hesitations, but kept talking for the right length of time.

4a **1** … both require sensitivity and tact.
2 … you must need a lot of patience.
3 … would survive without a good sense of humour.
4 … a doctor needs a lot of knowledge and technical skills … for a teacher of young children … creativity and energy
5 … a good listener … a teacher … a doctor.

4b I suspect, I imagine, I suppose

Use of English 2 p.18

1a Ask students to say what the person in the picture is doing then look at the example sentence. Elicit the way to complete the second sentence so it means the same as the first. Draw students' attention to the changes that were necessary; grammatically the change from passive to active and in vocabulary changing 'to assume' to 'to take something for granted'.

Students who have taken FCE will be familiar with Key Word Transformations but should note the difference that in CAE they should use three to six words (two to five in FCE). Take time to go through the task strategy. Highlight the fact that the questions are not related or even linked thematically and that they are designed to test both grammar and vocabulary.

1b Show students the help section before doing question 1 together as a class. Give students a few minutes to do questions 2–4 before checking the answers and highlighting vocabulary such as comparing *convert / exchange* and *take back / withdraw*.

2 Again give students a few minutes to complete the task, pointing out that there may be more than one possible way to complete each sentence, before they compare answers.

3 The task analysis should reinforce the changes required in Paper 3 Part 5. Highlight the fact that changes in vocabulary are often related to changes in register in the same way that changing active/passive is (e.g. take back – withdraw; get someone down – depress; get on – board).

This area of language is expanded in the following Language development section.

Key

1b 1 In the latest famine crisis there *are estimated to be over/more than* 10,000 refugees.
2 Our *offer can be withdrawn* at any time before acceptance.
3 Can you *convert these dollars into euros* for me?
4 I*'d rather you'd asked me* before you used my computer.

2 5 There's no doubt that *the harder we work the happier* we are.
6 Several men who had *been imprisoned for political offences were* released yesterday.
7 The weather *really gets me down/gets me depressed/gets to me* at this time of the year.
8 Your mobile phones *must be switched off before boarding* the aircraft.

3 1 active to passive: 1, 2, 6, 8
passive to active: Example
2 vocabulary: Example, 1, 2, 3, 7, 8

Language development 2 p.19

1 The section assumes students are familiar with how the passive is formed in different tenses and focuses on its use. Practice of forming the passive in different tenses is available in the Student's Resource Book.

Students could work in small groups to decide why the passive is used in each case before checking their answers by looking in the Grammar reference. Get students to match the sentences to the examples there. If necessary, start with a review of passive structures, either using some of the sentences from the Use of English exercise (e.g. *It will be assumed by your interviewers that …*) or with sentences from the Grammar reference. Elicit how the passive form is constructed (using a form of *be* + past participle) and how various tenses are formed.

Extra!
If you have time, you could follow up by getting students to put the sentences into the active to compare and see why the passive is more appropriate.

2 Start by getting students to notice that each pair contains the same information, but one uses the passive and the other the active. Encourage students to extend their answers by saying not only which is correct but why.

As an extension, see if students can think of situations where the other sentence might be used.

3 Students decide if the five ideas are true or false, with examples to support their opinions.

4a Use the points in Exercise 3 to explain why sentence 2 is more formal than sentence 1 (uses the passive).

4b The vocabulary should all be known to students at this level. It is noticing the contrast in register that is important.

4c Start by getting students to skim the text and identify why the letter needs to be more formal (it is a business letter to a prospective customer unknown to the writer).

5 Emphasise that there are a number of possible ways to paraphrase some of these sentences.

6 Remind students of the factors that make texts appear more formal (choice of language, use of passive, more distant/less emotional). Check that students are using passive forms correctly.

Note: There is further practice of passives in the Student's Resource Book, followed by another chance for students to test themselves through a Use of English task.

▶ **Student's Resource Book, pages 11–12**

Key

1 1 who said it (police) is known/understood
2 don't know who by
3 distance – avoiding attributing blame
4 general law – who makes it/who does it unimportant
5 to provide a link between publicity and posters
6 who sends them unimportant – part of an impersonal process
7 general widespread belief

2 1 b – it is less personal, more formal
2 a – friendly conversational style
3 b – who interviewed them is not important
4 a – appeal direct to customer

3 1 True 2 False 3 False 4 True 5 False

4a Sentence 2 is more formal.

4b 1 h 2 d 3 i 4 e 5 a 6 j 7 b 8 f 9 c 10 g

4c 1 We **organise** excursions to the following **destinations**.
2 We **guarantee** that you will not be **disappointed**.
3 I have **enclosed** prices until the end of the year and trust that is **satisfactory**.
4 We **anticipate** a huge **response** to the advertisement.
5 Customers are therefore advised to **reserve** a place now while there are **sufficient** places available.

5 **Suggested answers**
1 A lot of people like these tours.
2 We will tell you your seat numbers one week before we travel.
3 Children less than five years old can have a discount.
4 If we have to cancel the trip, we'll give you all your money back.

6 **Suggested answers**
1 On some trips, proof of age will be required.
2 We apologise for the lack of a guide.
3 The cheap hotel room was satisfactory.
4 The trip was cancelled due to insufficient numbers.

Writing 2 p.20

The Writing 2 sections are directly related to the exam tasks. They follow a process approach, getting students first to plan the organisation, language and content before they actually attempt the task.

This page focuses on a character reference, one of a number of possible options in Part 2 of Paper 2.

1 Use the discussion questions to establish the purpose, nature and content of character references.

2 Give students a moment to carefully read the task before looking at the questions together. It is worth spending some time going through them to get students familiar with the process of identifying style, register and content. Refer students to the Exam reference on page 169 and point out the factors that examiners are looking for (task achievement, correct register, etc.). Encourage students to highlight or number each of the key points that have to be covered.

Elicit the name of this type of job (au pair) and find out if anyone has any experience of being an au pair.

Background

The term *au pair* comes from French and means 'on a par' or 'equal to'; this denotes the fact that the person is considered not as staff but as living on an equal basis in a caring relationship within a family. Au pairs can be male or female and are usually young people from another country that come and live with a family for up to a year. In exchange for light housework and childcare duties, they receive board and lodgings, some 'pocket money' and a chance to learn the language.

3a Students look at the notes and start to group them into topics.

3b Apart from the first and last paragraph, the order of the other three is not that important.

3c The emphasis here is on being selective, only including relevant information and being able to complete the task within the word count.

4a When students have identified the tenses used, elicit why a range of tenses is important (for variety and to impress the examiners!).

4b There are clearly more expressions here than could be used in one reference. Go through them with the class, checking why, for instance, a particular tense is used, working out what the next word(s) is likely to be or, for alternatives, to fit into the gaps. Emphasise that the phrases are generative and can be used in any reference.

5 As all the preparation has been done, the writing would best be done for homework.

6 As this is the first piece of writing on the course, spend some time on the editing checklist. When marking the compositions, look for evidence of editing and return those that have clearly not been edited. One idea for early in the course would be to get students to edit each other's work before handing it in to be marked.

Note: There is further Reading practice in the Student's Resource Book on a topic related to the second part of the module (jobs).

▶ **Student's Resource Book, page 16**

Key

2 **1** Include five topics: relationship / how long you've known them / their character and personal qualities / attitude to children / relevant skills.
2 Anything that fits the points in 1. They do not all have to be positive.
3 Neutral
4 If the reader has a clear impression of the person concerned.

4a Present perfect for unfinished actions / time periods
Past simple for completed actions
Present continuous for a present temporary activity

5 **Suggested answer**

I have known Anna Kurtz for six years, both as a colleague and a friend. We first met at secondary school when we did two weeks' work experience together in a local kindergarten. Since then, I have got to know her very well and have come to appreciate her many talents. At present, we are both at the same teacher training college, learning to become primary-level teachers.

Whatever she does, Anna brings two outstanding qualities to her work – a sense of responsibility and her ability to handle new and unexpected situations calmly and sensibly. She has always proved herself to be honest and hard-working in whatever professional situation she finds herself in, and gives the impression that she is acting in the best interest of others.

With children, Anna inspires confidence. She has a strong sense of discipline, yet children find her fun. In schools where we work, she is much liked and respected.

I regret having to say anything negative about Anna, but in the classroom she is not always very tidy. As a colleague, this can sometimes be a bit irritating, and it is all the more surprising because her bedroom in college is always meticulously neat and clean. Nevertheless, it is not a major problem and doesn't really interfere with Anna's ability to do a good job.

For the reasons I have given, I have no hesitation in supporting Anna's application as an au pair. I believe she would care for your two young children very well.

(250 words)

Module 1: Review p.22

The reviews at the end of each module are designed to recycle, extend and reinforce both the vocabulary and the grammar covered in the module (including the listening texts). They could be used immediately after finishing the module or returned to sometime later. The sections could be done at different times, either in class or for homework. Additionally, they could be teacher-marked to assess progress or peer-marked to assist learning.

1 **1** D **2** B **3** A **4** C **5** B **6** C **7** D **8** A **9** B **10** C

2 **1** Sorry I **didn't make** the plane on Friday. I hope it didn't mess you around too much. I know you **had** already booked me a hotel room, but presumably you were able to cancel it. Unfortunately, my father **hasn't been** very well recently, and on Friday morning, while he **was cleaning** the car, he fainted and was rushed into hospital. Luckily, the doctors say he's likely **to be** home in a few days.
2 I got a job in Scotland about ten years ago, and I **have been** there ever since. Recently, I met this really nice guy and **we're getting** married on the 27th of next month. Of course, mother's scandalised because by the time we tie the knot, we'll only **have known** each other for two months! When **you get** some time off, why not come and see us? If you do, **I'll** organise a get-together with some old friends.

3 **1** I was encouraged to undertake further research.
2 My money has been transferred into my account.
3 You'll probably be given a grant.
4 The candidates are being interviewed right now.
5 I had to cut short my visit.
6 We were greeted with nothing but kindness by the locals.
7 Your proposal must be received by next week.
8 He must have been badly hurt / hurt badly.

4 **1** been unemployed **2** required; inform; arrival
3 received; accepted **4** need; have
5 pleased; recommend; position/post

Module 2 Seeing is believing

This module includes topics such as magic, mysteries, superstition, coincidence and luck.

Lead-in p.23

Start by getting students to focus on the photos and saying what they know about the images shown. Establish that the creature 'lives' in Loch (= lake) Ness, which is just south of Inverness in central Scotland.

Check students know what the letters UFO stand for (Unidentified Flying Object) and that in (British) English they are pronounced as an abbreviation, letter by letter, not as an acronym/single word.

Background

The photo of the Loch Ness Monster was taken in 1934 by Marmaduke Wetherell, a big-game hunter, who was commissioned by a newspaper to investigate the mystery. It was later revealed to have been an elaborate hoax. It is one of the most famous pictures of the monster ever taken.
The UFO photo was taken on 16 January 1958 from a ship off the coast of Brazil. The photographer, a Brazilian named Almiro Barauna, claimed to have seen a dark grey 'object' approach an island, fly behind a mountain peak and then turn around and head back the way it came, disappearing at high speed over the horizon. On board the ship, around 50 other crew members are claimed to have seen the object.
The ghost photo was taken around 1910, and is almost certainly a fake; the Victorians' interest in the relatively new science of photography resulted in a spate of experimental photos such as this.

Before students discuss the questions, check some of the vocabulary (*phenomena* (plural) / *phenomenon* (singular), *rationally explained*).

2A The inexplicable

One way to begin would be to keep books closed and ask students to brainstorm what they associate with the word 'magic'. They might come up with black magic, witchcraft, Harry Potter, etc., as well as magic as a form of entertainment. If possible, elicit *conjuror*, which they will need to know before reading the text.

Reading p.24

1 Make sure students only read as far as the first paragraph. Emphasise the importance of using titles, sub-headings and illustrations to get an idea of the content and style of a text. Skimming the first paragraph should confirm these ideas and add information, not provide a first impression. The photo shows the conjuror Steve Cohen performing a card trick for a small audience. For questions 2 and 3, ask students to justify their answers and focus on which elements of the paragraph convey the writer's attitude.

2 Give students a suitable time limit to read the complete text (e.g. two minutes) before they discuss if their ideas in Exercise 1 were correct. If necessary, explain that they need to read quickly at this stage, as they will need the time to do the task. Confirm that the purpose of the initial reading is to say in a few words what the topic of each paragraph is.

3a Read the task rubric and check understanding with some simple questions. Elicit the factors that students could use to match the paragraphs. Then refer them to the task strategies on page 168, reading through them together.

3b Do the first question together, getting the students to match the highlighted words in the text with those in the paragraphs.

3c They can then finish the task on their own or in pairs. Encourage them to read through the complete text once again to check that the paragraphs follow on from each other.

4 These tasks are both a way of finding out how students arrived at the answers they did (e.g. a guess, a hunch, using the lexical clues) and developing the right techniques for the future. Students will need to learn to spot the links themselves. Point out how this will also improve their own writing.

5 The discussion could be done in a number of ways (pairs, groups, whole class, etc.), depending on the time available. Encourage students to develop their answers by adding experience of magicians that they are familiar with.

Extra!
With extra time, they could also discuss the final quote in the text (*'Magic works for everybody ...'*); for example, is it equally popular with men/women?

Key

1 2 A performance of magic by a famous conjuror to a small group of people.
3 The writer is impressed by Steve Cohen (uses adjectives such as *astonishing* and *remarkable*).

3b/c 1 E *Steve Cohen = The young red-haired American = he* (line 10); *highly intimate / a tiny crowded room = small and personal / private functions*
2 A *I can't go and see it = I don't really believe the story, either; his mobile goes off = When it rang* (parallel phrase), *One member of the audience = he = the owner of the phone = the man; he is ... astonished = jumped out of his seat*
3 D *It's tricks like these* = refers back to the phone trick; *sell-outs/popularity = his success*
4 B *the language of eastern philosophy = his talk of spiritual energy; ... must surely do the trick = Well, up to a point*
5 F *Any magician ... ought to be able + Which is why tonight I have managed to sustain my scepticism; However* (contrast) *now things are beginning to get weird* (refers forward to two tricks with women's arms and wedding rings)
6 C *It's this last one* (refers back to the last of the three tricks mentioned in the paragraph before); *They want ... to believe in something = [they] don't want to know how it's done.*

Photocopiable activity

Activity 2A fits here. It is a jigsaw reading activity that focuses on words and expressions that give cohesion to a text.

Vocabulary p.26

1 All the words except *well-paid* and *close* are used in the text. Students need to work out the differences in meaning and use. Check understanding of *PIN* (Personal Identification Number).

Extra!
Having completed the exercise, students could be asked to find other contexts to use the sentences.

2a Students could start by comparing the meaning of the three words before completing the sentences, or work from the completed sentences to derive the differences in meaning/use.

2b Students could use a dictionary to check their answers. Check the pronunciation of words derived from *deceive* before they complete the sentences.

2c If students answer 'no' to all the questions, ask them how cheating in other contexts (e.g. exams) should be punished, or whether advertising tricks or deceives people.

3a/b Students could be introduced to the idea of componential analysis to compare the items by putting the words in a grid on the board and ticking columns according to whether they are firm/gentle, with palm/hand/finger/nail, long/short action, etc.

3c The story endings could be written as group work or individually for homework. If you ask them to use exactly 50 words, there will also be some focus on structure, as students add/delete words as necessary. Brainstorm any other 'touching' words that students know and compare.

Key

1 1 a lucrative b well-paid 2 a number b digit 3 a visualise b see 4 a intimate b close

2a 1 trick 2 deceiving 3 cheats

2b

Noun – idea	Noun – person	Adjective	Adverb
trick [C]	(trickster)	trick	
trickery [U]			
cheating [U]	cheat		
deception [U] (deceit)		deceptive deceitful	deceptively deceitfully

NB *Tricky* (adj.) usually means 'difficult, complicated or full of problems', and is therefore not related to these meanings of *trick*.
1 trick 2 deceptively 3 deceptive 4 trickery 5 cheat 6 deceitful 7 deception 8 Cheating

3a 1 rubbed 2 tapped

3b 1 pressed 2 holding 3 rubbed (*scratched* is also possible, but you are more likely to rub your neck when feeling tired and tense (as here), and scratch it when you have an itch or are thinking) 4 tapped 5 scratched 6 pushed 7 patted 8 stroked 9 touched (*felt* is also possible, but the verb *feel* is needed for gap 10, where it is the only possibility) 10 feel

Note: There is further practice of some of this vocabulary in the photocopiable activity that comes after Language development 2.

Listening 1 p.27

1a Ask students what the pictures show and whether they are lucky or unlucky. (NB Black cats are considered unlucky in many countries, but lucky in others, e.g. the UK.)

1b Students may need to be given examples or asked to think of what other people do, such as carrying certain objects (good luck charms), wearing special clothes, crossing themselves, using a familiar saying.

2 Remind students that more formal talks start with a clear introduction which summarises the main aim of the talk. This helps the listener to orient him/herself to what they are going to hear.

3a Give students a minute to read the five summaries and guess the order they will come in before they hear the main part of the talk. They can write their guesses beside the boxes, then number the points as they hear them in Exercise 3b.

3b After checking Exercise 3a, give students a minute to look through the points a–d and possibly give further examples. Stronger students could note the word/phrase used as they listen and match points to each section. Point out that rhetorical questions raise interest as well as summarising what the section will be about. You may need to play the recording a couple of times. Follow with some discussion on why it is important to hear these markers at the start of each section.

4 Here, students are practising the skill of listening for specific information. First, check the rubric and confirm what they have to do (e.g. How many words can they use? Do they use their own words or words they hear?). Get students to write as they listen.

5 You could discuss the following additional question with students:

How far do you agree with the speaker's conclusion about these types of superstitions?

Key

2 C

3a C, E, A, B, D

3b 1 c *The main explanation for this seems to be …*
2 b *Another reason is …*
3 d *What is hardest to understand, however, is …*
4 a *So what kind of people are most superstitious?*
5 b *Finally, we must …*

4 **1** 75/seventy-five **2** salt **3** touch wood **4** cross their fingers **5** set phrase **6** 'Bless you' **7** blame themselves **8** eats fish **9** putting on (his) socks

Use of English 1 p.28

1 As a lead-in, ask students what the picture shows and what they know about it but be careful not to discuss in detail, correct or give information that is contained in the text.

2a Students first read the title and skim the text to get a general idea of the content and to answer the two questions.

2b Give students a couple of minutes to read the task strategy before they start the task. Remind students that the correct word must fit both the meaning and structure of the sentence. Look at the example and then do question 1 together comparing the choices in each.

3 The task analysis draws attention to some but not all areas covered in Paper 3 Part 1.

4 As a follow-up question, ask students what other ancient monuments they are interested in or would most like to visit.

Key

2b **1** A **2** B **3** A **4** D **5** C **6** A **7** B **8** D **9** A **10** C **11** B **12** D

3 **1** choice of verb: Example, 6, 7, 8, 9, 10
2 adjective + noun: 1, 5

Language development 1 p.29

1 There are a number of ways to approach this review. Weaker students could start by reading the Grammar reference on page 175 before using it to find and correct the mistakes. Stronger students could attempt the task first and then use the grammar reference to check their answers. Alternatively, raise awareness by putting a pair of sentences (e.g. *Egyptologists, who have studied the Sphinx closely, believe it is 5,000 years old* – with and without the commas) on the board and get to students to compare and contrast them before doing the task.

2 This exercise looks at words such as pronouns, prepositions and quantifiers that are often used in more advanced relative clauses. Emphasise that one, two or three of the choices might fit each gap.

3a The exercise starts by giving four examples of how relative clauses can be reduced. Students rewrite the sentences using the complete clause.

3b Then students reduce clauses in the same way.

4 Emphasise that students should use a range of ways to combine the points. Also, establish what effect it will have on the text (reduce word count, avoid repetition, read better, etc.). If necessary, do the first one or two sentences as an example.

Extra!
As a follow-up, ask the students if they agree with Miranda, who believed her bad luck was more than coincidence. Could it have been the ring?

Key

1 1 A beautiful part of Britain is Wiltshire, where the ancient monument of Stonehenge **is**.
2 Stonehenge is a circle of stones which ~~they~~ date (*which* refers to *stones*) / dates (*which* refers to *circle*) back over 5,000 years.
3 The monument, **which** thousands of people visit each year, is 50 metres across. (non-defining clause must have the pronoun)
4 The original purpose of the monument, ~~that~~ **which** has not been discovered, might have been for sun worship. (Needs a non-defining clause; a better answer to avoid confusion would be *The monument, the original purpose of which has not been discovered, might …*)
5 June 21st, the longest day, is the day on **which** the stones line up with the rising sun.
6 Little is known about the people who built Stonehenge, or their beliefs. (needs a defining clause)
7 Some of the stones, **which** weighed up to 3 tonnes, were carried over 200 kilometres. (non-defining clause must have the pronoun)
8 Modern engineers, **whose** ~~efforts~~ to repeat this achievement have failed, don't know how the stones were transported. (needs a possessive relative pronoun – *their efforts*)

2 1 B/C
2 B
3 A (*pilots* needs *who(m)* not *which*)
4 C (with *two*, *both* is possible; *neither* and *none* are not)
5 B/D
6 A (= *it doesn't matter what* – can be used with/without *that*) / B (all they read – with/without *that*). (C would be possible if it was *all that* not *all what*)

3a 1 Tibet, **which is** situated between China and Nepal, is home to the famous yeti.
2 The yeti is a human-like creature, **which is** said to live in the high Himalayas.
3 People **who live** in the area say it is a common sight.
4 The first person **who catches** a yeti will become famous.

3b 1 Many years ago, people walking in the mountains claimed to have seen a tall, hairy figure in the distance.
2 However, there was no one carrying a camera who could take a photo.
3 A photo of a huge footprint, taken in 1951, remains the only real evidence.
4 The hunt for the yeti, also described as being like a giant bear, continues.

4 **Suggested answers**
1 Miranda Seymour is a well-known writer who has written many books, some of which are biographies. **2** In 1995, she wrote a book about a poet called Robert Graves who had travelled extensively in Egypt. **3** After the book was published, someone gave her an antique gold ring which had belonged to the poet. **4** She started to wear the ring, at which point strange things started to happen. **5** Her husband, to whom she had been married for 14 years, left her, after which, she was burgled. **6** The next thing to happen was that her mother, who had always been healthy, was diagnosed with cancer. **7** Then Miranda lost her teaching job, and finally her tenant, who had only just moved in, left. **8** Miranda looked at the ring which she was wearing on her finger. **9** That very day she gave it away to a museum which collects objects that belonged to the poet. **10** Immediately she found a new tenant who was perfect / a perfect new tenant. **11** Her mother, who hadn't had cancer after all, got better. **12** On top of this, she got her job back, which she had lost earlier.

Photocopiable activity

Activity 2B is designed to be used here. Students play a version of noughts and crosses to review relative clauses, expressions using relative pronouns and reduced relatives.

▶ **Student's Resource Book, page 18**

Writing 1 p.30

Many students are able to write reasonably good sentences, but are less able to construct and link good paragraphs. In an exam, they do not spend enough time planning what they are going to write. This section is designed to teach them an appropriate way to approach a piece of writing.

One way to start would be to give them all a piece of paper, set them a writing task with a time limit of 30–40 minutes and let them get on with it. Then stop them after ten minutes, collect in the papers and compare their approaches. See how many of them have started writing and how much they have written already. Compare with those that have produced a plan and what it looks like.

1 Discuss the questions as a whole-class activity. Remind students that an hour is more than enough time to write 220–260 words, so there is plenty of time to plan. Equally, it is hard to include a sufficiently wide range of structures, vocabulary and linking devices within the word count unless it is planned carefully. Many students will say that it is a good idea to use the time to write the piece out again neatly. Point out that:

- it is a waste of time to write the same thing out twice;
- if the writing is planned initially, there will be no need for major rewrites;
- neat crossing out is perfectly acceptable in the exam; it is the content that is being marked, not the presentation;
- students often make careless mistakes (e.g. missing words out) as they copy out a piece of writing.

2 The features listed are those of a formal text; students should be able to identify which of the pieces of writing are more and less formal.

3a Having read the task, students should identify the number of paragraphs needed. One way to show this would be to underline the different parts that need to be answered and then add a conclusion giving their overall view, such as a recommendation to see the film.

3b As they read the brainstorming notes, students may not know the words *spooky, eerie* and *hypersensitive*. They should notice that the two crossed out sentences are not directly relevant to the task.

3c/d Having decided on the best order, students complete the paragraph plan using the headings in Exercise 3c and the points in Exercise 3b.

4 If necessary, brainstorm a few recent or classic films of the genre that students might have seen. Encourage students to follow a similar plan for their reviews.

Extra!
The reviews could be written for homework and be displayed where students could read them before or after the following class.

Key

2 1, 2, 4

3c 1 Title/type of film/overall impression
2 Plot summary 3 Opinion in detail
4 Recommendation

3d A 'good' paragraph plan might be:
Paragraph 1: Introduction – Title/type of film/overall impression
'The Others' – spooky ghost story – keeps you in suspense
Paragraph 2: Plot summary
Simple plot – well structured. Grace – husband away at war – two children – mysterious illness (hypersensitivity to light) – three servants arrive from nowhere (▶ strange events)
Paragraph 3: Opinion in detail: acting, direction, music
Great acting (N. Kidman as Grace) – slow moving – eerie atmosphere
Paragraph 4: Conclusion – Recommendation
See it! Best film of the year.

2B It's only logical!

One way to begin would be to brainstorm words formed from the noun *luck – (un)lucky, (un)luckiest, (un)luckily* – and how the word is used. *Luck* is uncountable, so for specific events, we say *a bit/piece/stroke/element of luck*; we say someone *is lucky* not *has luck* unless we use other words before the noun, e.g. *have good/bad/no luck, did you have any luck? I had a bit of luck.*

Or use a good dictionary to identify idiomatic expressions with *luck* (including those in questions 1 and 2 of Exercise 1).

Listening 2 p.31

1 Before students discuss the questions, check that they understand *accident prone*.

2 The task type – sentence completion – should give students a big clue as to what they have to do here. Start by looking carefully at the task strategies on pages 170 and 171 and asking a few questions to check they fully understand what is required (e.g. How many words do you need? Do you need to change the words you hear?). Give students a minute to read the notes before they listen. At this stage of the course, students could be given a few minutes to compare answers before they listen for the second time to ensure that everyone is listening for more or less the same information.

3 Ask students if they remember any of the other signals/markers he used. If there is time, allow students to listen again, stopping after each section to identify the markers used.

4 Students could discuss the questions in groups. Remind them that there are no right or wrong answers and that they should get into the habit of trying to express an opinion rather that saying that they don't know. You could ask students who said at the beginning that they were unlucky if they think it is possible to change.

Key

2 **1** national magazine **2** diaries **3** think; behave **4** pictures **5** message **6** (more) relaxed **7** instinct(s) **8** positive

Speaking p.32

One way to start would be to brainstorm what students know about Paper 5 Part 1. Then get them to read the Exam reference on pages 171–2. In Part 1, students are expected to answer questions about themselves. As well as covering a wide range of grammatical structures, it tests their ability to use interactional and social language, hence the need to listen and respond appropriately.

1 One way to approach the exercise would be to play the two exchanges with books closed, asking students to compare them. Then play it again with books open so they can give examples of why the second is better. They should notice that there is some interaction in the first exchange. The speaker answers the questions he is asked. However the responses are much too brief and mechanical, and repeat too much of the question. Elicit the factors that make the second exchange so much better.

These include: substitution (*What do you like? – What I enjoy...*), expansion (*football – I play ... we've got a match ...*), linking phrases (*in fact*), short answers (*Oh, very.*), the use of linking words to create longer sentences (*Apart from... , I ...; but; as well as ... , I ...*) and the sense that the speaker wants to communicate something interesting about himself.

2a Get students to work together, deciding which is the best response to each question and why.

2b Give students a few minutes to look at the useful phrases, thinking about what questions they could be used to answer and various ways to use them. Remind them of the points they made in Exercise 1 before they practise answering the questions.

3 Start by getting to students assess the conversations they just had and what the strengths and weaknesses were. Read the introduction and point out that Part 1 of the speaking test has itself two separate parts (1 – The examiner asks basic questions such as 'Where are you from?'; 2 – The examiner asks further questions on topics such as work, leisure, travel and future plans.)

3a Stop the recording after the first part so students can discuss the question.

3b After the second part, give students a few minutes to answer the questions, particularly to think of suggestions for the second question.

3c Here, the focus is on dealing with difficult moments such as not being able to think of a suitable answer. At this point, the class could brainstorm other 'fillers'.

4a This would work best with groups of four, but would be possible with groups of three, with all three students acting as assessor at the end. If necessary, remind students of the key points in the task strategies before they begin. Ensure that only the 'interlocutors' look at the questions on page 205.

5 The 'assessors' could use the task strategy as a tick list to record each candidate's strengths and weaknesses on this part of the paper, and then compare their opinions with those of the 'candidates' themselves!

Note: In any future speaking activities, notice and comment on how well students are responding, and refer back to this section to remind them of the phrases they could have used.

Key

2a **1** A is a better answer as it is a personal response that reflects an aspect of studying. It is a more complex sentence. B repeats words from the question then just gives a list.
2 B is a better answer, expressing his likes with good varied vocabulary (*not that keen on, are addicted to*). A doesn't answer the question as it is not about his/her personal opinion.
3 A is a better answer. It is longer, more complex and has a richer vocabulary (e.g. *it will come in useful*). In B the sentences are short, simple and without any interesting detail.
4 A is better as it answers the question being a personal opinion about parts of the country. B is a generalisation focussing on the weather.
5 B is better as it includes interesting phrases (e.g. *Well, as a matter of fact*) and more interesting vocabulary (e.g. A – *I like to sleep* / B – *I try to catch up on my sleep*).

3a He answers fully, and gives relevant answers. He sounds relaxed and natural, whereas Cécile's answers are short and formulaic, as if she has rehearsed them.

3b **1** *That's a tricky question ... I'll have to think about that.*
2 She should have made one up.

Use of English 2 p.34

1 Go straight into the questions. Encourage students to expand their answers with examples.

2 The open cloze will be familiar to students who have taken FCE, but it would still be wise to spend a few minutes focusing on the best way to approach the task.

2a Give students a suitable time limit (e.g. 60 seconds) to skim the text to answer the first question and to scan the text to find the answer to the second question.

2b Check the rubric, then read through the task strategies on page 169, highlighting the key points. Do the first question together to demonstrate how to approach the task. Remind students to read through the text once more when they have finished to check that it all links together correctly. Point out that, in the exam, they would have about 15 minutes for this task.

3 The first question should raise awareness of articles before the following Language development section. The second question may help students to focus on areas of grammar that they find hardest.

Key

2b 1 *have* – auxiliary needed for present perfect passive, plural after *such events*
2 *each* – reflexive, two people, each one sits next to the other
3 *out* – expression = 99%
4 *in* – preposition before the noun in a fixed expression = *shared, known to both*
5 *the* – definite article, only one other
6 *all* – quantifier before number
7 *or* – fixed expression = *approximately*
8 *not* – negative in contrastive expression *not only A but B*
9 *One* – only/first reason of many given
10 *than* – comparative expression, after *more*
11 *them* – pronoun refers back to the people gathered
12 *if* – conditional clause
13 *be* – auxiliary needed in passive structure
14 *most* – superlative structure needed after *the*
15 *would* – modal before infinitive *be*, expresses a hypothetical point

3 1 articles: 5, 14 quantity terms: 3, 6, 7, 9, 10, 14

Language development 2 p.35

One way to begin would be to start by brainstorming when the various articles are used, and putting some rules on the board. Another approach would be to get students to attempt Exercise 1 in pairs first, then check their answers.

1a Students should work individually to complete the text before comparing their answers. Getting them to justify choices should help them focus on how well they know the subject.

1b Give students a few minutes to think about and plan their coincidence stories before they tell them.

2 This exercise focuses on nouns that can cause confusion because they are unusually singular, plural or uncountable. When going through the answers, get students to compare how the words are used in their own language. Errors may arise from translation. This section could be followed with students forming their own similar sentences from stems that focus on the required form (e.g. *At school, my favourite subject ... , In my country, the police ... , There ... a number of differences between ...*).

3 This exercise focuses on other common quantifiers and some of their more advanced uses. Again, students could refer to the grammar reference before or after attempting the task. If appropriate, ask about other games of chance students try.

4 Many of these quantifiers express the idea of 'none' or 'all'. However, only one fits in each gap. Students should identify the differences between them.

Key

1a 1 ø – *notice* = U
2 *a* – *holiday* = C
3 ø – before most countries
4 ø – *accommodation* = U
5 ø – *space* = U
6 *a* – fixed expression (*knowledge* is U, but *a good knowledge of something*)
7 *the* – specified earlier (France)
8 *the* – *the industry as a whole*
9 ø – *advice* = U
10 *an* – *area* = C + sing.
11 ø – *good weather* = U
12 *a* – *hotel* = C + sing. unspecified
13 ø – in general
14 *the* – specified (the food of that hotel)
15 *an* – one
16 *the* – specified
17 *an* – one of many
18 ø – fixed expression
19 *a* – *time* C = *moment*
20 ø – *contact* = U
21 *the* – specified
22 ø – plural
23 *the* – specific (news of people we knew)
24 *a* – expression
25 *the* – defined
26 *the* – defined
27 ø – before a number
28 ø – name of sport
29 *the* – before musical instrument
30 *the* – superlative
31 *the* – only one

2 1 The police **have** not charged the suspect because there isn't **any** evidence.
2 If you think you're getting ~~a~~ **the/ø** flu, **some** good advice is: stay in bed and drink lots of fluids!
3 Politics **isn't** a subject that most people enjoy~~s~~ studying.

4 At school, maths **was** my favourite subject and athletics **was** my least favourite.
5 On the flight home, **some** of my luggage**s** came open and **some** of my belongings are missing.
6 Four days **is** a long time to wait for an appointment.
7 Two per cent is a small pay rise, and I expect at least 80 per cent of the staff **are** going to go on strike.
8 A number of coincidences **have** been noted.
9 The number of lucky escapes **has** increased year on year.

3a 1 *Many* – people are countable 2 *Most* – in general; *much* = money (U) 3 *many* – need *of* after *lots* 4 *a few* – positive, a significant number 5 *little* – negative, dismissive

3b 1 *Each/Every* – used before plural noun (*both* not possible as more than two colleagues)
2 *Each* – before *of* (*every* not possible)
3 *both* – two games (*each/every* can't be used as pronoun – compare with *each/every one*)
4 *each/every* – Wednesday and Saturday of every week
5 *Every* – before plural time expression (*each/every week/month*, etc. but ~~*each*~~*/every two/few weeks*, etc.)

4 1 *Both* – two 2 *either* – one of two
3 *none* – negative of *many* 4 *not* – negative emphasis on *none* 5 *neither* – negative of two
6 *no* – negative uncountable 7 *the whole* – complete 8 *all the* – every one of them

Photocopiable activity

Activity 2C could be used here. It is a pairwork activity designed to practise the vocabulary covered in the vocabulary section on page 26 and in this Language development section.

▶ **Student's Resource Book, page 23**

Writing 2 p.36

For many students used to producing discursive or narrative text, this type of writing is very different from anything they have done before. They therefore need help in developing the necessary register, style and formatting.

1a First, check that students understand what is meant by a 'information sheet' (this particular type is often called a 'leaflet') and elicit examples of where you might find them. Then look at the two extracts and answer the questions.

1b Students talk about any clubs or societies (the two words have the same meaning, but in some contexts *society* is more formal) that they belong to or would like to belong to.

2 In order to write any composition well, the writer needs to have a clear idea of: why they are writing, who they are writing to/for, what they need to include. Start by getting students to read the task before answering the four questions.

3 Students start by planning the content and organisation of the information sheet. Remind them of the three areas that must be covered: background information, practical information and reasons for joining.

3a/b Students divide the points into the headings which best fit the points above.

3c Emphasise the need to invent any facts, figures or details for the information sheet.

4a This part focuses on a suitable style. Start by getting students to compare the two openings. Check understanding of the vocabulary (e.g. *budding* = just starting, *a grandmaster* = a chess player of a very high standard).

4b Draw students' attention to the fact that some of the points are in the less formal first person (*We are …, Our members …, we participate …*), some in the direct second person (*It doesn't matter whether you are …*) and some in the more formal third person (*The club offers …, The club meets …, Members have the opportunity to …*). Students combine phrases to form complete sentences that they could use in the information sheet.

5 Now that the information sheet has been planned in detail, give students 20 minutes to actually write it and another five to ten minutes to check it.

6 Remind students of the importance of checking their work for errors and simple mistakes – there is a checklist in the Writing reference on page 188.

Key

1a 1 Both aim to give information simply and clearly to the casual reader, who might be a potential member.
2 The Film Club information sheet is trying more to 'sell' the club directly to the reader (*the place for you*), giving special offers, and the language is more marked (*as little as/one of the best*). The Hillwalking Club information sheet is more neutral in approach and style. Both, however, make their points clearly and succinctly. Both start with an introductory message to the reader saying who the club is for, then proceed in a friendly tone giving the key points, almost in note form. An information sheet of this kind is often very brief.
3 Headings, subheadings (either neutral – *Walks, Activities* – or persuasive – *Join now and get*) and bullet points.

2 1 To tell people about a chess club with the aim of getting new members. It is addressed to local people.
2 Three main parts (1 background, 2 practical information, 3 reasons for joining), which can be broken down into sub-parts (e.g. 1 when founded, who the members are; 2 where/when meet, who can join, how join; 3 reasons for joining, other benefits)
3 Probably a mixture of neutral facts, a friendly tone addressed to the reader, and phrases intended to persuade.
4 If it persuades people to become members!

4a Text A is a fairly neutral opening and gives two key facts very simply (club history and membership). It is a better opening, as it is more appropriate for a council information information sheet focusing on the facts and information stated in the question. Text B is addressed directly to the reader in an attempt to draw him/her in.

4b **Suggested answers**
1 One of the most popular features of our club is ...
2 New members ...
3 We are ...
4 Members have the opportunity to/are able to ...
5 The club meets ...
6 We participate in/organise/run ...
7 Experienced players are often on hand ...
8 Anyone wishing to ...

5 **Suggested answer**

CITY CHESS CLUB
Who are we?
The thriving and popular City Chess Club was established in 1904, and has members from their early teens to near 80.
Where and when do we meet?
We meet at the City Leisure Centre every Monday, Tuesday and Thursday evenings, 7 p.m.–10 p.m. There is a large car park and easy access by public transport.
Who can join?
Anyone interested in playing chess. Our members include novices and county champions, men and women alike.
Why join the club?
• Atmosphere
The club has a relaxed, friendly atmosphere. Many members come for a nice, quiet game after work.
• Friendship
You will have the opportunity to meet members of all ages and form friendships with people from many different backgrounds.
• Improving your game
We offer coaching for those wanting to achieve the highest level of personal ability, and experienced players will often be happy to offer advice and assistance to beginners.
• Competitions
There is a range of tournaments: from quick-play tournaments (maximum 20 seconds a move) to weekend tournaments. We also participate in a number of live on-line tournaments.
What other benefits are there?
• Social events
We organise a regular social calendar of dinners and dances.
• Games Centre facilities
Members are able to use the Games Centre bar, and have access to other activities in the centre at reduced rates.
How do you become a member?
If you would like to join our club, you can phone the number below or simply turn up on club nights. You will always be guaranteed a warm welcome.

Module 2: Review p.38

1 **1** Her films are *deceptively* simple.
2 *As a matter of fact,* I'm quite superstitious.
3 *He's got a very good attitude* to his work.
4 Who *is responsible for* this mess on the floor?
5 There's always someone *on hand* to help.
6 Most people hear about the shows *by word of mouth.*
7 There hasn't been an accident yet – *touch wood*!
8 Many young people feel his films *strike a chord.* / His films *strike a chord* with many young people.
9 What do you *attribute your success to*? / *To what* do you *attribute your success*?
10 His mobile phone *went off* unexpectedly.

2 **1** cheating **2** origins **3** ability **4** up **5** round/over **6** digit **7** itching **8** in **9** prone **10** leap **11** explanation **12** legendary

3 **1** C **2** B **3** A **4** D **5** C **6** B **7** D

4 Reincarnation is the belief that, after death, some aspect of **each / every one** of us lives again in **another** body, **either/whether** human or animal. Indeed, most ~~of the~~ tribes avoid eating certain animals because they believe that the souls of their ancestors live in them. Reincarnation was once a belief mainly of ~~the~~ Hinduism and Buddhism, **both** of **which** are Eastern religions. Recently, however, a number of Western belief systems **have** started to incorporate it into their teachings. Perhaps the reason for this is that it seems to offer **an** explanation for a range of unexplained phenomena, such as the ability of people to regress to a past life under ~~the~~ hypnosis. Of course, reincarnation remains a belief, and there's very **little** chance that it **will ever** be proved.

Exam practice 1, TRB p.178

Paper 1: Reading
1 E 2 B 3 A 4 D 5 A/D 6 D/A 7 C 8 D 9 E 10 E 11 A 12 C 13 F 14 A/E 15 E/A

Paper 3: Use of English
Part 4
1 drop 2 depth 3 led 4 running 5 settled
Part 5
1 never been to a more 2 been going to dancing classes 3 it is unlikely that Yasmin 4 is said to be the most 5 book in advance are entitled
6 deadline for handing in your assignment
7 have/stand little chance of 8 is responsible for leaving

Paper 2: Writing
Suggested answers (NB Question 3 is designed for students who are taking the set-text option in the Writing paper and answers are dependent on the novel they have read.)
1

Dear Sir or Madam,
With reference to your newspaper advertisement of 25th May, I am writing to express interest in the post of temporary assistant at the art gallery for the summer period. I am 26 years old, single and graduated in Modern Languages last year. I have a valid driving licence.
I have a certain amount of experience as a guide, having spent two months last year working at a local folk museum. This job involved accompanying groups of foreign tourists around the museum. Two years ago, I worked part time as a courier for a local travel agency, taking tourists to areas of cultural interest in the region where I live.
My first language is Italian, but I am fluent in English and have a good command of French and German, plus a limited knowledge of Spanish. I believe I am the right sort of person to work as a guide, as I am genuinely interested in history and art and enjoy dealing with large groups of people.
Although your advertisement was quite informative, there are a few points I would like to clarify with you. First of all, what period would the job cover? Secondly, as the gallery is not in my town, would you be able to help me find accommodation, or suggest someone who could? And thirdly, what kind of salary would you be offering?
I hope my application is of interest to you. I am available for interview at your earliest convenience.
Yours faithfully

2

SOCIAL ACTIVITIES

Here are some ideas of things to do in the evening during your stay.

The town has an excellent, Olympic-size **swimming pool** with additional facilities, such as **sauna** and a warm-up gym area. Use of the swimming pool and gym costs £3 per visit, whilst use of the sauna is £2. It is not necessary to book.

There's also a **leisure centre** with indoor tennis, five-a-side football, bowling and many other activities. The centre charges £10 per visit, plus a book of tickets which you use for each activity. Tickets are £1 each. You need to book in advance, stating your choice of activity.

Just outside of town, there's a place where you can go **horse-riding**. This costs £15 per hour if you can already ride, while the cost of an instructor is £10 extra per hour. Horse-riding can be booked online or by phone (010-334-4457).

For those who prefer something less energetic, the town offers various **museums** and a **modern art gallery**, which are all definitely worth a visit. Highly recommended is the **Museum of Sculpture** which has works by many famous international artists as well as local sculptors. Entrance is free to local museums, but the art gallery charges £5 (group discount may be available).

Shopping is, of course, a favourite amongst visitors, and most shops are open until 9 p.m. The **Burrington Mall** is extremely popular, but not particularly cheap. In the small streets behind the mall, local craftsmen still produce hand-made jewellery and pottery. They are always willing to give demonstrations of their work.

Paper 4: Listening
1 website **2** 2,068/two thousand and sixty-eight **3** lucky charm(s) **4** Touch wood **5** ladders **6** science **7** teenagers **8** developing/evolving (*either order*)

Module 3 Values

This module includes topics such as the nature and pressures of fame/celebrity, autograph collecting, standing up for beliefs, expressing ideas and opinions, and raising money for charity.

Lead-in p.39

Start with books closed. Ask students to define the term *celebrity* (NB it can be uncountable (= the concept) or countable (= a person)) and to give examples of the types of people (actors, singers, designers, sports stars, etc.) who become celebrities. If appropriate, ask students who the biggest celebrities in their countries are at present.

Get students to look at photos on page 39 and discuss the questions and quotes below. Do not spend long on this, as some of the downsides feature in some of the following exercises.

Background

Julia Roberts, born in October 1967, is one of Hollywood's most successful actresses. She has appeared in numerous hugely popular films and was the highest-paid female star for four years (2001–2005). The first picture reveals the downside of being such a big celebrity: even when out shopping and wearing casual clothes, she cannot escape the cameras – her privacy is constantly invaded.
The second picture shows her winning a Golden Globe for her role in the film *Erin Brockovich* (2000), for which she also won a BAFTA (The British Academy of Film and Television Arts) and an Academy Award (better known as an Oscar). This amazing clutch of accolades pushed her career to even higher levels, and in 2003 she broke records for female earnings when she was paid US$25 million for her role in *Mona Lisa Smile*. In 2007 she was estimated to be worth around US$140 million – surely an illustration of the upside of being a celebrity!
You might want to tell students that they are going to read an interesting anecdote about Julia Roberts in the reading exercise on pages 40–41.

3A The burden of fame

One way to begin would be, again with books closed, to give students the phrase *the _____ of fame* on the board and brainstorm possible words that could fit in the gap (*nature, price, pursuit, rewards, pressure*, etc.).

Reading p.40

1 Get students to look at the title and subheading check the meaning of *limelight* (a situation in which someone receives a lot of attention) and *fierce glare* (harsh attention). Then get students to discuss the question. If students find it difficult, ask them to guess what some of the stranger effects of being well known might be.

2 Be careful here that students only skim the article. Point out the method suggested here of reading the first and last sentence of each paragraph and introduce the idea of topic sentences. In multiple-choice questions, it is especially important that students do not waste time reading the whole text in detail, as each of the five questions (seven in the exam) will only relate to a part of the text. Elicit the purpose of skimming here (to get a general sense of the text, its style and a rough idea of its organisation/layout so students know where to search for specific ideas).

3a/b Having read the rubric, students should study the task strategies on page 168. Then get them to cover the four answers and read the first question. Ask them to scan the article to find the anecdote about Julia Roberts (lines 12–22), then in their own words say what the writer's point was. Then they can uncover the four answers and find the one closest to their own. Remind students that, for each correct answer, there are three wrong answers and if they have time, they can check why the other three are wrong – often it will be that the information is just not stated.

3c Students should work on their own doing questions 2–5 in the same way.

4 They should then compare answers in small groups before checking the answers with the whole class.

5 The discussion could take some time and raise interesting issues such as the relationship between the press, celebrities and the readers, and what the limits are, if any, on what the press can/should report.

Key

3 1 C – *this casually dressed woman did not conform to the image he had in his head* (line 19)
2 A – *fame can engender distrust and isolation, meaning that nobody can be taken at face value.* (line 30)
3 C – *seem less well-adjusted … perceiving slights where none exist* (line 38)

4 D – *embody a wide array of archetypal traits* (combine various qualities) *which have current appeal* (are desirable) (line 56)
5 B – *the public have an insatiable appetite for seeing the famous toppled* (line 64)

Vocabulary p.42

1a Start by finding the nouns in the text and eliciting the verbs from which they are formed.

1b All the nouns needed are in the text, so students can search for them to check they are correct and to check the spelling. Focus on the endings *(-ness, -tion, -ity, -ance, -ence)* that are typical of nouns. Get students to identify patterns (e.g. the adjectives ending in *-ant/-ent* form nouns ending in *-ance/-ence*).

1c When students have identified the nouns, get them to decide which syllables are stressed. There is an opportunity for some learner training here. Get students to check how the word stress is marked in their dictionaries before they listen and check their answers. Focus attention on words that change the stress according to word class (e.g. *identify – identification*).

1d Pre-teach *soap (opera), minor* and *naive* if necessary. Point out that the text uses eight of the ten nouns in Exercise 1b. Finish by getting students to use the other two words (*devotion, imagination*) in sentences of their own related to 'celebrity'. Point out that *pay (more) attention to something* (question 3) is a strong collocation and link to Exercise 2.

1e The discussion is a quick focus on soaps, which are a worldwide phenomenon, and whether students watch and enjoy them.

Background

Soap operas, now better known as 'soaps', were so called because, like many classic operas, they focus on the complications of love in everyday life. Originally shown on daytime TV, they were aimed at housewives stuck at home, and funded by the soap manufactures that targeted them, advertising in the breaks. Then longest radio soap in Britain has been going every day for over 50 years, and the longest TV soap, *Coronation Street*, has been shown three times a week since 1960. (One actor has been in it since the beginning!)

2a Collocations exist across various word forms. Here, the focus is on adjective + noun. The collocations, but not the sentences, are all from the reading text. When students have matched the halves, they should underline or highlight the collocations.

2b Focus on some of the vocabulary before students write their own sentences. For example, *appetite* is not just for food; beliefs and convictions are very similar, so religious beliefs and *political convictions* are also strong collocations; a *household name* is someone famous; and *the popular press* refers to newspapers with the widest readership but which focus more on gossip and human-interest stories.

3a Again, the expressions are all in the text, so students could use it to check their answers and see the expressions in context. There is also an opportunity for some learner training here, as students find the expressions in a dictionary to see how idioms are listed. For example is *to play a joke on someone* listed under *play* or *joke*? They could then guess which word to check first in the others before looking them up in their dictionaries.

3b When students have completed the sentences, they could be asked to personalise them a little by giving examples from their own experience, for example of when they have played a joke or got on the wrong side of someone.

Key

1a 1 contradict 2 isolate
1b/c1 scrutiny 2 arrogance 3 rudeness 4 inconvenience 5 attention 6 devotion 7 imagination 8 neutrality 9 innocence
Patterns: stress on penultimate syllable for words ending in *-tion*, stress two syllables before the last on words ending in *-ence* or *-ance* and on words ending in *-iny* or *-ity*.
1d 1 inconvenience 2 attention 3 rudeness 4 neutrality 5 scrutiny 6 arrogance 7 innocence
2a 1 f (insatiable appetite) 2 c (vast array) 3 h (different perspective) 4 a (religious convictions) 5 g (casual clothes) 6 d (bad temper) 7 b (social skills) 8 e (household name) 9 j (public scrutiny) 10 i (popular press)
3a 1 joke 2 wrong 3 face 4 sorry 5 power 6 far
3b **Suggested answers**
1 take you far 2 get on the wrong side of 3 have power over 4 be taken at face value 5 playing a joke on me 6 feel sorry for

Photocopiable activity

Activity 3A could be used any time after this section. It is a groupwork activity to review vocabulary and aspects of text cohesion for reading and writing purposes. Students match sentences from different texts and complete with the missing words.

▶ **Student's Resource Book, page 28**

Listening 1 p.43

One way to begin would be to ask students to compare the words *signature* (anyone's name written a distinctive way to identify them or show they have read/written something) and *autograph* (famous person's signature given as a souvenir). They may translate into the same word in the students' language.

1 Ask students to look at the photo and say what they know about the person. Ask if they can read the signature and why they think that some people collect autographs.

Background

The photo is of Ray Charles (1930–2004) who is famous for being the 'father of soul music'. He went blind at the age of six, but went on to have a highly successful musical career.

2a Read the rubric together and ask students to predict the order of the four topics covered. It is reasonable to assume that the talk will follow a conventional chronological order, from how he started, to how it developed, to his feelings now. Multiple-choice questions test sections of a text in order. The text contains signals for each section, and in interviews, the signals are often in the questions.

2b Students listen for the gist. Play the recording once through without stopping and ask students to identify the content of each section. Get them to identify the words or phrases that helped them. They can relate any details they remember, but emphasise that the purpose at this stage was to get a general understanding and order of the content.

3a Students read the rubric and the first question with its three possible answers. Ask if they expect the speaker to use exactly the same words as in the answer. Hopefully they will realise that, as the section is identifying paraphrase, he will not. Elicit examples of how one or two of the answers could be paraphrased. (What would be another way of saying *to please/impress someone, to be like his friends*?)

3b When students have listened to the first part and answered the question, they should match the answers to the explanations and check the script.

4 Give students a minute to read the questions and answers before they listen to the rest of the interview. Ask them if they can remember the paraphrase used in each answer.

5 Ask students to brainstorm the pros and cons of Charlie's job on the board before going on to the second question. What is the most unusual job the class can come up with?

Key

2a 4, 2, 1, 3

2b The questions shape the structure of the interview (e.g. first question *What started you off as an autograph collector?* signals that he is going to speak about how he started collecting).

3a B

3b 1 C – He collected something else (model planes) to be like his friends.
2 A – His father (*my dad*) gave him his first autograph, but that's not why he continued to collect them.
3 B – The autograph impressed people (*made people … look up to me*).

4a 2 C – *Today, looking back, it makes me blush …*
3 B – *Once I realised there was a financial angle to it all, collecting became that much more entertaining.*
4 B – *I like the fact that … I don't want to do this today, I don't have to.*

Language development 1 p.44

1a As with previous Language development sections, the exercise could be done first as test to see how much they know before using the Writing reference to explain the mistakes. Alternatively, students could start by reading the Writing reference and then refer back to it as they do the exercise. Problems with capitals vary according to L1 influences. Generally, they are not a problem at the start of a sentence, but are often missed in names of days/months and for nationality adjectives. Adding the full stops should help students to think about how a piece of writing should be separated into distinct sentences.

1b Brainstorm other uses than those in this text or get students to find them in the Writing reference.

2a Students should answer the five questions by referring to the five sentences a–e. The main problems with apostrophes are usually confusing *its* and *it's* and including an apostrophe where none is required (e.g. *apple's 50p*). Point out that irregular plurals form possessives in the normal way (e.g. *children's, men's*).

2b The trickiest point in this text is probably the possessive form of *Dr Curtius*. Also watch out for 1800s – it is a common error to include an apostrophe here when it is actually a plural. The word *o'clock* requires an apostrophe because it is a contraction of *of the clock*; however, this is never used in its uncontracted form.

Background

The famous wax museum was founded by Marie Tussaud; therefore the apostrophe in *Madame Tussaud's (wax museum)* is correct. However, it is often seen without the apostrophe, even on the museum's own website and literature.

3 Commas are harder, as they are often required in complex grammatical structures, and so correct use of the commas comes with understanding the way the clauses of those structures function. There is more work on the use of commas in subsequent units, for example with conditional and cleft sentences. The use of a comma before *and* in lists is considered acceptable by some people (e.g. *London, New York, Tokyo, and Rome*).

4a Speech marks can be singular (') or double ("") but must be consistent throughout a piece of writing. The points to emphasise are: the use of capitals to begin a quote, the use of commas to separate the direct quote from the rest of the sentence and the fact that other punctuation comes inside the speech marks.

Background

Fred Allen (1894–1956) was a popular radio comedian in USA in the 1930s and 40s, moving to TV and films in the 50s. He has a long list of humorous quotes.

4b Get students to skim the dialogue for general understanding before they try to punctuate it. Hopefully they should realise that it is meant to be humorous!

5 These three punctuation marks are less common but useful nevertheless. The uses of dashes and semi-colons sometimes overlap. It would be helpful to draw students' attention to examples of their use in subsequent modules.

Extra!
An extension exercise would be to get students to complete these sentences from the board using their own ideas, taking care to use the correct punctuation.
1 My ambition is simple ...
2 The things that interest me are ...
3 My motto in life is ...

Key

1a People interested in celebrity should visit the popular London attraction Madame Tussaud's wax museum on Marylebone Road. It features wax images of people from all walks of life: from modern music superstars (e.g. Miss Dynamite) and Hollywood legends to scientists such as British astrophysicist Prof. Steven Hawking. Avoid Saturdays in July and August. The winter months are much quieter.

1b In Exercise 1a, capitals are used for: starting sentences, cities, names, titles, roads, nationalities, days, months.
They are also used for: countries, geographical features (rivers, mountains, etc.), some abbreviations (e.g. BBC, UNESCO) and, importantly, the pronoun *I*.

2a 1 contractions, possession
2 just use apostrophe, no extra *s*
3 one sister, more than one brother
4 *it's*: contraction of *it is; its*: possessive pronoun
5 *actors*: plural needs no apostrophe; *actor's*: possessive of *actor*

2b Madame Tussaud's is one of London's oldest attractions. One of its most popular displays is the Chamber of Horrors. The collection was started in Paris by Marie Tussaud's mother's employer, a Dr Curtius. Marie brought Dr Curtius' original collection of heads to London in the early 1800s, and it's been constantly updated ever since. It's only two minutes' walk from Baker Street tube station. But don't forget that the museum shuts at six o'clock!

3a 1 in lists (the 'serial comma')
2 to separate subordinate clause from main clause when the subordinate clause comes first
3 before question tag
4 after introductory adverb or adverbial phrase
5 non-defining relative clause

3b 1 *not necessary*
2 An autograph collector needs a notebook, a couple of pens, a camera and a lot of patience.
3 Tell me, what are you going to do next?
4 *not necessary*
5 I managed to get a ticket, believe it or not.
6 Pierce Brosnan, the actor who played James Bond, is Irish.

4b **A**: 'A table for two,' said the celebrity, 'in a quiet corner preferably.'
B: 'Sorry, sir,' replied the waiter, 'we're full.'
A: 'Do you know who I am?' asked the surprised celebrity.
B: 'No, sir,' said the waiter, 'but if you ask your mother, I'm sure she'll tell you.'

5b 1 To be successful, you need three things: talent, determination and good luck. (*to introduce a list*)
2 I'd like to see the show again – in fact, I'm going to book tickets tomorrow. (*adding extra information*)
3 Katie is a great actress; she has sensitivity and a good voice. (*closely linked points*)

▶ **Student's Resource Book, page 29**

Use of English 1 p.45

1 Elicit names of the people in the photo or get students to skim the text to find out. If the difficulties raised in question 2 have not come up in previous discussions, spend a few minutes talking about them now.

2a Students skim the text for general understanding. Remind them to focus on the text that is there and not to worry about the gaps at this stage.

2b Read the rubric together and then the task strategies. Remind them to check each gap for both the form of the word required and whether it is positive or negative.

Do question 1 together, using the Help point as an example, before giving students a suitable time limit to complete the task.

3 The task analysis should highlight the changes students are required to make in this task. The task tests formation of all types of words, especially nouns. Language development 1 in Module 4 looks at word formation in more detail.

4 You could also ask students if they know of any other lasting relationships, celebrity or otherwise.

Key

1 1 Paul Newman and Joanne Woodward
2a Because of the pressures Paul's fame had on their family life.
2b **1** unpredictable **2** stability **3** glamorous **4** difficulties **5** uncomfortable **6** popularity **7** professional **8** unacceptably **9** partnership **10** performance
3 **1** 1, 5, 8,
2 2, 4, 6, 9, 10

Writing 1 p.46

1 Start by asking students what, if anything, they understand by the term *coherence*. Then get them to read the Writing strategy notes. An understanding of coherence is also important for Paper 1, as Part 2 relies to some extent on understanding the sequence of information and how the paragraphs link together. Get students to analyse both texts, highlighting the good and bad points.

2 Students read the four ways of organising a text and identify which way was used in the first text. Different ways of organisation will be appropriate for different pieces of writing.

3 Rewriting the second paragraph could be done step by step, or the students read the four steps and then be left to rewrite the paragraph, comparing their attempts at the end.

4 One approach to this writing task would be to get students to plan it in class, possibly in pairs, leaving the actual writing to be done at home.

Key

1 A (the ideas are in a logical sequence with linking expressions)
2 The information is presented in time order.
3 **Suggested answer**
If you want to become famous, you need to think positively. You have to believe that you deserve success. Therefore, the first thing I would do if I wanted to become famous would be to decide how I was going to achieve it. Then I would set myself a small number of daily priorities and make sure I started to reach my goals. Finally, the most important message to myself would be 'Never give up!'

▶ **Student's Resource Book, page 31**

3B What I believe in

One way to begin would be to compare *to believe someone/something* and *to believe in someone/something*.

Introduce the topic of public protest by getting students to discuss the picture, which shows people campaigning against a third runway at Heathrow airport.

Listening 2 p.47

1 Start by getting students to skim the task to establish what it consists of: three unrelated extracts, with two, three-option multiple-choice questions per extract. Read the strategy for dealing with the task and give students a minute to read the questions before they listen.

One option would be to stop after the first extract to compare answers and review strategy.

2 Students could be given a copy of the audio script for the analysis to highlight how the information they were listening for was presented and how the distracters worked.

The paper tests mainly feeling, attitude, opinion, purpose and gist rather than specific facts.

3 You could start the discussion by asking students which of the three causes they would be most/least likely to support. Then ask for others. If appropriate ask students if they have ever participated in a public protest of any type.

Key

1 1 B – *we actively encourage a way of living in which no creature has to suffer needlessly.*
2 C – *we also offer free talks to schools ... It's a particularly worthwhile thing to do*
3 C – *it was just that one of my friends talked me into keeping her company*
4 B – *The actual arrangements left a bit to be desired*
5 A – *to protest about the firm not offering services in the Welsh language.*
6 A – *I still think that by fighting for the things you believe in, ... you can contribute to global justice.*

Speaking p.48

Following on from the discussion after the listening, if any of the class has demonstrated for/against something, you could start by asking them what exactly they did to protest. Try to elicit some of the items in the pictures.

1a When students have matched the vocabulary to the illustrations, spend a few minutes focusing on the five verb + noun collocations (*sign + petition, hand out + leaflet*, etc.). The five forms of protest are good examples of strong collocations. In the context of protesting, it would be easy to guess the nouns in each case (e.g. *sign a ... , hold a ... , write in to a ...*). Remind students to learn the vocabulary as single chunks.

1b Students could choose the correct preposition before checking their answers in a dictionary. Give a few examples of how the first seven can be used with things or people (e.g. *put pressure on the company/the boss, have an influence on the leader/decision, generate publicity for the candidate/cause, contribute to the appeal*), but the last three can only be used with things.

1c Students may have to use their imaginations here to think about reasons for and impact of protests. One approach might be to divide the class into five groups and give them an illustration each, asking them to come up with a scenario that answers the three questions. Then judge/vote on the most likely/imaginative/amusing, etc.

2 The missing words/phrases are from either Exercise 1a or 1b. Tell students that they might need to change the tense of the verbs.

3a/b Elicit or remind students of the format of Paper 5 Part 3. It is a collaborative task with the candidates discussing a task together and aiming at a consensus. Give students a few minutes to read the task strategy before they listen to the instructions. The emphasis is on the two parts of the task.

3c/d Play the recording a couple of times if necessary for students to analyse the task.

4a One way to do this would be to stop the recording after each sentence that begins with one of the phrases in the box and establish which of the five functions it has. Another approach would be to get students to number the phrases 1–5 as they hear them and let them compare at the end. Encourage students to think of the longer phrases as 'chunks' with lots of linking of the words (e.g. *As_a matter_of_fact*) to make them sound natural.

4b Students identify the option that can't be used and why.

5 Students use the same pictures to do the task themselves in pairs.

6 The task analysis should link back to the points in the task strategy. If the discussion went on a lot longer than four minutes, remind students that in the exam they need to be conscious of the time if they are to cover both parts.

Key

1a 1 B 2 D 3 C 4 E 5 A
1b 1 on 2 in 3 about 4 on 5 on 6 for 7 on 8 down 9 to; about 10 to
2 1 took part in a march 2 change 3 minds 4 back down 5 held a meeting 6 put forward their views on / express their opinion on 7 hand out 8 generate publicity for 9 sign a petition 10 put pressure on
3b 1 You have to (1) talk about the advantages and disadvantages of each method of showing your feelings about different issues; (2) decide which would be the most effective
2 About four minutes
3d They carried out the task very successfully. They discussed the first part of the task before coming to a conclusion. Each speaker participated fully, but did not dominate, and they encouraged each other to speak.
4a 1 qualifying – *Mind you …; Having said that, …*
2 emphasising – *Of course, …; As a matter of fact, …*
3 adding – *Besides that, …; Not only that, …; As well as that …*
4 disagreeing – *Well, actually, …; As a matter of fact …*
5 moving on – *Anyway, …; Anyhow, …*
4b 1 Anyway 2 Having said that 3 Actually

Photocopiable activity

Activity 3B is designed to work here. It is a whole-class discussion of hypothetical situations involving moral dilemmas, designed to practise discourse markers, getting a point across and paraphrasing.

Language development 2 p.50

Start by reminding students of the theme of the listening on page 47 and ask them to complete the sentence. 'Sometimes you ______ stand up for your principals.' (Answer: *have to*) Elicit other ways of expressing necessity, prohibition, advice and permission.

1a Explain that 'semi-modal' refers to structures like *have to* that are not true modals because they differ grammatically but express the same functions. Remind students that different modals can express the same function, and tell them that they should be able to find two different ways to rewrite each of the sentences. If necessary, do question 1 together first. Students can use the Grammar reference to check their answers.

1b Students use a suitable modal/semi-modal structure to complete the transformations.

1c Other questions that would practise the language would be:

What have you done recently that you didn't have to do?
What do you have to do before the end of the month?
Have you made any mistakes in the last few days?
What advice would you give a classmate about learning to use modal verbs?

2 This exercise focuses on ability and related meanings. The greatest difficulties tend to be in referring to the past. As students correct the sentences, get them to think of reasons why the sentences are not possible as written.

3 *Will* and related forms (*shall, won't, would*) is often over used as a future form and underused for talking about intention, volition and habitual activities. This exercise is designed to highlight some of those uses.

Students should identify the time reference and purpose of each sentence before using one of the forms to complete it.

4 Introduce the idea of expressing modality lexically by asking students how they express the meaning of *must*. From *obligation*, elicit the verb *oblige* and its passive form *be obliged to do something*. Then elicit other verbs that express the same concept (*required, made, forced,* etc.). Point out that many of the structures are passive, especially the more formal ones, as we are usually more interested in the action than who requires it.

Extra!
A follow-up here, especially in an English-speaking country, would be to get students to look for notices outside the classroom and note down how they express modal concepts.

Key

1a 1 You **can't/mustn't** demonstrate here.
2 I think **you should/ought** to go on strike.
3 We **must/have to** have a vote before we call a strike.
4 We **don't have to/don't need to/needn't** send all these letters today.
5 I think it's too late to protest now, you **ought to/should have protested** before.

1b 1 must/have to speak
2 don't have to be in a union
3 had to go on strike
4 needn't have closed / didn't have to close
5 didn't have to go
6 could have moved on
7 should / ought to have been back

2 1 Harry started collecting last year and so far **has been able to** raise £10,000 for charity. (use *be able to* for present perfect)
2 We couldn't get to meet the minister yesterday but eventually we **were able to/managed to** speak to him on the phone. (*could* is not possible for past ability at a specific time)
3 Jack **could have photocopied** the leaflets at the office yesterday, but he forgot. (to express something that was possible but didn't happen)
4 They say there **could/might** be nearly half a million people on the march tomorrow. (*could* or *might* for possibility)
5 Why not come with us on the demo? You **might** enjoy it. (possibility)
6 The damage might **have been** caused by the people who were demonstrating. (modal + *have* + past participle when referring to the past)
7 You must **have known** there was going to be trouble when you saw the crowds. (past deduction)

3 1 *will* (future intention/fact) 2 *would* (past habit) 3 *won't* (refusal) 4 *would*; *wouldn't* (characteristic annoying activity) 5 *shall* (offer) 6 *Would* (request) 7 *won't* (refusal) 8 *will* (annoying habit) 9 *would* (prediction) 10 *will* (obligation/order)

4 1 Visitors **are required to** report to reception on arrival. (formal)
2 You **are under no obligation to** answer the following questions. (quite formal)
3 **I'd better** phone home and tell them that I'm going to be late. (informal)
4 I **felt obliged to** invite my cousins to our wedding. (neutral)
5 All library books **are to be returned** by the end of term. (very formal)
6 **It is forbidden to use** mobile phones in this area. (formal)
7 **It is advisable to** take out insurance when travelling abroad. (formal)

Photocopiable activity

Activity 3C would work well here. It is a variation of the traditional game 'Bingo', designed to revise the form and use of modals and semi-modals.

▶ **Student's Resource Book, pages 33–34**

Use of English 2 p.51

The task will be new to most students so spend a while reading the task strategy to establish exactly what needs to be done. Establish that they must find one word that fits in the gaps of all three sentences.

1a Ask them first to think about what type of word is required: a verb in the infinitive after the modals *might*, *should* and *could*. Elicit possible words for the first sentence (*relieve, reduce, break, etc.*). Then ask which of them fits the second sentence (*break*) and confirm it also fits the third (*to break out* – to start suddenly).

1b Give students a couple of minutes to do the next part using the Help clues before checking the answers.

2 Give students three minutes to complete the task.

3 The task analysis should highlight the fact that Paper 3 Part 4 is testing various aspects of vocabulary such as homonyms (different meanings of words with the same spelling and pronunciation), idioms (e.g. *look on the bright side*), phrasal verbs (e.g. *break out*) and collocation (e.g. *running water* / *running battle*).

Key

1b 1 body 2 bright
2 3 drawn 4 running 5 personally
3 idioms: break bad news; bright ideas
phrasal verbs: break out; draw out

Writing 2 p.52

1 Start by getting students in small groups to come up with ways that anyone can help to raise money for a good cause. Get examples of activities they have been involved in.

Background

The photo on this page shows a Coinstar machine. These are often located in supermarkets; their main purpose is to enable people to convert coins into notes without going to a bank, but partnerships with charities also give users the option to donate their coins to charity, in this case Cancer Research UK.

2 Remind students that the Paper 2 Part 1 task is compulsory and that it involves combining information from more than one source. Give students a couple of minutes to firstly read the task, underlining the key points, and secondly look at the input material. They can then answer the six questions designed to help them plan the report.

3 Students could be left to work through the steps in a–e individually or there could be some feedback after each stage to check that everyone is heading in the right direction.

4a Elicit the fact that the formality and impersonal nature of such reports makes the use of the passive common.

4b Remind students that many of these sentence stems can be adapted for a wide variety of reports and that there are clearly too many to include in one composition.

5/6 Now that the report has been so carefully planned, 20–30 minutes should be plenty of time to write and check it.

Key

1 **Suggested answers**
Sponsorship, jumble sales / car-boot sales / garage sales (where used articles are sold); dances, dinners, talent shows where admission is charged; going round knocking on doors; doing errands for people (e.g. washing car, gardening, taking dog for a walk); auctions, etc.

2 **1** The director of a charity has written to you (a helper) asking you to write a report about a recent fund-raising day.
2 The Board of Governors will want a clear summary of what happened, whether or not the event was successful and any recommendations for the future.
3 An overview of the day, who was involved, how the money was raised, recommendations for the future.
4 The pie chart shows the relative percentages of how the money was raised (i.e. where it came from).
5 The style will be impersonal/formal.
6 Whether it is clear and succinct, well laid out (clear headings) with techniques (e.g. bullet points) which help the reader to read it.

3 Suggested answer

JULY FUND-RAISING DAY

Introduction

This report is intended to:

1 give an overview of our recent fund-raising day for your charity;

2 indicate who raised the money and how;

3 make recommendations for next year.

Overview

On 22 July this year, a substantial sum of money was raised for the charity. The total raised for disadvantaged children exceeded the sum achieved last year. In the main, the day was very successful.

Participation

Most of the 50 volunteers were students from the university.

How the money was raised

The largest part of the total raised (40%) came from visiting houses and knocking on doors. Collections were also carried out in the street, accounting for 30%, and we raised a further 20% from a jumble sale. The remainder came from a variety of sources.

Recommendations

In the light of this year's experience, I would make the following recommendations:

1 that we organise a wider variety of activities, including street parties (if we could make them work successfully);

2 since some members of the public thought our street collectors were not legitimate fund raisers, we should, in future, issue them with special badges to avoid misunderstanding.

Conclusion

To sum up, I would say that this year's fund-raising day was a great success and that we could make it even more successful next year.

(220 words)

▶ **Student's Resource Book, page 38**

Module 3: Review p.54

1 **1** A **2** C **3** C **4** D **5** C **6** A **7** B **8** D **9** A **10** C

2 **1** The **arrogance of some politicians** is breathtaking.
2 She was jealous of her **husband's popularity**.
3 Switzerland preserved **its neutrality** throughout both World Wars.
4 There were **many contradictions** in what he said.
5 Their **partnership has lasted** a long time.
6 Getting a fine is an **inconvenience**, but nothing more.
7 Her fame gives her a **feeling/sense of isolation**.
8 He was proud of his **(many) achievements**.
9 She'd never expected so **much devotion from** so many fans.
10 Can you **put your signature** on this form?

3 **1** must/should **2** Can/May **3** have **4** could **5** been **6** better **7** able **8** shouldn't

4 *All the President's Men*, directed by Alan Pakula and starring Robert Redford and Dustin Hoffman, is about two young reporters from the *Washington Post* who, after a lengthy investigation, discover that President Nixon had been lying to the nation about a break-in that occurred in the Democratic Party offices in the Watergate Hotel. Pakula's film, praised by everyone for its acting, won many plaudits from the critics, including Vincent Canby, the film critic of the *New York Times*, who said, 'In my view, no film has come so close to being such an accurate picture of American journalism at its best.'

Module 4 Life's rich tapestry

This module includes topics such as relationships, personality, decision-making, types of intelligence and memory.

Lead-in p.55

With books closed, put students into small groups to brainstorm different people they have relationships with (e.g. within family, at work – colleagues / boss, different types of friends, neighbours, teachers, shop assistants). See which group can get the most in two minutes. Then ask them in what ways the relationships differ.

Alternatively, ask them to draw some concentric circles with the word *me* in the centre and then add the names/titles of people they have relationships with to represent how important those relationships are to them. In pairs, they should choose a couple of examples to explain why.

Then ask students to open their books, look at the pictures and discuss the questions. Get the students to give as much information as possible. Are the two boys brothers, twins, identical/non-identical twins? Are the other two colleagues? Equals? Is she telling/showing him what to do? Or what she has done? etc.

4A Making choices

One way to begin would be, again with books closed, ask the students to think of any famous partnerships or double acts that they can think of (give hints as to areas of work: comedy, design, fiction, cinema, retail, etc.). Alternatively, give them half of the double act (choose suitable pairs according to where you are teaching, their experience or what is topical) and see if they can come up with the other partner, e.g. Laurel and … (Hardy), Simon and … (Garfunkel), Butch Cassidy and … (the Sundance kid), Bonnie and … (Clyde), Batman and … (Robin), Nureyev and … (Fonteyn), Lennon and … (McCartney), Rogers and … (Hammerstein), Sherlock Holmes and … (Dr Watson), Don Quixote and … (Sancho Panza), Fred Astaire and … (Ginger Rogers), Adam and … (Eve), Samson and … (Delilah), Starsky and … (Hutch), Tom and … (Jerry), Marks and … (Spencer), etc.

Reading p.56

1 Get students to look at the photos and headings. See if they know the saying where the main heading comes from (*Two's company, three's a crowd*) and predict the answers to the subheading.

2 Emphasise the importance here of just skimming. They should not be answering questions yet, just getting an idea of the style of the article as a whole and content of each section. If necessary, give a time limit to skim the article as a whole and then immediately ask for answers. Alternatively, do it section by section with 20–30 seconds to skim before asking what the important 'ingredient' is, and then quickly repeating with the next. Don't give students long to keeping searching for an answer after the allotted skimming time. Finding one point for each section is enough.

3 Start by reading the rubric. Then look at the task strategies on page 168 together. Check that students know what they have to do by asking one or two concept questions (e.g. Will the answers in the text use the same words as in the questions? – No, a paraphrase). Do question 1 together. Get students to underline the key words in the question (*have a good effect*) and the words in text D that express the same idea (*He can make me less impatient and I can make him less hesitant).* NB It might be worth pointing out that in the exam there is no example question. Give the students an appropriate time limit (15 minutes) to answer the questions before they compare and justify their answers.

4 To justify their answers, get them to underline the relevant sentences in the text. Spend some time focusing on the alternative ways to say the same thing and comparing the items of vocabulary that do the same (e.g. *important – crucial, trivial argument – squabble*) in terms of meaning, connotation or register.

5 Students could discuss the first question as a whole-class activity before doing the second one in small groups.

Background

Ian Hislop is the editor of the satirical magazine *Private Eye* and appears regularly on British TV.

Extra!
Other discussion questions could be

- Do you think you work better on your own or if you have someone else to bounce ideas off?
- Do you have an idea for setting up a business? Who would you like to have as a partner?
- If there is time, students could create a business plan for their idea and present it to the class a few days later.

Key

3 **2** B (line 33) *it's healthy to have a bit of confrontation*
3 A (line 14) *you've the right to say … without one of us feeling crushed by it.*
4 D (line 57) *we hardly ever see each other socially*
5 B (line 18) *the fact that we were meant to work together*
6 B (line 23) *Stefano dislikes the business side of things, and I would rather stay in the background when it comes to public relations.*
7 C (the paragraph starting at line 41)
8 A (line 11) *I sometimes appear to be the one chosen to do a particular job.*
9 C (line 51) a *very clear demarcation of the areas we're in charge of*
10 A (line 8) sq*uabbling about who left the top off a pen*
11 A (line 6) *it doesn't feel like work at all*
12 D (line 54) *it's crucial to have a partner who you can bounce ideas off*
13 D (line 61) *we help each other to be objective about things*
14 B (line 30) *We take a cocktail approach … red … white … end up pink!*
15 B (line 21) *we have a common purpose*

Photocopiable activity

Activity 4A could be used here or after Vocabulary. It is a pairwork activity in which students match short paragraphs to sentences that summarise the main meaning of the each paragraph.

Vocabulary p.58

1a Ask students to read the extract and identify idiomatic expressions. Discuss what effect using these types of idiomatic expression has on producing this type of writing.

1b As all the expressions are from the text, there are two possible approaches here. One would be to get students to scan the text to find the answers; the other would be to give them an opportunity to guess/remember the answers before checking them and finally scanning the text to see them in context. A number of follow-ups are possible, depending on the teaching situation. Students could be asked to come up with paraphrases for the expressions (e.g. 1 *big-headed*, 2 *argue*). Alternatively, ask them to compare them with similar expressions in their own language and decide whether they translate the same way. Additionally, discuss the usage notes in the key (see below).

2 The expressions highlighted here are used to add emphasis, to say how much something happens. Get students to compare a simple phrase such as *A is cheaper than B* with *A is a bit/slightly/much/ considerably cheaper than B* and the effect the adverbs have on meaning and style. It is unlikely that these will be new to students, but the idea is to highlight the usage.

2a Students should first see the phrase in context before they mark the meaning. See notes and alternatives in the key which could be elicited from the class. Students should be very familiar with absolutely before an extreme adjective, but will probably be less familiar with its use here at the end of the clause.

2b Students should skim the text in 30 seconds before they start. Get students to see that most of the sentences work with nothing in the gaps; the added words add emphasis and degree. The exception is question 1 where *barely* (like *hardly*) has a negative meaning and therefore totally changes the sense.

2c Focus here on the grammar of the expressions.

3 Start by getting students to look at the picture and skim the text. When they have completed the text with the correct prepositions, get them to identify which are prepositional phrases and which are dependent prepositions (following a noun or verb).

Key

1a *hit it off* – get on well , have a good relationship
from the word go – immediately, right from the start
the chemistry's working – there is a mutual attraction/rapport

1b **1** full of – To say someone is *full of X* is common in both positive (e.g. *hope/good ideas*) and negative expressions (e.g. *rubbish*)
2 have
3 take – Can be used in positive form with a negative connotation (e.g. *he takes himself/his work too seriously*)
4 toes
5 bed – Always used in the negative. The word *bed* refers here to a flower bed.
6 sense
7 eye – Note that *eye* is countable, whereas *vision* and *sight* are uncountable.
8 heated – Compare a *heated discussion* (= passionate) with *a heavy discussion* (= about a serious topic) and *a hot debate* (= about something topical).
9 out – Multiword verbs are types of idiomatic expressions

2a 1 not upset – not one bit (NB only used with questions and negative statements)
2 much – *also by far, by a long shot*
3 quite a lot – compare the difference when stress is on *fair* (cautious/guarded) to when it is on bit (positive/upbeat)
4 100% – *totally/completely* would work in the same way here, too.
2b 1 barely 2 quite 3 absolutely 4 a fair bit 5 by a mile 6 a bit of 7 at all 8 full of
2c The expressions are used in the text to modify verbs, adjectives, noun clauses and comparatives.
Used after a verb clause: *at all, absolutely*
Used before a verb clause: *barely*
Used with a comparative: *by a mile*
Used before an adjective: *quite*
Used before a noun clause: *a fair bit, full of*
3a 1 *on* – on a regular basis
2 *in* – in the business
3 *on* – an effect on something
4 *between* – rivalry between
5 *of* – full of something
6 *with* – deal with
7 *in* – in the background
8 *in* – in perspective

▶ **Student's Resource Book, page 39**

Listening 1 p.59

1 Start by eliciting the job title *sales manager* and what the job might involve (leading a team of sales reps, dealing with customers, thinking up new promotional ideas, etc.). Ask students to look at the list of words and say if they are positive or negative. Check the pronunciation, especially the word stress, before students work in small groups to choose the top four. Groups could then compare their list justifying their choices with examples.

2 Give students a minute to read the rubric and three questions before they hear the first part. The idea is that students use the introduction to get a general understanding and scope of the programme. This task is preparation for multiple matching, where the order of information in the questions is not the same as in the recording. There are four people to whom the information has to be matched, plus overlap and distracters. This exercise requires understanding gist meaning and paraphrase.

3a/b Remind students that in Part 4 they will need to identify and interpret information given by the speakers to complete a multiple-matching task. This exercise is designed to help practise that skill. They should read the rubric before listening to the second section twice, either listening for strong points the first time and weak points the second time, or answering both sections the first time and checking their answers on the second time.

3c This section highlights the parts of the recording that should have helped students to answer the questions. All the speakers use interesting idiomatic expressions to express the point summarised more succinctly and formally in the question. Get students to match the notes to the person who said it and to which point A–L it corresponds.

4a/b Depending on the class and time available, this could be a quick discussion or a longer, more detailed activity if they start to include their own ideas about what effect the loss of each one will have or what each of the managers is likely to do in the future.

4c Listen to the final part and see if any of the groups had the same combination. They could compare his decision with their own.

Key

2 1 Tom has to decide who to make redundant.
2 Four people are involved in his decision, and he has to make two of them redundant.
3 Corporate printing, packaging, digital supplies and labels

3a

Mike	Strong point: D	Weak point: K
Joanne	Strong point: F	Weak point: H
Jason	Strong point: A	Weak point: L
Carol	Strong point: C	Weak point: J

3c 1 Jason (L) 2 Carol (C) 3 Carol (J) 4 Jason (A) 5 Mike (K) 6 Joanne (F) 7 Mike (D) 8 Joanne (H)

Language development 1 p.60

The focus here is on word formation using affixation, an area that is directly tested in Paper 3 Part 3, but is clearly also relevant in other papers.

Start by looking at the summary box together. Establish that suffixes are added to the end of a word and usually change the word class, and that prefixes are added to the start of words and often change the meaning of a word. Also emphasise that changes might involve a combination of the four change types, including adding more than one suffix e.g. *fortune* (n.) – *fortunate* (adj.) – *fortunately* (adv.), and that the addition of affixes might involve other spelling changes, e.g. dropping the *e* in *fortunate* or adding an extra *l* in *travel* (v.) – *traveller* (n.).

1a The aim here is to highlight the range of suffixes that can be used for changes between different word classes and their associated spelling changes. Point out that sometimes two adjectives can be formed with different meanings or uses (e.g. *amuse – amused/amusing*) or that two nouns can sometimes be formed from a verb (e.g. from verb *teach* –

teacher for the person and *teaching* for the idea). Some suffixes can be used for more than one word class (e.g. *-ing* to form both nouns and adjectives).

Students should use dictionaries to find the words needed or to check the spelling of them, where necessary. Spend some time on the pronunciation of the words, highlighting those where there is a change in stress position.

1b Students should copy the table into their notebooks before using the words they found in Exercise 1a to complete it. Then identify any inherent meanings in the suffixes as in the example (e.g. if someone is *dependable* you are *able* to *depend* on them). Note how the verb participles (*-ed* and *-ing*) are commonly used to form adjectives from verbs.

2 The internal changes here all involve changes from verbs and adjectives to nouns, although in the exam the changes could be in either direction!

3 Look at the table together, highlighting the meaning of each group of prefixes and emphasising the range of word classes they are used with; for example, the negative prefixes with adjectives (*unbelievable*), but also with nouns (*non-fiction*) and verbs (*disobey*); the prefix *mis-* with the verb (*misunderstand*) but also with the noun (*misunderstanding*) and adjective (*misunderstood*). Students should then add the extra words to a larger version of the table in their notebooks. Point out that some of the words can be used with more than one prefix (e.g. *print – misprint, reprint*).

4 The emphasis here is on learner training. The two words given are just examples of how students could record a complete word family. Another way would be do it in table form. The point is to add as many words as they can to complete the group, to compare and contrast words of the same class and to organise the words in some way. As a follow-up, get students to do the same with other words such as *profession, connect* or persuade.

Key

1a **1** amused/amusing, collapsible, dependable/dependent, different, hesitating/hesitant, influential, pleasing/pleased, productive
2 affectionate, aggressive, dangerous, energetic, envious/enviable, funny, historical/historic, hopeful/hopeless, successful
3 amusement, confrontation, decision, defence, discovery, participation/participant, persistence, pleasure, safety/saviour
4 accuracy, cruelty, confidence, diversity, happiness, jealousy, loneliness, popularity, tolerance
5 beautify, deepen, generalise, legalise, popularise, strengthen, widen

1b **1** Suffixes: *-ing, -ible, -able, -ent, -ant, -ful, -ial, -ed, -ive*
Examples: *amusing, collapsible, dependable, dependent, hesitant, influential, pleased, productive*
2 Suffixes: *-ate, -ive, -ous, -etic, -c(al), -ful, -less*
Examples: *affectionate, aggressive, dangerous, energetic, historic(al), hopeful, successful, hopeful, defenceless*
3 Suffixes: *-ment, -ation, -ion, -ce, -y, -ness, -ant, -ence, -ure, -iour*
Examples: *amusement, confrontation, decision, defence, discovery, participant, persistence, pleasure, saviour*
4 Suffixes: *-cy, -ty, -ence, -ity, -y, -ness, -ity, -ance*
Examples: *accuracy, cruelty, confidence, happiness, diversity, jealousy, loneliness, popularity, tolerance*
5 Suffixes: *-en, -ise*
Examples: *deepen, generalise*
6 Suffixes: *-ify, -en*
Examples: *beautify, strengthen*

2 breadth, choice, death, flight, height, length, proof, strength, success

3
• disappear, unbelievable, non-fiction, non-conformist, unpopulated, irreversible, insecure
• misunderstand, misprint
• co-develop, co-education, co-exist, co-worker
• enlarge, endanger, enrich, empower
• replace, reappear, rearrange, redrawn, reprint, redevelop
• underpaid, undercook, underdevelop, underpopulated
• overpaid, overcook, overdevelop, overdrawn, overpopulated, overpower, overprint
• prearrange, pre-cook, prefix, predate, pre-exist, pre-school

4 **1** envy **2** enviable **3** unenviable **4** enviably **5** envious **6** enviously **7** hesitation **8** hesitancy **9** hesitant **10** hesitantly

Photocopiable activity

Activity 4B can be used at any time after this section. It is a board game activity to review affixation and word formation in which students identify and correct the wrong word forms.

Use of English 1 p.61

1 Start with books closed and ask the class a few questions about their shopping habits and their attitude towards shopping. Then ask them to briefly discuss the question together.

2 Find out what students know/remember about Paper 3 Part 3. Elicit that it always consists of a text with a total of 10 questions; that the base words will require changes to make them fit grammatically as well as to the sense of the sentence/text; that each word might require a number of changes; that at least one word in the text will require a prefix; and that in the exam they would have about 10 minutes to complete the task.

2a Students skim the text, ignoring the gaps and discuss what they recognise about themselves.

2b First, read the rubric, then the task strategy on page 169. Look at the example together and the process of identifying the correct word; before a noun (shoppers), the word is likely to be an adjective; it does not need to be made negative. Remind students to look out for negatives and plurals.

3 When they have completed the task, they could discuss the questions in the task analysis and which words helped them to decide the class of word required.

4 The discussion could include how people shop (e.g. at home, at night, online) as well as the increase in sites such as eBay where people buy from other individuals.

Key

2b **1** disastrous **2** exceptionally **3** curiosity **4** temptation **5** restricted **6** substantial **7** decisive **8** uncharacteristically **9** persuasion **10** introductory

3 **1** 4, 9
2 1, 6

Writing 1 p.62

One way to begin would be to ask students to discuss in groups whether their parents and teachers were strict or relaxed with them when they were younger and how much freedom they were given.

1a Students should start by looking at the two paragraphs and answering the three questions about each one to establish the style and purpose of each text.

1b Having identified the style, purpose and target readership of each paragraph, students can choose an appropriate opening for them.

1c The idea here is to identify certain techniques and the terms used to describe them. Students identify five types of sentence.

1d These are some other ways of making informal writing more interesting.

2a Students skim the second paragraph and decide which paragraph it follows on from and what features link the two.

2b/c Before students rewrite the second paragraph to fit the second text, quickly recap some of the features in Exercises 1a–1d that they should try to incorporate. Look at the phrases and elicit how they might be used, for example by asking what they mean (talk things through = discuss the consequences of something).

Key

1a **A 1** a formal article **2** to educate/advise/inform **3** other teachers
B 1 a less formal article/report **2** to complain/criticise **3** other teenagers

1b **A 2** – the other opening is too informal, not directly relevant and more illustrative /less factual.
B 1 – the other is too formal, too distant (written in third person) and less emotional

1c **a** – A2 **b** – B1 **c** – B2 **d** – A1 **e** – B2

1d Intensifying adverbs (*immensely*), adjectives (*unthinking*), mixture of short and long sentences (*Maybe we are.*), lively expressions (*easily led astray, it's hardly surprising if*)

2a A – formal in vocabulary (e.g. *belittle*) and style (e.g. use of passive – *Advice and guidance need to be given*), aimed at teachers about children

2c **Suggested answer**

Basically, when it comes to making choices, our parents need to give us more freedom than they often do. Perhaps we'll make mistakes. Fine, then we'll talk things through and try and learn from them. Maybe they should start by letting us make smaller choices first, to give us a chance to show that we can really be trusted to be responsible. And if our decisions turn out to be the right decisions, how good it would be if our parents told us so! Then maybe they could say to us, 'Next time there's a big choice to be made, we'll try leaving it to you!'
(102 words)

4B Human nature

Start by asking students to talk for a few minutes about any brothers and sisters they have and whether they are older or younger or their experience of being an only child. Elicit or teach the word *sibling*.

Listening 2 p.63

1 Remind students that Part 3 is a conversation lasting about four minutes; the task is either sentence completion or multiple choice. This exercise is the latter, with a choice of four options to complete sentence stems or answer questions. Students will need to understand specific information as well as opinions and attitudes.

Give students a moment to read the task strategy on page 171 and the rubric, but ask them not to read any further. Elicit a suitable strategy. One approach is to start by reading only the stems/questions before listening for the first time. As they listen, students should try to note down the answers in their own words. Then, before it is repeated, they look at the options, choosing the closest. Use the second listening to check the answers.

2 Give students a few minutes to compare and justify their choices before giving them the answers. Exercise 3b on page 43 gives examples of different types of distracters that they could identify here. If there is time, play the recording once more, stopping after each of the interviewer's questions and showing how they reword but signal the different questions in the task.

3 In the discussion, students can refer to people in their extended families (their own siblings or their cousins), as well as to what they know of their friends and how they relate to their siblings. On the basis that people are similar to their friends, find out if there is any tendency for older siblings to have best friends who are also first-born children and for younger siblings to have best friends with older brothers and sisters.

Key

1 1 B – *at the time of each birth, parents are in a new situation … they relate to the new child in a different way.*
2 C – *the oldest child typically dislikes and avoids changes and risk-taking*
3 C – *they seem to cultivate different personality characteristics or skills.*
4 B – *how great an impact … is mainly related to the gap between them*
5 D – *adopt specific patterns of behaviour learned from the people who take care of them*
6 A – *help people become aware of why they think and behave in the way they do*

Speaking p.64

The section starts with vocabulary development on the topic of families before using the theme to give practice in Paper 5 Part 2, the individual long turn.

1a Ask students to look at the photos for a moment before discussing the six sentences. Then comment on some of the phrases used. Note the dependent prepositions (e.g. *engrossed in*), the collocations (e.g. *a caring parent*) and the language of conjecture (*They obviously, They probably, It looks as if*).

1b Get students to note the dependent prepositions and emphasise that the topic is the relationship between the individuals as they try to use the expressions in complete sentences.

2a This looks at similar and/or confusing words and phrases. As well as asking students to identify the correct one in each case, ask them to give examples of how the other might be used.

2b Ask students to focus on the box containing additional idiomatic phrases (including some multiword verbs) that relate to relationships, and then to match them to the phrases in italics in the sentences below. Do the first one together as an example.

2c This discussion could be done immediately after the preceding exercises or the following day as a review/revision of the vocabulary.

3a Remind students that in Part 3 they need to react and respond to the pictures, not just describe them. Their comments should include speculation and hypothesis as well as description. Remind them of the comments in Exercise 1a.

3b Elicit what the first thing in the task will be (compare and contrast the pictures) before they listen for the second thing.

3c/d Students listen to her answer and discuss it first in terms of content then focusing on her task achievement. Tell them not to worry about grammar and vocabulary, etc.

4 Apart from missing one part of the task, it is a good attempt. Ask students to look at the extracts from what she said and predict what the missing words are. Play the recording again for students to complete the text and highlight the phrases she uses to speculate or hypothesise.

5a It would be useful to move students around at this stage to get them to do the task with someone they are less used to working with. One can be the examiner starting the other off and stopping him/her after one minute.

5b This reflects the way Part 2 moves into the two-way discussion in Part 3. There is obviously no correct answer; it is the flow of ideas and the interaction that are important.

6 Encourage students to be honest about themselves and their partners.

Key

1a **Suggested answers**
1 A, B, C **2** A, B, C **3** D **4** B **5** A, B, C, D
6 A, B

2a **1** *tightly-knit* (= closely connected) – compare *intimate* (private);
strong bond (= emotionally) – compare *tight bond* (physically);
we relied on each other – compare *each relied on the other*;
have interests in common – compare *share* ***the*** *same interests*
2 *extended* (definition beyond parents and siblings) – compare *increased* (bigger)
only child – compare *single parent*
3 *of*
get on (= relationship) – compare *get by* (survive)
4 *expectations* (= beliefs) – compare *aspirations* (desires/ambitions)
conscientious (= diligent/hard-working) – compare *conscious* (aware/concerned)
disappoint them – compare *let them down* (separable phrasal verb so pronoun goes in the middle)
5 *sheep* – an idiom meaning *the rebel*
from the start – compare *from the word go*
do as + clause – compare *do anything that* + defining clause
have cross words (= argue) – compare *speak in any way*
healthy – a positive thing

2b **1** takes after **2** get their own way
3 looked up to **4** see eye to eye
5 runs in the family **6** lost touch
7 got on with **8** fell for

3b **1** compare and contrast the pictures
2 say how important the relationships are and how they might change

3c She talks about pictures A, B and C.

3d The candidate was a bit vague when saying how important the relationship is. She could have used clearer signposts for the examiner.

4 **1** there's definitely **2** might **3** obviously become less **4** bound to **5** may well
6 could affect

Photocopiable activity

Activity 4C can be used after this section; it is a pairwork activity to review useful vocabulary for talking about family relationships.

▶ **Student's Resource Book, page 43**

Language development 2 p.66

Probably the best way to begin is to analyse a noun clause on the board to show how a group of words can function like a noun. Then read the explanation of noun clauses in the summary box and look at the Grammar reference for examples of the various types. One of the main problems with noun clauses is confusing them with other clauses, such as relative clauses. Get students to compare:

1 *It is amazing that she knew the answer* – noun clause
2 *She was the only one that knew the answer* – relative clause
3 *I liked what she said* – noun clause (*what she said* = *it*)
4 *I didn't like the way that she said it* – relative clause (can omit *that*)

1 Ask students to skim the text for general content and say how teachers can use MI theories in their work. They should then study it more closely, looking for examples of the three basic types of clauses.

2 Ask students to look at each sentence and decide which type of clause is possible before choosing the correct word in each pair. Before checking the answers, give students time to compare and justify their choices.

3 Start by asking students to skim the text and say what the writer's problem is and what he/she plans to do about it (*keeps forgetting things, would like to take a dietary supplement*). It is worth pointing out that it's slightly unnatural to have quite so many noun clauses in a text of this length. Secondly, as well as giving practice of noun clauses, it is also good practice for Paper 3 Part 2 (the open cloze), as these type of words are often tested there.

4 The exercise uses transformation to compare using different noun phrases to express similar ideas. After students have completed the exercise, ask them to compare each pair in terms of emphasis and degree of formality.

5 Students could spend some time writing the sentences in class or do it for homework. Elicit how to follow on from the first stem and give an example (*to* + infinitive + clause, e.g. *It's easy for me to understand why some students find noun clauses confusing*).

Key

1 1 a **evidence that** more people are becoming aware
b **encouraging that** more teachers are adopting
c **suggests that** people learn differently
2 a **What is good for one learner, How you teach**
b One feature ... **is how it identifies**
c depends **on who/what you are teaching**
3 a **Using words, To achieve the best results**
b If the goal **is to help** learners
c it is **important to adopt**, no **desire to change**

2 1 which – *wh-* clause as object of the verb, *which* because of limited choice. Question word clauses are related to questions (*which areas* not ~~*What areas*~~ *of the brain are associated with intelligence*?)
2 that – following an adjective
3 that – *that* clause as object of a verb, not related to a question
4 *both* – *that* clause after adjective (*likely*) or noun (*likelihood*)
5 *both* – degree (*how far*) or simply *yes/no* (*is intelligence affected by diet*).
6 *both* – *to* + infinitive (more formal) or -ing clause (less formal) as subject
7 what – *What have researchers been trying to discover?*
8 What – *what* clause as subject (*What we learn now*)

3 1 *how* – method 2 *where* – location 3 *who* – person 4 *which* 5 *whether* – with *or not* 6 *that* – after adjective 7 *what* – what time 8 *if/whether* – indirect question 9 *whether* – choice, before to + infinitive 10 *if/whether* – both possible when *or not* comes later in the phrase (but *if* would avoid repetition) 11 *that* – after adjective 12 *what* – what was the ending? 13 *How/Why* – as subject 14 *how* – after preposition 15 *that* – after noun + *be* 16 *what* – as subject (*what is it called?*)

4 1 parents to want 2 How parents; bring up 3 to be confused 4 Knowing who to turn to 5 It is important to understand 6 for parents to ignore 7 Why some people have children

▶ **Student's Resource Book, pages 44–45**

Use of English 2 p.67

Key word transformations were covered in Module 1B. Start by eliciting what students know about the task: eight questions, requiring candidates to complete the second sentence so the meaning is the same as the first, using between three and six words including the word given. The task tests both grammatical structures and vocabulary.

1a Look at the example and the two possible answers. Using concept questions (e.g. Are both answers grammatically correct? – yes; Do they use the word given? – yes; Do they use 3–6 words?) to establish that the first answer is correct and that the second is incorrect as it uses too many words.

1b Give students a suitable time limit (8–10 minutes) to complete the task before they compare answers.

2 Use the task analysis to review noun clauses and the grammatical (e.g. active – passive in example) and vocabulary (e.g. dependent prepositions in question 5 *prevented from*) areas covered.

Key

1a 1 How far our intelligence *is affected by environment is unclear* to scientists. [Adding *our* makes it seven words.]

1b 1 It *struck me that Jessica* is looking well these days.
2 Having a pet *to take care of makes* some people very happy.
3 There's a *strong likelihood/possibility that the concert* will be cancelled.
4 The *discovery of the letter was* an incredible surprise to us.
5 As a result of his back injury *he was prevented from playing* in the game.
6 Whether or not *you attend the meeting doesn't* really matter to me.
7 I'm sorry. I *take back everything/what I said* about your sister.
8 We *should cut down (on) the amount* of salt in our food according to Nick.

2 1 Noun clauses: Example, 1, 2, 3, 6
2 Vocabulary: 1, 2, 3, 5, 7, 8

Writing 2 p.68

1a The task here is a competition entry, so start by asking students of their experiences of entering/winning competitions of any type. Hopefully at least one student will have a success story of some type.

1b Within the broad description of a competition, there are a number of possible writing styles. This exercise highlights the range of format and styles and how they might differ.

2 As with all writing tasks, it is important to analyse the question carefully, to identify the content, purpose and style of the piece. When students have looked back at the task strategy, give them a few minutes to read the task, highlighting the important parts and to answer the five focus questions.

3a Students should follow the steps here and the advice on page 30 to plan the content and organisation of the article.

3b Both plans would be appropriate in this article.

4a Having planned the content and organisation, remind students that the next thing to do is plan the style and some of the language to include. Elicit ways of making informal writing more interesting. If necessary, refer back to the Writing strategy on page 62.

4b Students compare the openings and justify choices.

4c Spend a moment highlighting how noun clauses can be used to introduce the topic, give examples and make suggestions.

5/6 As with previous compositions, now give students an appropriate time to write the article (20–25 minutes) and 5–10 minutes to check it.

Key

1b
- a story: strong narrative coherence, interesting story (use of adverbs)
- an article: interesting points, coherent argument, helpful/useful for the reader
- a review: clear idea of what is being reviewed/clear (and balanced) opinion, will it help the reader decide to read/see it (opinion language)
- a description: does it bring the person/place to life, vivid picture (use of adjectives)

2 **1** Two main parts (importance of good memory vs. bad memory; suggestions)
2 To inform and entertain; the competition judges, and then the readers of the magazine
3 article
4 informal style (but not too informal)
5 engaging the readers and choosing examples/situations the reader can relate to; making useful suggestions

4b **1** b Asking the reader a question directly is a good way of opening an article.
2 a A clear statement is used as a 'topic' sentence. b) is probably too informal
3 b This is more of a topic sentence. a) sounds like the middle of a paragraph, not the topic sentence.
4 a *In short* helps connect the paragraph with the rest of the text and indicates to the reader the purpose of the paragraph.

5 Suggested answer

THE IMPORTANCE OF A GOOD MEMORY

Are you one of those lucky people with a good memory, someone who can remember everyone's name at a party and never forgets an appointment? Or do you have the misfortune to remember little of what you read and hear, and are generally absent-minded?

Having a good memory is useful at all stages of our life. Being able to remember facts and figures helps us sail through our exams at college, and when we are at work, we are superbly efficient. On the other hand, if we suffer from a poor memory, we struggle with our studies, writing out notes for ourselves hoping this will help the information to 'stick'. And of course at work, our office will be in a state of chaos, with little reminders and forgotten bits of paper all over the place.

But the idea that there is nothing we can do about a bad memory is wrong. For a start, when we want to remember something, for example someone's name, we should try to concentrate on it and use it as often as possible in conversation. But by far the most useful technique is to build up a visual picture around the name to create an amusing mental image of it. In other words, we can learn to remember the name by association.

In short, we might have a naturally poor memory but there are ways of improving it and remembering things we never thought possible. After all, once something is in our memory it takes an awful lot to forget it!

(261 words)

▶ **Student's Resource Book, page 47**

Module 4: Review p.70

1 1 C **2** B **3** A **4** D **5** C **6** A **7** C **8** D **9** A **10** B **11** C **12** D

2 **1** I was impressed with/by the breadth of his experience.
2 He was an only child.
3 It was a really tough decision.
4 There is a very close bond between all the brothers.
5 She was made redundant because of the cutbacks.
6 He is very/so full of himself.
7 Why have they fallen out?
8 In the end, I always get my own way.

3 **1** depth **2** energetic **3** productive **4** influential **5** actors **6** happiness **7** unhesitatingly **8** non-conformist **9** endanger(ed) **10** belief **11** confrontation **12** responsibility **13** enrich **14** lives **15** discoveries

4 **1** **Deciding** who to live with is the most important choice we make in life.
2 Some people really know how **to** dress in order to show their personality.
3 If you want to understand a person, look at **what** their life choices are. / … look at their life choices.
4 It's always difficult to decide what **to** spend our money on.
5 I know **that** if our choices reflect our personality, they turn out to be better choices.
6 It's interesting **that** decisions based on the opinions of others are usually wrong.
7 The question is **how far we** really have a choice in any particular situation. / The question is **whether** or not we really have a choice in any particular situation.
8 It's easy **to say** that we don't have a choice.

Exam practice 2 TRB p.184

Paper 1: Reading
1 D 2 C 3 A 4 D 5 A 6 B 7 C

Paper 3: Use of English
Part 2 **1** set **2** their **3** to **4** the **5** which **6** of **7** make **8** in **9** on **10** whose **11** of **12** What **13** bring **14** Although **15** were
Part 4 **1** draw **2** gain **3** slow **4** landed **5** point
Part 3 **1** daily **2** unexpected **3** unaware **4** embarrassment **5** volunteers **6** spontaneous **7** debatable **8** publicly **9** convincing **10** attraction

Paper 2: Writing
Suggested answer

Report on Information Day
By: Antonia Marinova

I have compiled this report using feedback from families who attended, plus my personal impressions. Although things generally went well, there is clearly room for improvement in some areas.

Timetable
The start time of 10.30 was probably too early, as not everyone had arrived in time for the informative opening speech. I suggest beginning at 11.30 in future. Also, we must be careful to limit the amount of time devoted to closing speeches. I noticed many people leaving before these had finished. In addition, the session on practical details might come better earlier in the day.

Video diaries
In general, these were popular, but we received complaints about sound and picture quality, and some people felt that one would have been enough. I suggest we update our equipment and, in future, check videos for quality and length in advance.

Question-and-answer session
Next year, we must be much more careful when choosing the families. Although it's important to show both sides of the situation, we must avoid families who had particularly negative experiences.

Conclusions
My impression was that the event was successful. If we make the changes I have suggested, it could go even better next year.

Paper 4: Listening
1 D 2 B 3 A 4 A 5 A 6 C

Module 5 Global issues

This module links together related topics such as the rise of slow cities (cities reacting against uniformity and other aspects of modern life), the development of a global brand, sensitive tourism, island communities and environmental concerns.

Lead-in p.71

Start with books closed and introduce the topic of globalisation. One way to do this (which also acts as a review of word formation in the previous module) would be to firstly elicit the noun *globe* (the world), then from that the adjective *global* (around the world), then the verb *globalise* (to spread around the world) and finally the fairly new noun *globalisation* (having business activities around the world).

Alternatively, give students a gapped definition and elicit the missing words, finishing by asking which word the definition represents, e.g. *The* [1]______ *of making something such as a* [2]______ *operate in a lot of* [3]______ *around the* [4]______ . (Answers: 1 process, 2 business, 3 countries, 4 world)

However it is done, they need to know the word and its meaning before looking at the pictures and discussing the three questions. Students might need some help with some of the concepts in the quote (e.g. growth of a material kind, a finite environment, evolving).

The photos show a woman in Kenya gathering the bean crop (left); the check-in area at London's Heathrow Airport (top right); and a monkey in Thailand drinking from a can bearing one of the most recognised logos in the world.

Background

E.F. Schumacher was born in Germany in 1911 and moved to Britain in the 1930s. By the 1950s, he had become a government economic advisor. While in Myanmar (Burma), he developed the principles of what he called 'Buddhist economics', based on the belief that good work was essential for proper human development and that 'production from local resources for local needs is the most rational way of economic life'. He became a pioneer of what is now called appropriate technology: earth- and user-friendly technology matched to the scale of community life. His bestselling and highly influential book *Small is Beautiful* was published in 1975. He died in 1977.

5A In the slow lane

With books closed, spend a few minutes eliciting examples of what students like or dislike about city life. Encourage them to think of big cities in general and make sure they know the word urban.

Reading p.72

1 Stress that students should only read the title and subheading at this stage before they discuss the three questions, which are designed to create interest in the text and to help them develop the skill of prediction.

2 Give students a suitable time limit to read the text quickly (e.g. two minutes), trying to find answers to the three questions above. They do not need to look at the missing paragraphs at this stage.

3 This is the second time students have looked at Paper 1 Part 2, as it also featured in Module 2 – the difference here is that, like the exam, there is one extra paragraph. Quickly look at the task strategy on page 168 together to remind students of the best approach.

3a Get students to read the paragraphs on either side of question 1 looking for linking words or ideas. Look at the highlighted words and elicit suggestions as to what they refer to. Find the most suitable paragraph and find the links within it.

3b Get students to follow the same process with the other five questions using the Help section if needed, before comparing answers. Remind them to read through the complete text once again to make sure it all fits together properly.

4 Ask students to go through the text highlighting and joining all the linking words, phrases and ideas.

5a Ask students to scan the complete text looking for examples of what Slow Cities are doing or have pledged to do.

5b Put the list on the board and check some items of vocabulary (e.g. *linger, frenzy, drift by*) before the discussion. Finally, ask students where they would prefer to live!

Key

3a/b 1 B ***They*** *were inspired by* refers back to *Bra and three other Italian towns;* ***The movement*** *was first seen* – refers back to *they gathered to the 'Slow' banner in droves.*
2 G *the enjoyment of food and wine … conviviality = long lunches*
3 C *To win* ***it*** = *the designation* Città Slow / *mark of quality; the … manifesto pledges* (line 28) refers back to the list in paragraph C
4 F *Again in keeping* refers back to *traditional dishes; a more traditional way of looking at life* = *Despite their longing for kinder, gentler times*
5 A *a balance between …* (line 46) = *this ideal combination*; the negative point after *Nevertheless* (line 48) contrasts with the successes in paragraph A
6 D *At the very least* contrasts with *it is hard to make people …* (line 54)

5a closed some streets to traffic, banned supermarket chains and neon signs, given the best sites to small family-run businesses, subsidised renovations that use regional material, serve traditional dishes in hospitals and schools using local produce, cut noise and traffic, increased green spaces and pedestrian zones, closed food shops on Thursdays and Sundays, opened the City Hall on Saturday mornings, encouraged use of local produce, promoted technology that protects the environment, preserved local aesthetic and culinary traditions, fostered spirit of hospitality and neighbourliness.

5b **The Slow City way of life**
linger over coffee
watch the world drift by
enjoyment of food and wine
fostering of conviviality
promotion of unique, high-quality and specialist foods
small family-run businesses
handwoven fabrics and speciality meats
materials typical of the region
traditional dishes
traditional architecture, crafts and cuisine
green spaces
pedestrian zones
The globalization of urban life
high-speed frenzy
frenetic way
processed meals
noise pollution
people shouting into mobile phones
speed through the streets
culture of speed
stress of urban living

Background

'Slow Cities' were inspired by the Slow Food movement to preserve their unique character. According to the Mayor of the Tuscan town of Greve, 'The American urban model has invaded our cities, making Italian towns look the same. We want to stop this kind of globalisation.' The Slow City programme involves enlarging parks and squares and making them greener, outlawing car alarms and other noise that disturbs the peace, and eliminating ugly TV aerials, advertising posters, neon signs. Other priorities include the use of recycling, alternative energy sources and ecological transportation systems. The movement rejects the notion that it is anti-progress and holds that technologies can be employed to improve the quality of life and the natural urban environment.
To qualify, a city must be vetted and regularly checked by inspectors to make sure they are living up to the Slow City standard of conduct.
The movement hopes eventually to build a worldwide membership inspired by the notion of 'civilized harmony and activity grounded in the serenity of everyday life by bringing together communities which share this ideal. The focus is on appreciation of the seasons and cycles of nature, cultivation of local produce and growing through slow, reflective living.' For more on the movement, including their charter of principles, see www.cittaslow.org.uk

The movement currently has about 100 cities in ten countries.

Vocabulary p.74

Knowledge of idiomatic phrases will help students in various parts of the exam. The text on pages 72–3 is rich in idioms; this exercise focuses on some of the more common ones that students should be able to use themselves.

1a Students identify and explain the idioms in the extract; one way would be by rewriting it using other expressions.

1b After students have matched the halves and identified the idiomatic expressions, they could find each one in the text and explain the meaning.

1c This is only meant to be a short discussion to reinforce the expressions and their meaning.

2a The correct words are all used in the text. The words to choose from in each question have a similar meaning or appearance, but only one is correct in the context. Remind students that they should be looking out for collocations, distractions and prepositions as well as nuances of meaning.

The exercise is good practice for Paper 3 Part 1. When they have done the exercise and compared answers, ask them why the other two words in each group are wrong or how they could be used.

2b This could be a quick discussion or expanded to include other aspects of government control of the freedoms of individuals.

3 The text contains a number of words/expressions derived from the word *live*. Students could start by finding them in the text – *to live* (lines 57/58), *good living* (line 47), *standard of living* (paragraph E), *life* (paragraph E), *quality of life* (line 22) – and explaining their meanings. Students could use dictionaries, either to help them answer or to check their answers when they have finished. Encourage them to think of the differences in meaning or use of each item.

If there is time, ask students some questions to reinforce the vocabulary. For example, if they are studying abroad, ask how the cost/standard of living compares with their country, how their lifestyle has changed since arriving, etc.

4a When students have looked at the extract and the meanings of the two verbs, discuss their connotation. Point out how *gulp* has a negative connotation, whereas *sip* is very positive; also how this creates attitudes towards the two drinks – coke and wine – and adds to the overall feel of the text.

4b Students could either think about the meanings of the verbs before they try to match them to the complements or use the complements to help them deduce the meaning after they have matched them.

Extra!
Expand the exercise by eliciting other foods or situations they would use the verbs with or by getting students to mime some of them.

Key

1a turning the clock back (also turn back the clock, put the clock back) = to return to a good situation experienced in the past.
strike a balance = to give the correct amount of importance/attention to two things

1b **1** f – get away from it all (line 2)
2 a – disturb the peace (line 51)
3 h – have a long way to go (line 48)
4 c – swim against the tide (paragraph G)
5 b – slowly but surely (line 27)
6 d – quality not quantity (line 37)
7 g – come in droves (also with other verbs, e.g. *gather*, *arrive*, etc.) (paragraph B)
8 e – to strike the right balance between A and B (line 46)

2a **1** promote (+ positive) = encourage – cf. provoke (+ negative), proceed (continue)
2 pledged = promised (what and when) – cf. decided (an idea), wished (not possible in present perfect)
3 ban = prevent/prohibit something – cf. bar someone, censor books/films, etc.
4 implement (a policy) = make changes (collocation) – cf. carry off out
5 curb = control/limit – cf. bring down (a number/amount), crack down on
6 epitomises = is a typical example of – cf. equalises (make the same), expands (make larger)
7 cut = reduce – cf. slice (cut thinly), chop (cut into pieces)
8 preserve = keep/maintain as is – cf. serve (help/assist), reserve (keep for the future/a particular purpose)

3 **1** **a** cost of living – the expense
b standard of living – the quality
2 **a** way of life – of a community over a longer period
b lifestyle – of an individual, easy to change
3 **a** livelihood – source of your income
b living – what you earn to live on
4 **a** alive – predicative (after a verb, not used before a noun)
b living – attributive (only used before a noun)
5 **a** lifelong – (adj.) continuing/existing all through your life
b lifetime – (n.) period of time that someone is alive
6 **a** live out – to live until the end of your life
b outlive – to live longer (than someone else)

4a gulp: to swallow something quickly
sip: to drink something slowing, taking small mouthfuls

4b **1** e – noisily
2 c – using muscle of mouth through a small hole
3 b – drink large mouthfuls, especially from a bottle
4 g/h – eat quickly, in big pieces because you are very hungry
5 a – taking very small bites
6 g/h – eat small amounts because you are not hungry or don't like/want it
7 f – (negative) eat or drink a lot of something quickly and eagerly
8 i – with teeth
9 d – empty the glass because you are thirsty

▶ **Student's Resource Book, page 50**

Listening 1 p.75

1 Start by looking at the advertisement and eliciting which company it is for (Nike /'naɪkiː/). Ensure students know the meaning of *logo* and *slogan*. To check, ask for other examples. It doesn't matter if students don't know much about the company, as they will hear all they need to in the listening exercise. It might be interesting to do a quick survey to ask everyone in the class to estimate how many items of clothing, footwear or equipment they have made by Nike to give an indication of the company's size and influence.

2a Ask students what they can learn from the first sentence (1 It is a talk therefore a monologue, not a discussion; 2 It is about the company Nike; 3 It will be a man's voice). The first time they listen is to identify the four parts of the talk and to summarise them in three or four words. The idea here is to raise awareness of text structure and of signals (e.g. time references, discourse markers) which indicate this.

2b Again, emphasise how these four sentences divide the listening up and clearly mark the content of each section.

3a Check that students are clear what is meant by *stressed* before they listen to the first part of the talk again. Note that it is repeated on the recording and that although all the words listed are used in the recording, students only need to identify those that are stressed. This is to draw attention to the fact that, in the exam, the words they need to do the task are likely to be stressed, as they provide key information in the text.

3b/c Having heard the piece twice, students should have no difficulty answering these questions. The point to emphasise is how knowing what the pronouns and other references refer to may be the key to getting the correct answer in the exam.

4 Give students a moment to read the notes before they hear parts 2–4 again and complete questions 3–6.

5 This could be done by dividing the class in two and getting one half to think of positive points and the other half negative points.

Photocopiable activity

Activity 5A can be used after this section or anywhere else in the module. It is a simulation focusing on the theme of big-business ethics and globalisation.

Key

2a **Suggested answers**
1 The history of Nike
3 Accusations against Nike
4 How Nike has/have responded

2b **1** d **2** c **3** a **4** b

3a **1** Blue Ribbon Sports **2** Nike **3** logo

3b **1** it/the company = Nike **2** it/it = the logo

3c **1** Nike **2** (Nike) logo

4a **3** footwear **4** working conditions **5** child labour **6** Opportunity International

Use of English 1 p.76

1 Brainstorm the positive and negative effects of tourism on both large and small communities. It might be better to focus on the positive points, as many of the negative points are covered in the text.

2a Students should skim the text to get a general understanding and answer the two questions.

2b Look at the example together and establish why B is correct (it is the availability of cheap travel that has lead to mass tourism, not its suitability or convenience, and it collocates with *readily*). Give students 10–15 minutes to complete the task.

3 This is designed to reinforce the idea that collocation is important in this task.

4 Have students answer *yes* or *no* to the three questions in the survey on page 206. Using a show of hands, estimate the percentage of *yes* answers. Compare with the results at the bottom of page 207. Note: The survey was probably carried out with a much wider age range than in the class.

Key

1 **Suggested answers**
Positive: creates work (many jobs needed in tourist industry), good for local economy (tourists spend money and brings foreign exchange into the country), educational (people learn about other places, languages, etc.), helps to preserve and protect monuments, traditions and cultures, as that's what tourists like to see)
Negative: increases local prices (tourists willing to pay more than locals), uses valuable resources (water, land, food, etc. go to tourists instead of locals), creates pollution, disturbs wildlife, destroys local culture (everywhere becomes the same), jobs are seasonal and low-skilled, tourists don't respect the places they visit (can offend local people)

2a Problem: Tourism damages local communities
Solution: Involve local people and respect their rights
2b **1** A **2** D **3** B **4** B **5** A **6** D **7** C **8** B **9** D **10** C **11** A **12** C
3 **1** 7, 8, 9
2 1, 2, 5

Language development 1 p.77

This section reviews the concept of gradable and ungradable adjectives and the adverbs that can be used to modify them. However, it also links to the previous section in that much of time the combinations are due to collocation.

1a Start by looking at the extract and the three questions. Students should establish that there are two types of adjectives, which use different adverbs to modify them. If students have difficulty with the concept, spend some time going through the Grammar reference on page 180.

1b When students have corrected the mistakes, give them a moment to check their answers against the Grammar reference before giving them feedback.

2 Remind students of some of the adverb + adjective collocations from the English in Use text on the previous page (*readily available, poorly paid*). Ask them to skim the text and answer two questions: *What is the situation on the islands? What do the local people want to do?* (Oil has been discovered, they want to get rich). Students might find dictionaries helpful for this exercise. For example, in the *Longman Exams Dictionary*, under the definition of *slow*, it highlights the collocation *painfully slow* in the examples.

Background

São Tomé and Príncipe form a country in the Gulf of Guinea off the west African coast with a population of about 175,000. Uninhabited until 1471, it was a Portuguese colony until independence in 1975. The people and culture are a mix of African (Angolan) and European.

3a Students should start by looking at the adjectives in the second box and identifying any that have similar meanings (e.g. *furious* + *irate*) and any that are extreme versions of another (e.g. *unusual* + *unique*, *similar* + *identical*, *pleased* + *delighted*). Then when they have skimmed the passage, ask them to complete each gap with a suitable adjective only. Finally, get them to go through it again, adding adverbs to each gap to modify the adjectives. To drive the point home, ask them to compare the versions with and without the adverbs and say in what ways the fuller text is better.

4 This exercise could be done in class or at home. Stress to students that they should try to stick to the theme of the module when they complete the sentences in ways that are true for them.

Photocopiable activity

Activity 5B is designed to be used after this section. It is a pairwork activity to practice modifiers with gradable/ungradable adjectives. Students search for adjectives in a word grid and match them to a suitable modifier.

Key

1a **1** to intensify the adjective
2 *very, really*
3 used only with ungradable adjectives
1b **1** 3
2 *very/vitally important or absolutely essential*
3 *absolutely unique* (ungradable)
4 3
5 *very/extremely rich* (gradable)
6 *absolutely furious* (ungradable)
7 *utterly/entirely opposed* (more formal)
2 **1** B **2** A **3** C **4** B **5** C **6** A **7** C **8** B **9** C **10** B **11** A **12** B **13** C
3a **1** absolutely/completely/pretty furious
2 pretty/quite/fairly cheap **3** virtually empty
4 rather/really pleased **5** fairly/quite/pretty similar **6** almost/virtually identical
7 quite/slightly/somewhat unusual
8 totally unique

► **Student's Resource Book, pages 51–52**

Writing 1 p.78

Start by looking at the picture and finding out if anyone in the class has been there or knows anything about it.

Background

Uluru, known for many years to English-speaking Australians as Ayers Rock, is 450km south-west of Alice Springs, the largest town in central Australia. Sometimes described as the world's largest monolith (block of stone used for religious purposes), it is 335m high and 9km around its base.

1 Students did a Paper 2 Part 1 task in Module 3. As it is a compulsory question, it is worth spending some time on. This section deals with the sub-skills of selecting the key relevant information and ordering it logically to answer the question.

1a When students have read the three pieces of input, comment on the different styles of the three types of writing. Check for any unknown vocabulary (e.g. *sacred*, *vandalism*).

1b Get students to find the answers to the questions by highlighting relevant parts of the input.

2a/b Divide the points into three or four paragraphs and produce a plan.

3 Students use their plans to produce a letter of 180–220 words.

Key

1b **1** It's been a sacred site for them for over 20,000 years.
2 Benefits: provides work, educates visitors
3 Discourage the current type, encourage respectful tourism
4 To increase visitor numbers.
5 For the Anangu people not to restrict tourism.
6 More control for local people in tourism and no climbing.

2a Notes that could form basis of sample answer:

Paragraph 1
saw your recent editorial … writing to say that I cannot agree with your criticisms of the Anangu policy … important site
Uluru is a sacred site for the A, who have been there for thousands of years.

Paragraph 2
Tourism is important, but needs to be managed so that these sacred sites are respected

Paragraph 4
If government wants to increase tourism to Uluru, they must do this together with A; A are trying to do this by encouraging greater understanding … through Culture Centre.
In my opinion, A should have greater control over tour promotion so they get a fairer share of profits. At the moment, only get 24%.

3 Suggested answer

I saw your recent editorial about the Uluru National Park and I am writing to say I cannot agree with your criticisms of the Anangu policy of restricting tourism. Uluru is a sacred site for the Anangu, who have been there for over 20,000 years, and while 400,000 tourists a year may not be huge numbers in national terms, it is still a lot.

I agree that tourism is important to the Anangu, but it needs to be managed so that its sacred sites are respected. For example, the Anangu themselves never climb the rock, and in my view, tourists should be prevented from doing so out of respect for the Anangu's traditions.

I also think the Anangu should have a much greater role in how the site is promoted and in tour organisation. At the moment, most Anangu are only able to get low-paid, seasonal employment, and the profits from tourism go to outsiders. However, they are trying, through the Culture Centre, to encourage a greater understanding of all aspects of their culture, not just the rock.

If the Government wants to increase tourism to Uluru, which seems to be its aim, it must do so together with the Anangu and implement measures to reduce environmental damage.

(208 words)

5B A fight for survival

The sub-skills focused on in this section will help students with the Part 1 tasks in Paper 4 of the exam.

Ask the class if they have ever studied history, whether they enjoyed it and what people can learn from studying the past.

Listening 2 p.79

1a Find out if any of them know where Easter Island is or anything else about it. Read the rubric and spend a few minutes getting possible answers to the two questions.

1b The sentences don't give direct answers, but students may be able to infer information.

Background

Easter Island, or Rapa Nui (also the name of the people and their language), was discovered by Europeans on Easter Day 1722 by the Dutch sailor Roggeveen. The island is now part of Chile, its nearest inhabited neighbour, 3,500km away, and has a current population of about 1,900, having dropped to a few hundred after young men were taken to America as slaves and European diseases ravaged the population. Apart from a small area reserved for the indigenous people, it is mainly used for sheep and cattle grazing. There are about 100 of the statues, which are between 3 and 12 metres high, standing.

2 More formal talks start with a clear introduction which summarises the main point of the talk. This helps the listener to orient him/herself to what they are going to hear. Give students a moment to read the introduction and check what they have picked up (e.g. Is it a monologue or a conversation?, Is he talking about Easter Island today?). Look at the task strategies on pages 170 and 171. Remind them they will hear the recording twice and that the information for the answers will come in the same order as the questions; but the sentences summarise the information in the recording, so students should listen for similar ideas. Look at the first question and elicit what type of word students should be listening for (a two or more syllable adjective after *the most*).

3 Remind students that well-organised talks are clearly divided into sections, and each section will be signalled by some kind of discourse marker. When the students have heard the recording twice, give them a short time to check their answers in groups before giving them feedback. Pick up on some comparisons of how the ideas were expressed on the recording against the summary (e.g. *intertribal competition – competition between rival tribes*).

4 Allow some time for students to discuss the questions. Students might know of other events from the past that serve as a warning or that people could learn from.

Key

2 **1** isolated **2** coast **3** protection **4** potatoes; birds **5** competition **6** transport **7** population **8** caves

3 **1** The talk is divided up into clear sections containing one main idea, with one question per section being the norm. Signals that help students to identify when the information they need is going to come up include sentences that introduce the sections, and parallel phrases to those in the questions.
Question 1 Introduction describing Easter Island: *this tiny Pacific island*
Question 2 *The focus of social life were the stone platforms called* ahu *… , these platforms were constructed …*
Question 3 *the fact that they all face inwards …*
Question 4 *At first, the islanders had no problems finding food …*
Question 5 *Because of easy access to stone, the statues were always at least five metres high, but …*
Question 6 *The amazing thing is how these huge statues got from the quarry where they were carved to the stone platforms …*
Question 7 *By this date, many of the trees …*
Question 8 *the absence of trees also led to soil erosion, so that plant and animal species became extinct, … When the first Europeans arrived in 1722 …*
2 The words were mainly nouns. This is a sub-answer of Q3

Speaking p.80

This continues the theme of environmental problems from the listening and contains a lot of related vocabulary before practising it in an exam format.

1a Ask students not to look at the vocabulary in the box when they first talk about the pictures. The idea is just to get the concept of each, i.e. climate change leading to flooding, dumping of rubbish leading to pollution, vehicle fumes leading to air pollution and then to global warming, etc.

1b Then give them sufficient time to work through the list and match the items to the pictures, using dictionaries where necessary. Highlight both individual words and collocations (e.g. *trigger allergies, build-up of greenhouse gases*). NB a *GMO* is a *genetically modified organism*; *DNA* stands for *deoxyribonucleic acid.*

1c Students should first try to match the captions to the pictures before completing them with suitable expressions.

1d There is obviously plenty of scope for argument here. Equally, it is too vast a subject for many students to have firm ideas on. Another question might be *Which of the problems do you think it would be easier to tackle?*

2 Remind students that, in the exam, Part 4 is an extension of Part 3, a three-way discussion developing the theme of the collaborative task. This section gives practice in both.

2a Start by looking at the task strategy for Paper 5 Part 3 on page 172.

2b Let students listen to the examiner giving the instructions for the task. Point out that much of it is fixed (*Now I'd like you to discuss something between yourselves ...*) and that the important part is signalled with something like *Talk to each other about ...*

2c/d The students should comment first on what the candidates said, then on how successfully they achieved the task. Did they speak about both parts? Did they reach a conclusion? Did they keep the conversation going? Were there any long pauses? Was there much interaction and turn taking/giving?

3a Ask students to complete the notes as they hear the recording again. Some of these structures (e.g. *the more ... the more ..., hotter and hotter*), are focused on later in the review of comparative and superlative structures in Language development 2 of Module 8.

3b Don't let this discussion go on too long, as it is only a warm-up for when they do the exam task themselves.

4a/b It would be a good idea to give students new partners at this point to do the exam task. They should feed back in one or two sentences.

5 Look briefly at the task strategy for Part 4 on page 172. At this point, the interlocutor (examiner) joins the discussion and extends it. Ask them to look at the questions and notice how they move into areas more directly related to the students' experience (*where you live, ... do you think ...*). Give them four or five minutes to continue the discussion, again concentrating on keeping the discussion going and turn-taking.

Key

1a/b **Photo A: Waste disposal/recycling**
bio-degradable materials
dispose of / dump (waste/rubbish)
incinerate/recycle waste
use up/run out of/conserve resources
Photo B: Air pollution
contaminate water supplies
destroy the ozone layer
give off/emit carbon dioxide/toxic fumes
pollute the atmosphere
trigger allergies
Photo C: Climate change
build-up of greenhouse gases
global warming
lead to/run the risk of famine/drought/flooding
Photo D: Species loss/deforestation
bio-diversity
deforestation
extinction
loss of natural habitat
wipe out/kill off (animals/fish/birds)
Photo E: Genetically modified crops
become pest/disease resistant
DNA technology
genetically modified crops
health hazards
improve flavour/nutrition
spray crops (with pesticides)

1c **1** E 1 recycled 2 bio-degradable 3 conserve 4 resources
2 A 5 give off 6 pollute
3 D 7 build-up 8 global warming 9/10 famine/flooding
4 B 11 genetically modified 12 pesticides 13 health hazards
5 C 14 Deforestation 15 habitat loss/loss of habitat 16 extinction

2b **1** Talk about relative importance of problems
2 Decide which is most urgent to address

2c air pollution

2d They cover both aspects of the task. Student 2 is better at turn-taking, responding to what student 1 says and asking questions. Student 1 tends to dominate, cuts student 1 off and does not invite responses from her.

3a **1** the more **2** and more **3** anywhere near **4** hotter and hotter **5** pretty much **6** nearly **7** by far the most

Use of English 2 p.82

Remind students what they have to do in Paper 3 Part 4.

1a Look at the first sentence in the example and establish that all three words (*strange/odd/peculiar*) would fit here. Then elicit which of the words fits in the second sentence (*odd*) and what it means in this context (not a matching pair). Finally elicit whether it fits in the last sentence and what the word means in that context (occasional).

1b Give students five minutes to complete the task and then let them compare answers.

The two verbs used for diseases referred to in the Help for question 2 are *catch* and *contract* (more formal). The three verbs used with hold are: *catch, take* and *get*.

2 The task analysis again draws students' attention to types of collocation and idioms tested in this part of the paper.

Key

1a odd

1b 1 sense 2 caught 3 date 4 beaten 5 company.

2 Idioms: Question 1 – not much sense in, a sense of panic; Question 2 – catch hold of something; Question 5 – to be in good company

Language development 2 p.83

This section assumes students at this level are familiar with the idea and basic forms of conditionals and focuses on more advanced uses and forms. If some students find the first part difficult, it might need to be supplemented with extra practice from elsewhere in the four basic types.

1a This is a review of the basic types of conditional sentences ('zero, first, second and third' or 'real, unreal (present) and unreal (past)'). However, the examples use less typical structures than would normally be found in grammar books (e.g. in question 1 *might* not *would*, in question 2 past tenses not present, in question 4 present continuous not *will*). Give students a chance to match the examples to the explanations before comparing and looking at the Grammar reference on page 180. Use some concept questions to check their meaning. For example:

1 Are the farmers happy? Has it rained enough? Is it likely to?
2 Did it happen once or many times? Does it happen now?
3 Is the town flooded? Why not? (It stopped raining.)
4 Am I definitely staying at home tomorrow? Why will I? Is it likely to snow?

1b Advise the students to look carefully at the whole sentence and to think about both the time reference and whether the situation is real or not before they choose an answer. Allow some time for discussion and justification before giving them feedback. Manipulate the six sentences to compare other forms with different meanings.

2a Tell students that these two sentences give examples of the two types of mixed conditional that are most common. Ask students to identify how each is made up and why.

2b They will need to identify the time of each part as they complete the sentences.

2c Encourage students to use their imaginations to complete the sentences in either thought-provoking or amusing ways and in various time combinations.

3 Start by asking students if they enjoy air travel and eliciting some negative aspects of it (cost, noise, pollution, space needed for airports).

3a Look back at sentence 1, question 1 in the Use of English exercise on the previous page. Ask students how it could be reworded and establish that *unless* is an alternative to *if*, in this case *if not*. Do question 1 together and draw students' attention to the other changes that were necessary (removing *don't*) before they work through the rest, referring to the Grammar reference if necessary.

3b Also ask them how they would feel if a new airport was built near where they lived.

4a When the students have compared the first pair of sentences and established that b is more formal, ask them to complete the exercise and justify their answers in pairs.

4b When students have completed the three sentences, reinforce the idea of when to use the structures by eliciting where they might have come from (e.g. sentence 2 could be from a notice in a hotel room), establishing that they are all quite formal.

Key

1a 1 c – second conditional for unreal situations in present or future
2 a – zero conditional for real, repeated situations that are always true
3 d – third conditional for unreal situations in the past
4 b – first conditional for real situations in the future

1b 1 If it **snows** again this week, the match on Saturday **could** be cancelled. (real possibility)
2 *If it **rains** during the night* (future), *the ground **might be*** (possible) *too wet to play*. (cf. If it ***had rained*** *during the night* (past) the ground **might have been** too wet to play.)
3 *It was a good holiday* (past), *but if it **had been*** (unreal past) *sunnier, I would **have enjoyed*** (unreal past) *it more*. (cf. It is a good holiday (present still there), *but **if it was** sunnier **I would enjoy** it more* (but it isn't).)
4 *If there is an avalanche warning* (future real possibility), *I **won't** go near the mountain* (result – 1st conditional) (cf. *If there is a warning, I **don't** go near the mountain* (regular timeless event).)
5 *If the typhoon **had hit** the island* (unreal past – it didn't), *everything **would have been** destroyed* (it wasn't). (cf. *If the typhoon **hit** the island* (unreal future), *everything **would be** destroyed*.)
6 *If a giant hailstone **hit** you* (past tense for unlikely future possibility), *it **would** hurt* (second conditional)! (cf. *If a giant hailstone **hits** you, it **hurts*** (general truth)! or *If a hailstone **had hit** you, it **would have hurt*** (unreal past)!)

2a **1** 2nd for the condition (unreal present – I'm not afraid and never am) + 3rd for the result (unreal past – I did go out)
2 3rd for condition (unreal past – we didn't listen) + 2nd for result (unreal present – result of past inaction is that we are in trouble now)

2b **1** *had not destroyed* (unreal past); *would/might still live/be living* (unreal present)
2 *were not* (always); *would not have built* (past)
3 *would not be* (present); *had not erected* (past)
4 *did not use* (present); *would not have killed off* (past – already)
5 *had invested* (past); *would/might not be* (present)

3a **1** Air travel will continue to grow *unless* economic conditions deteriorate again.
2 The total number of flights will *not* decline *unless* fares rise steeply.
3 Many people will continue to fly *whether* the price of tickets goes up *or not.* / *whether or not* the price of tickets goes up.
4 Targets for reducing atmospheric pollution can be met, *provided that* air travel is dramatically reduced.
5 Most people are in favour of new airports, *as long as* they are not built near their own homes.
6 A new airport might already have been built, *but for* opposition from local groups.
7 People should protest, *otherwise* the new airport will go ahead.

4a Sentence (b) in each case omits *if* and inverts subject and verb. Additionally, in sentence 3 *was* changes to *were*. It is formal, whereas sentence (a) is neutral in register.

4b **1** *Had* (formal letter) **2** *Should* (formal notice) **3** *Were* (formal letter/report)

Photocopiable activity

Activity 5C can be used at any time after this section. It is designed to recycle and practise language from across the module. Students can work either individually or in small groups.

▶ **Student's Resource Book, page 56**

Writing 2 p.84

Start by establishing the difference between *litter* (unwanted items dropped in a public place) and *rubbish* (UK), *trash* (USA) and *refuse* (/ˈrefjuːs/) (unwanted items put in 'official receptacles').

1 Give students a few minutes to discuss the two questions without looking at the task.

2 Check the Writing reference on page 192 if necessary. Get students to read the task, including the input, before they answer the six questions.

3a If there is time, let students look back at the notes on planning and coherence on pages 30 and 46 before they start. First, they highlight the key points.

3b Students should think about the number and scope of each paragraph.

3c Match the points highlighted in Exercise 3a to the paragraphs in Exercise 3b.

3d Discuss the pros and cons of each title and ask if anyone has a better one.

4a The focus here is on style. The six sentences demonstrate different ways to achieve an informative, lively, informal style.

4b As before, point out that they will not be able to use all the structures in one article.

5/6 It might be useful to let students write the article in their own time, but then spend some time in class going through the checklist with them before they hand in their work for marking.

Key

2 **1** Student representative on the Health and Safety committee; article for student newspaper
2 To inform students about Clean-up Day and the campaign
3 What happened on Clean-up Day; what the effect has been; what the principal's most popular new policy is; which of her measures have been unpopular
4 The context: who *you* are, what has happened, what you've been asked to write and why; from the poster: volunteers, bags/gloves provided, leaflets distributed, free supper; from the notes: successes and failures
5 Informal, conversational: you are a student writing to other students.
6 Is it persuasive, lively, interesting, well organised?

3d B (A is too formal, C is not about the litter day and the article is not about threatening litter louts but encouraging them to change)

4a **1** c **2** a **3** d **4** e **5** f **6** b

5 Suggested answer

The big clean-up campaign!

Have you ever wondered what our college would be like if it was cleaned up? If so, you should have been with us after Clean-up Day last Saturday.

The idea was that we would tidy it up and the principal would introduce a number of new policies to keep it tidy. On the day itself, we issued rubbish bags and rubber gloves and set you to work, while others distributed leaflets about how to keep the college clean. As a reward, everyone was given a delicious buffet supper free of charge!

And what a transformation! For the first time that any of us can remember, the college is litter-free, and doesn't it make everyone feel more positive about the place?

The most popular of the policies has been to ban smoking, even among smokers who, it seems, hated seeing cigarette ends around the place! Also, it's clearly been a good idea to put extra bins around the place.

The idea of the leaflets, however, was counterproductive, since it only created more litter! Also, students feel that the new fines for dropping litter are unworkable, particularly as they are too high.

So now it's down to us. If we all like a cleaner college, which it seems we do, let's try to keep it that way!

(218 words)

Module 5: Review p.86

1 **1** C **2** B **3** A **4** D **5** C **6** B **7** A **8** D **9** B **10** C **11** D **12** B

2 **1** catch on **2** wiped out / dying out **3** given off **4** away from **5** runs **6** swim **7** disturbing **8** over **9** dependent **10** go

3 As people are becoming more weight-conscious, the major fast-food chains are now offering healthy options as an alternative to traditional burgers and chips. However, if customers **hadn't complained**, there probably wouldn't now be salads on the menu at all. Some customers have been **terribly/extremely/really/very pleased** *or* **absolutely/highly/really/quite delighted** with these new additions, while others have been **bitterly/deeply/very/extremely/really/terribly disappointed** *or* **deeply/really/terribly upset**. **If you're / Should you be** one of the ones deeply attached to burgers and chips, don't worry, they're still on the menu.

But for this change of strategy, the fast-food chains' profits **would decline / would have declined** even further. The effect on profit margins has been **entirely/extremely/highly/very beneficial**. What people don't realise is that when you eat a Caesar salad, **you often consume** just as many calories as you would if it **was/were** a plate of burger and chips! It's **painfully/fairly/totally/extremely/very/perfectly/quite obvious** that if people simply **ate** less, there wouldn't be the current obesity crisis.

4 **1** Provided we make a concerted effort to fight racism, it will decline. (... to fight it, racism will ...)

2 Whether or not genetic modification of food is safe, many people are suspicious of it. (Whether or not it's safe, many people are suspicious of genetic modification of food.)

3 Unless something is done about climate change, there'll be a severe water shortage in a few years.

4 Had the Soviet Union not collapsed, globalisation might not have spread so quickly.

5 As long as we carry on burning fossil fuels, there will always be pollution.

6 Should the population carry on declining, there'll need to be more immigration into western Europe.

7 But for the Internet, we wouldn't know what's going on in some countries.

Module 6
Looking forward, looking back

This module reflects the focus on aspects of the future and the past. Topics include health and appearance in the future, changes in lifestyle, medical discoveries, personal experiences, visiting museums and the content of guidebooks.

Lead-in p.87

Start with books closed. If possible, take in some photos of some well-known people or models from magazines, both men and women, who are considered beautiful. Alternatively, elicit names of some and list them on the board. If the class is roughly equally divided, ask the male students to decide on the most attractive man, and the female students to decide on the most beautiful woman. Then see if the other group agrees! Ask if any of the people would have been considered beautiful 100 years ago and if any are likely to be considered beautiful that far in the future.

Then look at the photos on page 87 and discuss the questions. Ask if beauty is about body shape, physical appearance, hair, skin tone, fitness or anything else. The photos show three different examples of ideals of feminine beauty: a 17th-century French aristocrat (left), a catwalk model from the present day (top right) and an Egyptian queen from ancient times (bottom right). The discussion does not, however, need to be limited to feminine beauty.

6A Health and fitness

Photocopiable activity

Activity 6A could be used to introduce the module or at any stage during it. Students work in pairs to do a quiz about health and fitness, which pre-teaches or reviews a lot of vocabulary related to the topic.

Again with books closed put the words *health* and *fitness* on the whiteboard and ask students to compare and contrast the meanings. Elicit related words and collocations (e.g. *fit, get fit, physical fitness, fitness fanatic, heal, healthy, healing, healthily, health centre/ club/ farm/ food*, etc). Put the expressions/collocations *Health and fitness* and *Fit and healthy* on the white board and ask students why the word order changes. (In binomial expressions – two connected words joined with *and* – where the words have similar importance the shorter word precedes the word with more syllables.)

Ask the class if they are concerned about their health and fitness and what they do about it. Or ask if they are more concerned with their health or appearance and which they spend the most time and money on.

Reading p.88

Remind students that Paper 1 Part 1 consists of three texts each followed by two, four-option multiple-choice questions. The texts are related thematically but vary in type.

1 Give students a moment to determine the function of each text from the headings alone. Note that *K* in *10K* is an alternative to the more common *km* abbreviation of *kilometre*.

Background

Cancer Research UK is the world's leading independent organisation dedicated to cancer research, supporting the work of more than 3,000 scientists, doctors and nurses across the UK. It raises and spends over £250 million each year on scientific research.

2 Allow 90 seconds for students to skim the texts to get a gist of their content and function.

3 Remind students that in multiple-choice questions only one answer is correct and that the other three answers must be incorrect. If they have time, it is useful to establish why the other three choices are wrong as a way of checking that their choice is the right one. Give the class a suitable time limit (15–20 minutes) to complete the task before getting them to compare and justify their answers.

4 Conduct the discussion in small groups. Include other sponsored activities that students have entered.

Key

3 **1** A – All routes pass through beautiful locations, the other points are all mentioned but do not apply to 'every route'
2 A – *online only, with no entries accepted on the day*
3 C – *the change in body weight ... most impact on those under 20*
4 C – *scientists develop ways of giving us wrinkle-free complexions*

5 A – it is the only one that fits the criteria of being of universal appeal, positive and able to benefit everyone.
6 B – empirical (based on scientific testing) evidence is required in *the research questions, data collection, analysis and interpretation of results*. Not A as the paper can be submitted by students, not C as it can be *published* or *unpublished*, not D because presentation style is not mentioned in the article.

Vocabulary p.90

There is a lot of vocabulary in the section, so it would be useful to tackle it in stages.

Ask students to keep their books open on the reading text. Start by asking them to look through the second text for expressions that express;

- optimism, e.g. *it is likely that …*
- pessimism, e.g. *my main concern is …* , *there is not a shred of evidence that …*, *I fear that …*
- doubt, e.g. *I'm sceptical about…*

Now ask students to look at the table on page 90. Show that column A contains ways of expressing opinions about the future. Column C contains time references and column B predictions for the future. Note the use of *they* to refer to scientists in *they will find a cure*.

1a Give students time to prepare some sentences, then put them in groups to discuss them. It might be useful to revise some ways of agreeing or disagreeing first.

1b Leave students in groups to discuss their opinions of when the second list of events will occur (if ever). 'An eat-anything pill' is a drug that allows you to eat what you want without putting on weight. Finally, ask if they have any other predictions that they would like to share.

2a This exercise picks up some interesting vocabulary items (some of which are from the text) and contrasts them with similar words. Remind students that they need to choose the verb that fits the context according to its meaning or collocation. When they have finished, ask the students how they would use the other verbs.

2b If necessary, give prompts, such as which of the following can be affected by diet: hair, avoiding coughs and colds, fighting major illnesses, energy levels, etc.

3a The sentence is from the second text. Elicit the fact that *a wrinkle* is a noun, but *wrinkle-free* is an adjective, so in this case adding the suffix changes the form as well as the meaning.

3b The exercise gives common examples of this suffix.

3c It is a suffix that is common on food and drink labelling e.g. *alcohol-free, caffeine-free, additive-free*. Other examples are *stress-free*, *risk-free*, *accident-free*, etc.

Extra!
There is an opportunity for some learner training here. Encourage students to keep lists of different suffixes in their notebooks, constantly adding to them as they come across further examples.

Key

1 **Suggested sentences**
1 I'm confident/optimistic that a cure for cancer will be found in the not-too-distant future/within a decade/at some stage/one day.
2 I predict that no one will look their age a century from now/in 100 years' time.
3 I fear that obesity will reach epidemic proportions in the near future.
4 It seems far-fetched to think that we will be able to change our appearance at will a century from now/in the near future.
5 I don't hold out much hope that hair loss will be cured in the short term.
6 I can't see any evidence that we are going to achieve the perfect body in the near future/in the short term.
7 I think/believe that living to 100 will have become the norm one day/at some stage/within a generation.
I believe that hair loss will be cured in the long term.
8 In my opinion, it is highly unlikely that exercise will become unnecessary in 100 years' time.
In my opinion, it is highly likely that laser technology will make glasses and contact lenses history in the not-too-distant future.
9 I'm sceptical of claims that exercise will become unnecessary one day.

2a 1 deteriorate – qualities deteriorate; numbers decline, standards slip
2 pumped – pump money into something
3 pinpoint – to discover/explain exactly; diagnose an illness, place an object or person (remember where you last encountered them)
4 replicate – get the same result again; imitate = copy, indicate = show
5 point to – point to a fact or link
6 reversing – reverse a trend = move in the other direction, restore = return to a former state
7 applying – apply a cream, etc. to skin; install machinery
8 weigh – weigh up = consider both sides carefully

3a Without wrinkles. The suffix *-free* is added to nouns to form adjectives with a positive connotation.
3b **1** debt-free **2** rent-free **3** interest-free **4** sugar-free **5** trouble-free **6** tax-free **7** lead-free **8** fat-free

▶ **Student's Resource Book, page 61**

Listening 1 p.91

The aim of this section is to practise skills for Paper 4 Part 1 where candidates need to recognise feelings, attitude and opinions.

1 The exercise is designed to create interest in the topic. Keep it short and general as students have a chance to say what they do to keep fit at the end. It could be done in small groups or as a whole class activity brainstorming ideas on the whiteboard.

Suggestions include; Food – eat more organic/less fatty food; Exercise – take the stairs not the lift, cycle to work/school; Other – get lots of sleep, move to the countryside.

2 Get students to read the questions in a and b before they listen for the first time.

3a Give students a moment to read the opinions. Point out that these are summaries, not the words used in the extract. This time they listen for meaning.

3b Play the extract for a third time. This time they are listening for language used to express the opinions in Exercise 3a. Remind them that this part is not necessary in Part 1 of the exam.

4a Give them a moment to read the opinions. Point out that it is not a multiple-choice activity but for each of the points A–E they should decide if the speakers agree or not.

4b After they have listened again and identified the expressions used, elicit other possible ways of indicating agreement and disagreement..

5 Ask students if it is easier to exercise well or eat well. Ask if they prefer to spend a little extra to buy organic foods.

Key

2a A
2b suggestions: which way to turn, when to change partners, tripping over my own feet, keeping in time to the music
3a 1 S 2 B 3 M 4 B 5 S
3b **1** I realise that **2** You and me both **3** Come off it
4a A, D, E
4b Agreeing: *I don't deny; You do; You might have a point there*
4b Disagreeing: *So how come…?; Yes, but hang on!*

Use of English 1 p.92

1 If there is time, start with books closed and expand the discussion, getting students in groups to discuss their sleeping habits. They could discuss ideas such as: what time they like to/usually go to bed/get up, how they sleep, if they ever oversleep, if they have ever fallen asleep when they shouldn't have, the strangest place they have ever slept, the longest time they have been without sleep/have slept, etc.

2a Give students 30 seconds to skim the text and to answer the two questions.

2b Remind students of the task type by looking back to page 34. Brainstorm the types of words that student will need to complete the text or look at the task strategy on page 169. Remind them to look at the words both before and after each gap. Give them 10–15 minutes to finish the task.

3 Students should now go back through the text and identify what type of word was needed each time and what clues told them so.

4 Examples for the discussion could include: people who work at night, people with really boring jobs, people who work where not much happens, museum attendants, security guards, exam invigilators.

Key

2a **1** A medical condition which causes sufferers to fall asleep without warning at any time of day.
2 It would help narcoleptics and other groups of people stay awake; it can improve other aspects of mental functioning.
2b **1** their (plural possessive to match *sufferers*)
2 that (cleft sentence)
3 One (before a non-defining clause; *A man* would need to be followed by a defining clause)
4 himself (reflexive pronoun)
5 nobody (cf. *no one* fits grammatically but is usually written as two words)
6 all (before plural noun *sufferers,* cf. *nearly every sufferer*)
7 the (definitive article before noun – only one conclusion)
8 due (*is due to*, cf. *is because of*)
9 What (cleft sentence)
10 no (before noun + *at all*)
11 of (noun + *of* + noun
12 such (before *a* + noun)
13 like (= *such as*)
14 out (multiword verb: *carry out a test/trial*)
15 in (*in addition*)
3 **1** 6, 7, 10, 12
2 2, 4, 5
3 1, 9

Language development 1 p.93

Cleft sentences are one of the few bits of grammar that are not really studied until an advanced level. Students will therefore be much less familiar with them than with much of the other grammar reviewed in this course.

You could start by asking students to close their books and guess where and when the first human heart transplant took place.

Background

The recipient of the first transplant, 55-year-old Louis Washkanskey, died of pneumonia only 18 days after receiving the heart of a 25-year-old female donor after the anti-rejection drugs weakened his immune system. Barnard's second patient lived much longer.

1a Look at the four sentences and establish that the basic information in each is the same, but the emphasis changes. Then spend some time looking at the Grammar reference starting on page 181.

1b Referring back to the table or the Grammar reference, students rewrite the sentences.

1c This is not meant as a serious discussion, although it could become one! Encourage students to complete the sentences in a humorous way if they cannot do them for real.

2a Students should look back at the table at the top of the page to do this, noticing that it is just reversing the two parts. Sentence 2 could also be transformed in this way, but would involve an additional change and would be very formal and rarely used (*Performing the first heart transplant was what Christiaan Barnard did.*).

2b Emphasise that this is more a test of general knowledge than grammar, the idea being that the sentences give further examples of the structures in Exercise 2a.

Background

- Crick and Watson's discovery of the structure of DNA as a double helix also explained how genes were coded and how DNA could divide and reproduce. With Maurice Wilkins, they were awarded the Nobel Prize in 1964.
- Gene splicing is the manipulation of the genetic material of an organism to introduce genes from other organisms to alter the characteristics of the first. It has led to medical advances such as the production of artificial insulin and the development of super tomatoes.
- The first of well over a million born worldwide, test tube baby, Louise Brown, was born in Oldham, UK, in July 1978 as a result of the pioneering work of Patrick Steptoe and Robert Edwards. She worked for a while as a postal worker, leading to plenty of jokes about 'special deliveries'.
- Dolly the sheep, said to be named after the singer Dolly Parton because the cells that cloned her came from her mother/twin sister's udder, was born in 1996. She went on to have lambs naturally herself, but died at the young age of six in 2003. It is not known whether it was because she was cloned.

3a To focus on form, the content of the examples is the same. Elicit other ways that the structure could be used (e.g. *it was on* + day/date, *it was at* + time, *it was by* + method). Elicit and practise situations when the structure is commonly used, such as when correcting information (e.g. '*The first transplant was performed in 1977.*' '*No, it was in 1967 that …*') or responding to questions (e.g. '*Was the first transplant performed in the USA?*' '*No, it was in South Africa that …*') or emphasising a difference (e.g. *It was in South Africa, not the USA, that the first …*).

3b Point out that the structure can be use in the past, present or the future before students attempt the task.

3c This time, encourage students to complete the sentences in a way that is true for them.

Extra!
Get the class to look back at the Use of English text on the previous page and identify the two cleft sentences (*It is not just their own safety that is at risk. What this discovery has led to is …*).

Key

1a 1 the subject (*who*) 2 the action 3 the time 4 the place

1b 1 The country where the most cosmetic surgery is performed is Brazil.
2 The reason many people have cosmetic surgery is to improve specific facial features.
3 What I'd really like to change is the shape of my nose.
4 What you need to do is take more exercise.
5 The thing that companies are most keen to develop is a drug that prevents obesity.
6 All my sister has ever wanted is to look like a movie star.

2a 1 Dr Christiaan Barnard was the person who performed the first heart transplant.
3 1967 was the year when Dr Christiaan Barnard performed the first heart transplant.
4 South Africa was the place where Dr Christiaan Barnard performed the first heart transplant.

2b 1 DNA 2 1973 3 1978 4 Edinburgh

3b 1 It was the Danish geneticist Wilhelm Johannsen who coined the term *gene*, not Gregor Mendel.
2 It is obesity rather than cancer that is now the major cause of death in the UK.
3 It wasn't until the 1950s that the link between cigarette smoking and cancer was identified.
4 It was three years ago that/when we last had a holiday abroad.
5 It was only when I got to the airport that I realised I'd left my passport behind.
6 It's because you work so hard that you're always stressed out.
7 It is only by identifying its causes that we can overcome stress. / ... identifying the causes of stress that we can overcome it.

▶ **Student's Resource Book, page 62**

Writing 1 p.94

This section looks at aspects of cohesion: the relationships based on grammar or vocabulary between parts of sentences or across them that hold text together.

1a When the students have identified what the links refer back to, ask them what the sentences would be like without them (repetitious). Ask if references always refer back. (No, they can also refer forward, e.g. ***It** used to be considered just sport, but now for many people **exercise** is an essential part of their life.*) Spend some time looking at the Writing reference on page 200, which lists examples of logical, grammatical and lexical links.

1b Get students to skim the box and answer a gist question such as *What does the writer believe is the best way to remain young?* Then give them time to complete the text.

2a/b Students should have no problems finding examples of repetition in this piece of writing. Later, they could look for repetition in a piece of their own writing and try rewriting that, focusing on the cohesion.

3 It could also be a sports club or health spa if that would be easier for students.

Photocopiable activity

Activity 6B is designed to practise the cohesive structures covered in this section. Students work in groups, playing a game like dominoes, matching two sentences together and completing the second sentence with a word or part of an expression that provides cohesion to the first.

Key

1a 1 One method > several ways; Another > several ways/method.
2 there > a gym; then > last month; which is why > been going there every day since then; her > Tania
3 those people > people who eat a lot of meat; it > eating a lot of meat; so > you are one of the people who ... ; millions > those people; such > eating lots of meat
4 neither do > doesn't believe in plastic surgery

1b 1 whatever – no matter what
2 What's more – in addition to people believing in the need to use creams etc.
3 their – unscrupulous food manufacturers
4 such claims – food supplements can reverse the ageing process
5 neither do – don't believe the claims
6 the ones – the scientists
7 prevent ageing – = stay looking young
8 this aim – prevent ageing
9 First – of several ways
10 which – eating sensibly
11 Secondly – of several ways to achieve the aim of preventing ageing
12 such as – an example of exercise
13 Finally – of several ways
14 That's why – in order to keep mind active

2b **Suggested answer**

INTRODUCTION
We appreciate that the last thing that many of you have time to do is to keep fit. However, physical exercise is the key to maintaining a

healthy body and mind. That's why we've created a luxury club that offers a convenient and enjoyable way to achieve your personal fitness goals, whatever they may be. What's more, we pride ourselves on offering a welcoming atmosphere, one in which you can feel relaxed and at ease.
FACILITIES
Once inside this urban paradise, you will find a range of facilities, all of which are free to members. Why not take a dip in the heated pool or enjoy a snack by the poolside? Or perhaps you'd prefer a strenuous workout? If so, the Club offers a large variety of state-of-the-art exercise machines which will challenge the fittest among you.
We think our club is the best in the city and, we are pleased to say, so do our members.

▶ **Student's Resource Book, pages 64–65**

6B Unveiling the past

Start with books closed and ask students to define the verb *to unveil*, and elicit two related meanings: 1 to remove the cover (the veil) from something, to uncover (e.g. a new statue or plaque); 2 to show or tell people about something for the first time (e.g. to unveil new plans.)

So 'unveiling the past' means both to uncover and to tell someone about the past.

Listening 2 p.95

Remind students that Paper 4 Part 1 consists of three short extracts, usually dialogues, with three multiple-choice questions on each.

1 Give students a moment to skim the introductions to the texts for the gist and match the pictures to the extracts.

Ask students to read the questions and find the connection between questions 2, 4 and 5. (Answer: feelings – 2 *How did Liz feel* ...; 4 *How did he feel* ...; 5 *How does the caller feel?*)

Either deal with the extracts one by one, stopping after each to check the answers, or closer to exam conditions playing them all without stopping before checking the answers.

2 Give students a few minutes to discuss their choices. If possible, follow by giving them the audio script to check their answers.

3 Students could be asked to think of their earliest memories as well as vivid ones, either happy or not!

Key

1 1 C – *I was just so mesmerised by the sheer volume of people there*
2 A – *so overwhelming ... that I was in tears for most of the time*
3 A – *we were able to match a single hair found on the scene all those years ago with one from our current investigation*
4 C – *I'd been rather hoping to spend my years coming up to retirement doing something relatively undemanding*
5 B – *I lie awake all night wondering whether or not I should let her know.*
6 B – *If you speak to the teacher without saying anything to your friend, you run the risk she'll find out you've gone behind her back*

Speaking p.96

On way to start would be, with books closed, to ask students to talk about the last time they went to a museum/art gallery, where they went and what they thought of it. This should create some interest in the topic and bring up some of the vocabulary.

1a The photos feature an animated dinosaur that moves, roars (and smells!) at the Natural History Museum in London, and a guide who dresses and acts in character as a period solider at the Imperial War Museum, as well as more traditional museum displays. Students can use the adjectives given, as well as any others that they think appropriate.

1b There are a number of possible answers here. For example, are the two girls really into their drawings or bored stiff?

2a Students can use dictionaries or pool knowledge for this task. Encourage them to think of examples, not just definitions. Pay attention also to the pronunciation also (e.g. stress in *ex**hib**it* and *exhi**bi**tion*).

2b When students have completed the text, they could write related sentences to use some of the other words.

3a Students should by now be used to listening for two parts to the task.

3b He talks about photos A, B and C, and says that museums are more interesting when visitors interact somehow with the exhibits or staff. If necessary, refer back to the task strategy to have some points to assess his performance against.

3c He answers both parts of the task. He compares the way the exhibits are brought to life when the visitors interact with them in different ways, and he talks about how successful exhibits of this type (but not these specifically) might be. However, he includes some comments on the negative effects of queuing, which could be seen as irrelevant.

4a Establish that, to make what they say more interesting, students will need to add emphasis to important parts. This reviews three ways of doing so. When students have categorised the five examples, remind them that sentences 3 and 5 are cleft sentences.

4b Allow students to look back at page 93 if they need help to transform the sentences. Check they know what *fossils* are.

5 Remind students to answer both parts of the task. The instructions for student B in Task 2 are above the pictures on page 206. What is introduced here is the type of questions students will be asked to follow up their partner's long turn.

6 Encourage students to analyse their performance seriously and also to think about ways they could improve.

7 The discussion is a perfect opportunity for students to focus on their weaker areas.

Key

2a **1** admission fee: fixed amount of money paid to get in
voluntary donation: variable amount of money given by choice
season ticket: pass for a fixed amount of time that gives admission (probably at a reduced rate)
2 an exhibit: an object in an exhibition
an exhibition: a collection of objects on show to the public
an exhibitor: the person who puts the object(s) on show
3 a guide: someone who shows visitors around and gives them information
an attendant: someone who works for the institution and provides a specific service (e.g. security, information on request, cloakroom)
a curator: someone with specialist knowledge who looks after the objects
4 an event: a special performance/exhibition/etc. organised for a specific time period
an incident: a single, noteworthy happening
an experience: something that happens that influences or affects you
5 a catalogue: a detailed list of every object
a guidebook: a description of the most interesting objects bound into book form
a brochure: a short guide to the exhibition in leaflet form
an audio guide: a description of the objects in audio format (e.g. cassette, CD)
6 wander (around): to walk slowly, in a relaxed way, without specific aims
trudge (along): to walk heavily, as if tired and bored
stride: to walk quickly and purposefully

2b **1** exhibition **2** season ticket **3** event
4 wandering **5** catalogue **6** audio guide
7 attendants

3a **1** Compare and contrast photos
2 Say how successful the places might be in encouraging young people to take an interest in the past

3b He talks about photos A, B and C.

3c Yes, he dealt with both parts, but also included something irrelevant (the comment about queues).

4a **1** A stress on *doing*
2 B use of intensifier *much*
3 C change of normal word order: emphasis using cleft sentence structure (see Language development 1)
4 C introductory expression
5 D use of auxiliary verb *do*

4b 1 Why so many people are interested in fossils I just don't know!
2 It was the cost of admission that put me off going to the exhibition.
3 What I particularly enjoy is doing hands-on activities.
4 I'm not at all/in the least bothered about going on the museum trip.

Language development 2 p.98

Start by asking the class if anyone has ever tried to trace their family tree and if so, how far back they went and what they discovered.

1a Students could look through the Grammar reference either before or after doing the first exercise. Establish that the choice of whether to use *-ing* or *to*-infinitive is determined here by the verb before (e.g. *volunteer/guarantee/agree* + *to*-infinitive, *contemplate/involve/fancy* + *-ing*).

1b First, elicit how each verb should be followed (e.g. *contemplate/imagine* + *-ing*, *pretend/refuse* + *to*-infinitive). To help make the interaction more interesting, encourage students to form questions that they would really like to hear the answers to.

2 Here, the focus is on verbs that are followed by an object before the second verb. When correcting the mistakes, students should notice the form that each particular verb follows. Some verbs (examples are given in the Grammar reference) can be used with or without an object before the following verb.

3a Ask students to compare the two sentences:

1 *I remembered to write everything down.*
2 *I remembered writing everything down.*

Establish that both are grammatically possible and elicit the difference in meaning. In sentence 1, the meaning is 'I didn't forget to do what I had to' – the remembering preceded the writing. In sentence 2, the meaning is 'I had a memory of the action of writing it down' – the remembering came after the writing.

The text contains a number of the most common verbs that can be followed by either form with a change in meaning. Students should identify the differences and choose the correct form for each gap before checking in the Grammar reference.

3b Give students suitable thinking/preparation time before they complete their sentences.

4 This exercise combines a number of differences between *-ing* forms and infinitives, determined either by the form of the preceding verb or by the nature of the *-ing* form, which suggests repetition or duration.

Key

1a **1** finding **2** to help **3** to find **4** to trace **5** spending **6** to share **7** drawing **8** to discover

1b **1** *contemplate* + *-ing* **2** *imagine* + *-ing* **3** *be forced* + *to*-infinitive **4** *consider* + *-ing* **5** *pretend* + *to*-infinitive **6** *refuse* + *to*-infinitive

2 **1** *We chose Alice **to represent** us at the investigation.* (*choose* + object + *to*-infinitive)
2 ✓ *I found out* (cf. *I never expected him to find out* = he found out)
3 *I arranged **for** the library to send me all the information.* (arrange for someone to do something)
4 ✓ *I miss Amanda moaning* (verb + object + *-ing*) or *I miss Amanda's moaning* (verb + possessive + gerund (= more formal))
5 *I didn't expect **them** to discover anything very exciting.* (They discovered it.)
6 *Can I recommend ~~you~~ visiting* (or *you to visit*) *the industrial museum in town?* (*recommend* + *-ing* or *recommend* + object + *to*-infinitive (more formal))
7 *I was made **to** go first because I was known to have experience.* (*made* in passive + *to*-infinitive (cf. *made* in active + object + infinitive: *They made me go …*)
8 ✓ (formal) (cf. *like* + object + *-ing*: *I didn't like them suggesting*)

3a **1** *visiting* – thinking back to an earlier action **2** *to spend* – intention **3** *to learn* – reason/purpose (we stopped travelling in order to learn) **4** *seeing* – thinking back **5** *to set* – didn't do what we had to, future action **6** *waiting* – resulted in **7** *to find* – objective **8** *hitching* – method **9** *worrying* – this activity ceased **10** *to have* – change to another action **11** *exploring* – continued **12** *missing* – feeling sorry about the past

4 **1 a** ii I like to pay my bills at the start of the month. (I don't enjoy it, but feel it is the best thing to do – thinking of specific occasions with future reference.)
b i I like paying less than other people for things. (I enjoy the activity in general.)
2 a i I prefer reading about an exhibition before I see it. (present simple + *-ing* for a generalisation)
b ii I'd prefer to read about the exhibition before I see it. (*would prefer* + infinitive for a specific case)
3 a i I saw the driver drop his ticket before he went into the museum. (single short action)
b ii I saw the driver dropping people off outside the museum all afternoon. (repeated action)

4 a ii I can't bear to go to a museum today. (specific case)
b i I can't bear going to museums when I'm on holiday. (generalisation)
5 a ii I watched him painting a portrait; he only did the nose while I was there. (part of unfinished action)
b i I watched him paint a portrait; it only took 30 minutes. (complete finished action)
6 a ii I left Duncan waiting for the gallery to open and went home. (duration: he continued waiting while I went home)
b i I left Duncan to wait for the gallery to open while I bought us some coffee. (purpose: his purpose was to wait for it to open)
7 a ii I don't want you to go there tonight. (specific case)
b i I don't want you going there every night. (generalisation)

Photocopiable activity

Activity 6C can be used any time after this section, as it reviews language from the Vocabulary section and both the Language development sections. It is a grammar auction, with students working in groups to find mistakes in the sentences they are given.

▶ **Student's Resource Book, page 67**

Use of English 2 p.99

Start by eliciting requirements and tips for Key word transformations.

1a Establish why the answer given in the example is wrong, referring back to the Language development section if necessary. (*imagine* + object + *-ing*)

1b Either do the exercise in sections, with students using the Help section, checking answers as you go, or if the class is more confident, do it under exam conditions with a time limit of 8–10 minutes.

Pick up on any areas of grammar that cause problems referring students to sources of help where necessary. Areas might be the noun clauses in question 3, Future perfect tense in question 5, reported speech in question 7, cleft sentences in question 8.

Note that in question 3, Ruby (female) would like to be an *actor*, not *actress* as the female form of the noun is becoming less common and the word *actor* becomes non-gender specific.

2 The task analysis reviews *-ing* forms and infinitives as well as some structures used for giving emphasis..

Key

1a I could *never imagine him refusing* free tickets.
1b 1 He told *me not to forget to* lock the door when I went to bed.
2 The speaker's jokes went *down badly with* the audience.
3 What Ruby would like (*to do*) *is* (*to*) *take up* acting as a career.
4 What was it (*that*) *persuaded you not to* accept the job offer?
5 Next our parents *will have been married* for 25 years.
6 If you don't turn *up soon, it will mean* (*that*) we'll have to leave without you.
7 Chloe wanted to *know why Tom had acted so* irresponsibly.
8 It *was only when Emily arrived* that everything started to go wrong.
2 1 Example, 1, 3, 4
2 3, 8

Writing 2 p.100

1 In the discussion about guidebooks, ask what a good one should contain, e.g. historical information, enthusiastic descriptions, balanced/strong opinions, suggestions on where to eat, hotel reviews, etc.

2 The format is similar to previous modules. Students read the task, highlighting the key points, then look at the six questions to focus on content and style.

3 At this stage, students just make notes on the two places of interest and why they are worth visiting, before deciding how many paragraphs they will need. Three should be enough: one to describe each attraction and one to compare and contrast them, but they could possibly include a fourth to introduce the attractions or sum them up.

4a Give students plenty of time to read the three extracts, highlighting words and phrases that give each its style and catch the reader's interest.

4b Go through the list of expressions, highlighting how each would be continued and noting its style.

5 Students could write and check their guidebook entries, and then give them to classmates to comment on the content, organisation and general readability.

Key

2 **1** To try to give the reader a 'picture' of the attractions and clear idea whether they are worth visiting. The target reader is an overseas tourist.
2 Three. Describe two attractions, saying why they interesting and what can be seen there, then compare them.
3 Both: facts are important, but so is an evaluation of the place. Note that in this question, you are not asked to give practical information such as times, prices, etc.
4 Descriptive/semi-formal/impersonal style: language which creates a picture, brings the places to life and engages the reader.
5 Yes and no. A guidebook is independent and does not have the same aim as a brochure, which tries to sell a place. However, guidebooks need to make a place sound interesting.
6 If they can decide whether or not to visit the place. To know what sort of person it is most suitable for (e.g. children).

3 **1** The talk is divided up into clear sections containing one main idea, with one question per section being the norm. Signals that help students to identify when the information they need is going to come up include sentences that introduce the sections, and parallel phrases to those in the questions.
Question 1 Introduction describing Easter Island: *this tiny Pacific island*
Question 2 *The focus of social life were the stone platforms called ahu … , these platforms were constructed …*
Question 3 *… the fact that they all face inwards …*
Question 4 *At first, the islanders had no problems finding food …*
Question 5 *Because of easy access to stone, the statues were always at least five metres high, but …*
Question 6 *The amazing thing is how these huge statues got from the quarry where they were carved to the stone platforms …*
Question 7 *By this date, many of the trees …*
Question 8 *the absence of trees also led to soil erosion, so that plant and animal species became extinct, … When the first Europeans arrived in 1722 …*

5 Suggested answer

Located in the centre of the city, you'll find two museums quite unlike the sedate museums elsewhere in the city, both with a gruesome past.

The Bishop's Prison Museum, used originally in the 12th century as a prison for disobedient church people, later became a dumping ground for the city's lowlife. Not surprisingly, it became a much-hated institution and was frequently attacked by rioters in the 18th century. The exhibition features several scenes from prison life and dwells on the torture and grim conditions inside. The museum doesn't take long to walk through, but several audio stories from 'inmates' make you linger by the waxworks.

The Dungeon Museum, on the other hand, is a large Gothic horror show and one of the city's major crowd pleasers. Throughout its history, until the middle part of last century, this was the place where royalty, politicians and ordinary criminals were imprisoned and put to death. Inside today, medieval torture has become a spectator sport with all the ketchup-splattered authenticity of a horror movie. At every turn, the spine-chilling exhibits give you some nasty surprises. You are assaulted by rats, attacked by fire or put in a boat and sent backwards down a dark river. All the while, the music is ghoulish and the lighting spooky. Although teenage kids love the place, there is plenty to offend, so it's best avoided by young children.

In fact, both museums are really best avoided if you are of a nervous disposition or prone to nightmares. The National Museum is a much calmer experience!
(257 words)

▶ **Student's Resource Book, page 69**

Module 6: Review p.102

1 1 D 2 B 3 A 4 C 5 B 6 D 7 A 8 B 9 C 10 D

2 1 There has been a startling **rise** in the number of anti-ageing products on the market.
2 Do you believe the world will be **trouble-free** in the future?
3 I'm a pessimist. I don't hold out much hope **for** a better world.
4 First published in the 1930s, the data remains an absolutely **indispensable** resource.
5 Most health-food companies have online **brochures** of their products.
6 He has a very **selective** memory – he only remembers what he wants!
7 Can you **pinpoint** the site on the map?
8 The government's refusal to hand back the relics caused a major diplomatic **incident**.
9 What is the price of **admission** to the exhibition?
10 We need many **alternative** forms of fuel.
11 We need to **ensure** there are fewer nasty weapons about.
12 He **leads** a very solitary life.

3 1 I really regret not being able to come to the museum with you.
2 The shuttle is expected to reach / It's expected that the space shuttle will reach its destination by next Tuesday.
3 It is considered to be the most daring space mission yet.
4 What they did was make us work 12 hours a day.
5 The archaeologist has stopped talking to his assistant.
6 Do/Would you mind me sharing your programme?
7 It is the director who decides who goes on the mission.

4 1 where 2 were 3 What 4 was 5 making 6 reason 7 that 8 It 9 who 10 things 11 is 12 to 13 be 14 did 15 Whatever/Despite

Exam practice 3 TRB p.191

Paper 1: Reading
1 B 2 F 3 A 4 C 5 G 6 D

Paper 3: Use of English
Part 1
1 D 2 A 3 C 4 D 5 A 6 B 7 D 8 A 9 D 10 C 11 B 12 A

Part 5
1 if/when there is no alternative 2 had the aim of promoting 3 stop burning fossil fuels 4 soon be a shortage of water 5 suggested that we should 6 a book is preferable to 7 reminded Tom to post (her) 8 have the choice of staying

Paper 2: Writing
Suggested answer

Dear Sir,
I am writing on behalf of my class in reply to the article in the college magazine about the service provided by the company Gymwise which holds the franchise for the college gym.
As it seems there is some confusion about whether the gym has improved or not, we decided to conduct interviews and a survey amongst college students. We discovered the following:

- almost half the students we contacted were wholly satisfied with facilities, and another third were generally satisfied. That makes less than 20% who were dissatisfied, with a mere 5% completely dissatisfied. I think this speaks for itself;
- many students felt that the gym had greatly improved in the last year, with new staff who are helpful, and new equipment which is much appreciated;
- students also like the extended opening hours and Internet booking system, although we registered complaints about the fact that the gym gets very crowded at peak times.

Overall, the gym has improved, and those who have been there have recognised this. Unfortunately, a few students who have not been back since the improvements were made think that nothing has changed, and it seems to have been their views which were reported in your article.
I hope this letter will be printed and that Gymwise will continue to provide this invaluable facility.
Yours faithfully

Paper 4: Listening
1 A 2 C 3 A 4 B 5 B 6 C

Module 7 Breaking the mould

This module incorporates features on various people that have dared to be different in some way and have struggled against adversity. In Part B, the focus shifts to breaking the habit, where the link is various types of addiction.

Lead-in p.103

Start by asking students to look at the title *Breaking the mould* or by giving it to them on the board and eliciting what a *mould* is and what the expression means (to change a situation completely, by doing something that has not been done before).

Then look at the photos together and discuss the ways in which the people are breaking the mould. The skateboarder is attempting higher and harder jumps/tricks; the kite-surfer is trying to make surfing more unusual and exciting by using a kite or small parachute to catch the wind; the BASE jumper has just jumped from the world's second tallest building (the Petronas Twin Towers in Kuala Lumpur).

Background

The term *BASE jumping* comes from *Building, Antenna, Span, Earth*, which summarises the type of objects (tall buildings, radio masts/towers, bridges and cliffs/mountains) that BASE jumpers leap from with nothing but a parachute.

Discuss the other three questions, encouraging students to use suitable language when giving opinions or making suppositions. For the last question, they could consider: accidents in the home (a very high source of accidents), journeys to/from work, working life as well as hobbies, dangerous or otherwise.

7A Against the odds

Reading p.104

With books closed, ask students to talk in groups about something they have done which they considered very difficult, demanding or dangerous; a time they were successful 'against the odds'. Alternatively, ask them to think of any feats of endurance that they have read or heard about that impressed or inspired them in some way. Ask what the person/people did, why it was so unusual and why it impressed them so much.

1 Ask students to look at just the title and subheading of the article. Check that they understand what is meant here by *demons* (a powerful force – not the devil – that causes problems) and ask them how wide they think the Atlantic is (3,000 miles/4,800km) before discussing the questions.

Background

The photo shows the author of the article, Jonathan Gornall, with his fellow crew members, Pete Bray, Mark Stubbs and John Wills, training for their attempt to beat the record for rowing across the Atlantic in 2004.

2 Remind students how to skim and give them a minute to look for the answer to the question. They could look out for any words or ideas related to motivation or personal qualities.

3a First, look at the rubric which is the same for all multiple-choice questions.Then elicit a suitable task strategy (e.g. reading the questions, but not the four possible answers, before finding the part of the text that answers it; answer in their own words, then choose the option which is closest). Remind students that in Cambridge exams, no marks are lost for incorrect answers, so if they don't know an answer, they should attempt to rule out one or two wrong answers and then guess.

3b/c Do question 1 together as an example, following the questions that analyse each option.

3d Give students 10–15 minutes to finish the task.

4 The analysis should help focus attention on the types of question asked and the best way to answer them. Ask the students to look at each question and decide if the answer was contained in specific details or in the sense of the paragraph as a whole. They should notice that individual phrases will suggest a particular answer, but they will need to understand the general sense of the paragraph and other points that rule out the incorrect options.

5 Spend some time on the vocabulary in the discussion questions, drawing students' attention to both the connotations and the meanings of the words. In the first question, *shallow*, *selfish* and *destructive* are all clearly negative, but in the second question there might be some disagreement as to the connotation of *eccentric* (often positive in UK – see line 7) or *macho* (negative in UK – see line 50). For the third question, emphasise that a personal challenge could be large or small; it's a question of overcoming our limitations, doubts and fears. For example, for some people, picking up a spider could be as much a challenge as an Atlantic crossing!

Key

3d 1 C – (line 9 *in pursuit of what?*)
2 C – (line 16 *before succumbing to mental and physical pressures*)
3 B – (line 32 *We all have strange compulsions and we all choose to prove ourselves in different ways*)
4 A – (line 47 *we need to be free of our comfortable lives at times; to push ourselves …*)
5 D – (line 53 *her husband … left her to it* and line 56 *she finished the competition*)
6 B – (line 68 *spare a thought for a man haunted by his sense of self-worth and wish him well*)

Vocabulary p.106

The first two parts of this section explore the vocabulary from the text by comparing it to similar expressions, as paraphrasing is such an important element of CAE.

1a Start by asking students to look through the text for words or phrases connected to the topic of motivation. Then get them to compare their list with the items in column B. As they match the expressions, they should note how the words are used. For example, *search* and *quest* have similar meanings and are both followed by *for* + noun, whereas the adjectives *able* and *capable* are similar, but followed in very different ways (*able to do something*, *capable of doing something*). Give as much information as possible about how the words are used, or alternatively get the students to look up the words in a dictionary and then ask them questions about them. For example *grip*, like many verbs of emotion, is usually in the active to describe things (e.g. *it was gripping, it gripped me from start to finish*), but in the passive to describe personal feelings/reactions (e.g. *I was gripped*).

1b Having matched the expressions, it should be quite easy for the students to complete the second text. Point out that they might have to make some small changes to make the words fit.

1c There are five other similar parallel expressions to be found.

Background

Erik Weihenmayer from Denver, USA, was born with a condition that meant he was partially sighted through his childhood and totally blind from 13. A great athlete, he also runs, cycles and scuba dives. In addition to climbing and writing, he works as a high-school teacher and wrestling coach and is one of about 100 people to have climbed the 'seven summits', the highest peaks on the seven continents. There is a documentary film of his ascent of Everest called *Farther Than The Eye Can See*.

2 If students are unsure of the meaning of the expressions or how they are used (i.e. what follows them), they should find them in the text on the previous pages. For example, *no allure* is preceded by *have* or *hold.*

3a The text was about motivation but also about risk. The box gives some of the dictionary phrases using the word. To reinforce various aspects of vocabulary, ask the students to look at the box and give an example of *risk*:

- as a noun (*take a risk*)
- as a verb (*risk life and limb* = everything)
- as an adjective (*risky business*)
- in an adjective + noun collocation (*calculated risk*)
- in a verb + noun collocation (*run a risk*)
- in a noun + noun collocation (*a health/fire/security risk*)
- in a prepositional phrase (*at risk*)

The aspect of learner training to focus on here is that, at CAE, it is not enough to know a word like *risk*; the students will need to be familiar with a range of fixed expressions that use the word. They should therefore get into the habit of building up word groups like this (see Exercise 3c).

3b When they have completed the six sentences, ask them to make a few more with some of the other expressions.

3c Students could start by noting down any phrases or collocations that first come to mind before checking in a dictionary. Many dictionaries will list collocations or at least use them in examples. A good dictionary like the *Longman Exams Dictionary* groups these types of expressions. Students could record the expressions in whatever way works best for them, whether that is as a list, a table or whatever.

Key

1a **1** e **2** d **3** f **4** a **5** h **6** g **7** b **8** c
1b **1** succeeded in **2** quest **3** driven
4 prove himself **5** gripped **6** compulsion
7 pushes himself **8** capable of
1c despite great difficulties = against the odds
to give up = to abandon
problem = setback
to give in to = to be defeated by
to show = to prove
2 **1** There is no chance of beating this record.
2 The fact that someone else has already climbed Everest is beside the point.
3 The compulsion to succeed against the odds is far from being a male preserve.
4 This compulsion to take risks may come down to a basic human need for challenges.
5 The routines of everyday life have/hold no allure for people who crave risk.

3b **1** health/security **2** life and limb **3** element of risk **4** risky business / calculated risk **5** at your own **6** run

3c Competition: hold/run/launch a competition; enter/win/lose a competition; knock someone out/be knocked out of a competition; be in competition with; fierce/stiff/intense competition; not much/little competition; foreign/international competition; a competition winner
Doubt: have (your/no) doubts; raise/express/voice doubts; cast/throw doubt on something; be beyond/without/open to/in doubt; there's no/little/some doubt; serious/grave doubts; nagging/lingering doubts; an element of doubt; to doubt someone's word
Success: be a big/huge/great success; the secret/chances of success; a box-office/overnight success; a success story; without success; have success in doing something

▶ **Student's Resource Book, page 72**

Listening 1 p.107

Start by asking students to think of different types of common disability, such as being blind, deaf, physically handicapped or with learning difficulties.

1 Identify the disabilities the people have and discuss the question in small groups.

Background

The man is Stephen Cunningham, who has been blind from the age of 12. He was the first blind person to fly around the UK in a light aircraft. The woman is Paula Craig, a former police officer, who lost the use of her legs in a road accident in 2001.

2 Remind students that in Paper 4 Part 2 the answers will fit grammatically as well as contextually. Elicit the type of word students are listening for in question 1 (*the* + noun + *of*). Remind students to note down the answers they hear the first time they listen and then to check them the second time. Give them a moment to check their spelling before going through the answers.

Background

Evelyn Glennie, who gives about 100 performances a year, has played with a long list of top orchestras around the world. She has never considered herself as anything other than normal and prefers people to enjoy her music and forget her deafness.

3 Get students to think of different areas in which people could succeed such as the music business, the arts, business, entertainment, etc. Examples include:

- Beethoven: continued to compose despite his deafness
- Franklin Roosevelt: longest-serving US president despite severe physical handicap following polio
- Goya: great Spanish artist was deaf from his mid-40s
- Stephen Hawking: remarkable physicist despite being in a wheelchair and needing a computer to speak.

Key

2 **1** originality **2** delicately **3** clothes **4** flower(-)pots; spoons **5** 600 **6** music colleges **7** research centre **8** 'role model'

Language development 1 p.108

This section focuses on spelling. Remind students that this is important in most papers, but is particularly important in word formation in Paper 3 Part 3. It is designed to show common patterns and words that are often spelt incorrectly.

Start by asking students which English words they commonly misspell. (You might want to give some of your own problem words.)

1a The mistakes in the section all break one of the (loose) rules or patterns of English spelling. Students should correct the mistakes in each sentence and say why each one is wrong. Then let students check their answers by referring to the spelling guide in the Writing reference on page 202.

1b This time, students have to first identify the incorrectly spelt words before correcting them. Tell them that there are one or two mistakes in each sentence. Note that in question 4, *Yours truly* is a more common ending in USA; in question 5, the UK spelling is *centre,* but in the USA it is *center*.

2 The sentences in the section contain typical student mistakes. The idea is to train students in spotting mistakes in their own written work. Encourage students to get into the habit of checking their own work in the same way. They should by now have a few pieces of written work to look back through.

From these, they could produce a list of the words they spelt incorrectly (in the corrected form!) and then in future check specifically for them.

3 The words here are examples of those that are often confused. As students do the exercise, they should think about the difference between the words in each pair, considering word type (noun, verb, adjective), pronunciation and meaning.

3a The first part gives students examples of how they could keep a record of example sentences in their own notebooks. Draw students' attention to the collocations and phrases used (e.g. *lose your head, stationary vehicle, complementary medicine, to have a big effect on someone*).

3b Students write their own sentences. Encourage them to use collocations or other words that link to the meaning.

Extra!
The exercise could be extended with further confusing pairs (e.g. *site/sight, fair/fare, break/brake, peace/piece, hole/whole, accept/except, steel/steal*) or as warmers in future lessons.

4 Start by asking students when hyphens are used; they should know that they are common in compound adjectives (e.g. *open-minded, face-to-face*), especially those formed from participles (e.g. *heart-broken, hard-working*), and used to avoid confusion with certain prefixes. They are seldom used now in compound nouns.

When students have tried adding the hyphens, get them to check in the Writing reference on page 202, noticing how they help avoid confusion with prefixes.

Draw students' attention to question 6, which gives an example of how a hyphen is sometimes used when nouns are formed from multiword verbs. Another way is by reversing the order of the verb and particle.

5 It might worth pointing out that spelling is much harder for some students than for others (although all will have some difficulties), and these students will have to put more effort into learning spelling patterns. Different techniques will work best for different people. Mnemonics to specifically remember individual words such as 'don't get sep**ara**ted from your **para**chute' are as effective as the general ones. Remind students that the '*i* before *e*' rule only applies to (some) words with the /iː/ sound.

Key

1a **1** boxes; heroes **2** wonderful; visiting **3** swimming; entered **4** beginner; success; surprising **5** beautiful; prettier **6** regretted; cancelling **7** panicked; arriving **8** niece; unwrapped **9** humorous; practise (*v.*)

1b **1** cities; families **2** relieved; receipt **3** moving **4** truly **5** neighbour; opening **6** advice (*n.*)

2 **1** accommodation **2** necessary **3** immediately **4** medicine **5** separate **6** pronunciation **7** professional; skilful **8** assistance

3a **1** a loose b lose (compare pronunciation and meaning with *choose/chose*)
2 a stationery b stationary
3 a complementary b complimentary
4 a effect b affect

3b **Suggested answers**
1 Their coats are over there.
2 I can't hear what you're saying – please come over here.
3 He has a nasty cut on his heel that won't heal properly.
4 He's bone idle – he just sits on the sofa all day watching his idol, the TV.
5 The miners had a minor crisis down the coal mine.
6 They had to give their boots a thorough clean after walking through the mud.

4 **1** The BBC offers a **24-hour up-to-the-minute** news service.
2 The band split up in 1998 but **re-formed** after a three-year break.
3 She still sees her **ex-boyfriend** quite regularly.
4 His **co-star** in the film was a very **down-to-earth** person.
5 The economy has enjoyed a **consumer-based** recovery.
6 There have been several **break-ins** in our neighbourhood recently.

Photocopiable activity
Activity 7A provides further practice in identifying easily confused pairs of words. Students work in pairs, identifying and correcting the wrong word in each sentence.

▶ **Student's Resource Book, page 73**

Use of English 1 p.109

1 Start by looking at the picture and discussing the questions. The difficulties she had to overcome could include racism and being rejected by both the white people that controlled the sport and her own people for wanting to compete.

Background

Cathy Freeman, born 1973, was trained initially by her stepfather before getting a professional coach. At first, she had trouble being accepted as a serious athlete before going on to win the 400 metres in the World Championships of 1997 and 1999. She now campaigns for aboriginal rights.

The word aborigine comes from the Latin 'ab origine' meaning 'from the beginning' and is used to signify a region's indigenous inhabitants. The Australian Aboriginals have existed there for over 40,000 years. The current number of 220,000, down from about 300,000 when Europeans first arrived in 1788, makes up less than 1.5% of the population. Nearly 40% of them still lead a traditional way of life in the Northern Territories. The majority of the rest live in towns and cities and have much lower employment and life expectancy than other Australians.

2 Remind students that Paper 3 Part 1 is the multiple-choice cloze. It is always a single text with 12 questions. Each question has a choice of four words always of the same type (all adverbs, nouns, participles, etc.) so all fit grammatically. However, only one of the words fits, either in meaning, collocation or surrounding structure (such as to form a phrasal verb).

2a Give students a minute to skim the text and answer the questions.

2b Allow students a suitable period of time (10 minutes) to complete the task before they compare choices and reasons and then check the answers.

3 In the task analysis look for phrasal verbs as well as collocations.

4 Follow with the exercise with a discussion. .

Background

Students might be interested to hear of the great Olympian Steve Redgrave who won gold medals in five successive Olympics from 1984 –2000 in rowing, one of the most physically demanding sports, and therefore maintained his level of commitment and training for 20 years.

Key

2a 1 She became the first Aboriginal to be selected for the Australian Olympic team.
2 She is having to come to terms with not being successful in athletics now.

2b **1** A **2** B **3** D **4** C **5** D **6** B **7** C **8** A **9** B **10** D **11** C **12** B

3 Examples of collocations: Question 1 – fulfil an ambition; Question 2 – running track; Question 4 – a role model; Question 6 – take a break from something; Question 7 – actively support; Question 9 – troubled by injuries;

Writing 1 p.110

This writing skills section looks at ways of rephrasing ideas (or expressing them in other ways!). It is a skill that is specifically required in Paper 2 Part 1, but students should also be able to see similarities with the register transfer task in Paper 3.

Start by asking students what they know about exchange trips and if anyone has been on one. If so, ask them to talk a little about it and what they gained from it.

1 Ask students to read the rubric and look at the map showing where the Orkney Islands are. Ask them what they think life would be like there.

Background

The Orkney Islands, with a surprisingly mild climate, being warmed by the Gulf Stream, comprise about 70 islands, of which 16 are inhabited. The vast majority of the 20,000 inhabitants live on the largest island, confusingly called the Mainland. The population, originally of Scandinavian descent, was in decline for a number of years, but is increasing again, as the islands' remoteness becomes more attractive.

1a/b Get students to read the input material, marking the parts that they consider will be most useful.

2 Ask students to read the two extracts and establish: a) that they contain the same information as the input material, and b) that it is expressed slightly differently. Get students to find examples of parallel expressions and to note different techniques used to rephrase the input.

3 Elicit where extracts A and B would come from (the beginning and the end) and give students a suitable time to complete the middle section, including all the parts they highlighted in Exercise 1.

Key

2 The paraphrases use a variety of changes:
- in vocabulary (e.g. *partner someone up with / match someone with, began to understand/realised*)
- numerical (e.g. *four-week / 30 days*)
- positive–negative (e.g. *not there long / there only a short time*)
- change of word class (e.g. *to lead a happy* (adj.), *life* (n.); *in safety* (n.), *to live* (v.) *happily* (adv.) *and safely* (adv.); *a four-week* (adj.) *exchange, for four weeks* (n.))
- register change (e.g. *have the same aspirations / want the same things*)
- other paraphrases: *to welcome / to host; my own age / the same age as you; became one of the family / as part of the family / made me feel at home; learning about how people in other countries live / get to know people from a different culture; who we are or where we live / it doesn't matter where people come from*

3a Suggested answer

I have just returned home after the best four weeks of my life. For 30 days, I lived on one of the Orkney Islands, just off the coast of Scotland, as part of a student-exchange programme. The organisers partnered me up with a boy my own age and, although I was there only a short time, I became one of the family. Next year, I will be welcoming their son, Tom, into my own home.

The family I lived with were marvellous. Although they'd never met anyone from my country before, they gave me a warm welcome. They were very friendly and pleasant and showed me around. And Tom was great, we got on so well because we shared many of the same interests.

The whole programme was extremely significant for me because not only was this the first time I'd ever been abroad, it was the first time I'd ever been away from my family. At first I was very frightened, but I soon learnt to look after myself and I think I've become a much better person. What I also enjoyed was learning about how people in other countries live. I began to understand that no matter who we are or where we live, we all want the same thing: to live happily and safely.

(217 words)

▶ **Student's Resource Book, page 75**

7B Kicking the habit

Listening 2 p.111

Start with books closed and elicit the word *obsession*. Alternatively, introduce the word by playing a quick game of a variety of hangman (e.g. put the nine dashes on the board to represent the letters, telling students it is a noun, and draw ten chocolates. Tell the students that they have to guess the word before you eat all the chocolates, eating one/rubbing one out each time they choose an incorrect letter). Either way, elicit types of obsession.

1a In the task rubric, get students to note the important words (*habits which they are unable to control*) and to look at the list of eight possible habits in Task 1, before discussing the questions.

1b The class could be divided into groups, with each group brainstorming words and expressions for one of the eight problems before sharing them. The idea here is for them to start thinking about what the topic might include and some of the language that might be used.

2 Remind students of a suitable strategy for this type of matching; they can either answer both parts the first time they listen and check them on the second time or (a better idea) answer Task 1 the first time and Task 2 the second time. Remind them that they hear all five speakers once before the whole piece is repeated. Give students a chance to read through the complete task and the Help points before they listen. Check the meaning of any difficult words with paraphrases (e.g. *needless* = *unnecessary*, *deteriorating* = *getting worse*).

3 Before the role-play, check that students can use the expression correctly by eliciting the type of words that come next (e.g. *Have you tried + -ing?*, *If you didn't/were …* (2nd or 3rd conditional), *Would you consider + -ing?*).

Key

2 1 G – *gym membership … it's taken over my life. I'm now working out five nights a week.*
2 H – *torn myself away from the screen … One link leads to another.*
3 E – *a ready-meal on a tray … the wrong kind of stuff*
4 F – *Half the time, I don't need the stuff*
5 A – *I'm just ringing for the sake of it*
6 B – *none of them ever calls any more*
7 F – *my parents think … and I let them think that*
8 G – *My girlfriend's always telling me … We had a big row about it*
9 H – *it infuriates my parents*
10 D – *my last bill was astronomical*

Speaking p.112

1a Start by getting students to look at the words in and below the box. Establish that they are all adjectives and that the majority, like most adjectives of emotion, are formed from a noun (e.g. *thankful, regretful*) or from the past participle of verbs (e.g. *astonished, stunned, relieved*).

1b The box contains idiomatic expressions of emotions, some of which are similar in meaning to the words in Exercise 1a. See if students can match them before rewriting the sentences (*prey on your mind – worried; come as a huge surprise – surprised; be a weight off your mind – relieved; be over the moon – pleased*).

1c Focus on the example and point out to students the difference between using the vocabulary in Exercises 1a and 1b and just saying he looks happy/unhappy.

Extra!
There is a chance to expand the use of modals of deduction. For example, have students look at the example again and find the modal (*he must be feeling really thrilled about it*) then go on to ask where he is from (*He must be in an English-speaking country because the certificate is in English. He can't be British or Australian because he's in a left-hand drive car. He could be American or Canadian*).

Background

The athlete is Kelly Holmes who, at the age of 34, won gold medals in the 800 metres and 1,500 metres in the Athens Olympics in 2004.
The man is celebrating his retirement. The legal age for retirement in the UK is 65, although there are currently moves to raise this to 70.
The couple with the cheque are winners of the British national lottery, a twice-weekly draw with a top prize of many millions of pounds.

2 There are various possible answers here that students could argue as they wished (e.g. 1 – b, e, then d). Note use of second (unreal) conditional structures.

3 As in previous modules, students first identify the task, then they listen for content, and finally they comment on performance.

4 Allow students to listen again, this time noting down specific functional phrases.

5 Emphasise the need for fluency and task completion within the minute available.

6 Remind students of their answers to Exercise 4 and ask them if they managed to use any of these expressions in their answers to Exercise 5. Emphasise to them the importance of these aural signals to help their listener follow what they are saying.

7 If students have not experienced any of the situations in the pictures (it could be winning any race/competition – it doesn't have to be an Olympic gold!), spend more time on the second question where they can select other important/emotional events in their lives.

Key

1a 1 PLEASED: thrilled, elated
2 RELIEVED: thankful, grateful
3 WORRIED: anxious, apprehensive
4 SURPRISED: astonished, stunned
5 SAD: regretful, dejected

1b 1 Now the exams are over, it's a weight off my mind.
2 The news came as a huge surprise.
3 My parents were over the moon when they heard I was getting married.
4 The thought of losing my job has been preying on my mind for months.
5 Even though I didn't win, I have no regrets about taking part.

2a Most likely answers: **1** e **2** c **3** b **4** a **5** d

3a 1 Compare and contrast the photos.
2 Say what impact these events might have on these people's lives.
3b She talks about photos C, D and E.
3c She achieves the task well. She talks about all the photos and the impact of each. She compares the similarities (lifestyle changing, no going back) and mentions some differences (less freedom for the couple, more for the man retiring). She keeps speaking, structures her answer well and ends with a summary.
4 1 Right, well …; So …; Well, …; What about …?
2 both … and … are …; but in a differnt way because … whereas …; On the other hand …
3 I think …; perhaps …; maybe …; I'd say …

Photocopiable activity

Activity 7B is an extension of Exercise 7. It reviews the vocabulary of emotions and gives practice in describing situations and their implications. It also practises the use of paraphrasing techniques.

▶ **Student's Resource Book, page 77**

Use of English 2 p.114

Start by putting the class in groups to discuss their attitudes to chocolate. They could discuss how much they like it, whether they have an addiction or allergy to it, suffer cravings and what type of chocolate they like best.

1 Then get them to look quickly at the four questions.

2a They then skim the text to find the answers to the questions. Ask them not to look at the words on the right at this point but to read just the words in the text. Remind them that in the exam they should follow the same process to get an idea of the content and style of a text before they attempt the questions.

2b Remind them that will have 8–10 minutes to complete the ten questions and that they should check for plurals and negatives as well as focusing on changing the word class.

3 The analysis should again help focus on the type of changes to look out for in the exam. If time, focus on some of the spelling patterns. For example; please–pleasure, (drops the *e*, add *-ure* to form noun), profit–profitable (add *-able* to form adjective), innovate–innovation (change *-ate* to *-tion* in verb to noun).

4 Here the idea is both to demonstrate the range of word forms that can be derived from one stem and to give students a chance to extend their vocabulary in an organised way. Look at the example together: price can be both a noun and a verb, from which the participles *priced* and *pricing* can be used, in addition to the adjectives *priceless* (too valuable to estimate) and *pricey* (expensive). Allow them to choose the words that they think are most useful. Alternatively, divide all the words between the students, give them a few minutes with a dictionary to do the research and then get them to teach the class all the words that can be derived from the root, by coming up to the whiteboard and producing a word tree.

Key

1 **1** a **2** b **3** b **4** b
2b **1** undrinkable (negative adjective)
2 countless (adjective) **3** recognisable (adjective)
4 distinctive (adjective) **5** profitable (a adjective)
6 Unbelievably (adverb) **7** beneficial (adjective)
8 innovations (noun plural) **9** manufacturing (uncountable abstract noun) **10** pleasure (noun)
3 **1** undrinkable, unbelievably
2 innovations – needs to be plural as there is no indefinite article and is followed by an example proceeded by *such as*

Language development 2 p.115

1 The first part is designed to give an overview of the grammar, so it might be a good idea to get students to do it, to see how much they understand, before looking at the Grammar reference on page 183. Encourage students to think of the meaning of the complete sentence.

2a Here, students can choose the correct options and then check the answers in the Grammar reference, or refer to it as they are doing each question. Stress that in some questions, only one answer is possible, but in others, where there are two possible answers, they should try to identify the differences.

2b Students move from recognising the correct answer to producing it. Suggest that they look back at Exercises 1 and 2a for help.

3a This part deals with other useful structures that also come up in the exam at this level. They are alternative ways of expressing unreal situations.

3b Emphasise that there are a number of possible answers for many of these questions.

4 The missing words in the text cover everything on the page. Students need to complete it by selecting the most suitable tense for the verbs.

5 One variation here would be to get students to write sentences that were true for them, collect them in and read them out for the rest of the class to guess who wrote them.

Key

1 **1** is – *wish* + past for unreal present
2 watch TV – *If only* = *I wish*
3 not optimistic – *wish* + *would* for actions beyond our control that are unlikely to change
4 know – *If only* / *wish* + past perfect for unreal past
5 doesn't – *he lived* = past tense for unreal present (compare *he acts as though he lives there* = *he might*)
6 should leave now – unreal past: we haven't left yet
7 shouldn't smoke – *I'd sooner* + past = polite request
8 might – *Supposing* = *If* in a second conditional (it's possible)

2a **1** hope – real future possibility (compare *I wish* = unreal, *I want to pass …*)
2 could – *wish* + past form to express unreal present
3 was – *If only (I wish)* + past form to express unreal future
4 wishes – 3rd person present needs *-es*; ongoing wish in the present
wished – single act in the past – he no longer wishes it)
5 didn't – past simple for present habit, he does it regularly
hadn't – past perfect for single act in the past
6 had to – past obligation for unreal present
7 could – for unreal present ability (compare *would* cannot be used to refer to oneself/things, *should* for real situations)
8 you/he – *wish* + *would* possible with other people, but not with *I* as it expresses things beyond our control
9 you only / only you – *If you only* is a more emphatic variation of *If only you* (compare *I wish you knew* or *If you knew* = conditional)
10 had paid / had been paying – *wish* + past perfect simple or continuous (the reality is *I haven't paid …* or *I haven't been paying … recently*)

2b **1** *I wish I had a personal trainer* – about present
2 *I wish I hadn't eaten so much last night* – about past
3 *I wish he wouldn't keep sending me text messages* – another person's annoying habit that I don't expect him to stop
4 *If only I could get into these clothes I had five years ago* – own present ability
5 *I wish I had paid / been paying my girlfriend more attention* – regret about past (repeated) action

3a **1** *to stop / I stopped* – *it's time* + *to* + infinitive / *it's time* + subject + past
2 *you changed* – only use subject + past after *it's high time*
3 *stopped / would stop* – *I'd prefer* + past like a second unreal conditional; *would stop* is more distant/formal
4 *was / were* – past because I'm not stupid
5 *get / got / had got* – *get* = real likely possibility; *got* = unreal or unlikely possibility; *had got* = unreal past
6 *doesn't / didn't* – *doesn't* = present for real possibility; there's a good chance it won't / *didn't* = past for unreal; there's a very small chance it won't work
7 *know / knew* – *know* = real and timeless – they still know; *knew* = real past – they did at the time

3b **Suggested answers**
1 It's time to go / I went / I was going back to work.
2 Supposing I was/were to join a gym, how much would it cost per month?
3 I'd rather/sooner you didn't spend all your time watching TV.
4 He looked as if he had been sitting in front of a computer all night.
5 I'd rather / I'd sooner you hadn't told me what the film was -about.

4 **1** did **2** hadn't done (*unreal – I have done some*) **3** had been **4** exercises (*he might*) **5** was/were **6** didn't try **7** could get

Photocopiable activity

Activity 7C reviews expressions with hypothetical meanings. The class works in groups, playing against each other, to see how many correctly matched sentences they can form in a given time.

▶ **Student's Resource Book, page 78**

Writing 2 p.116

Start by asking students in groups to think about something that they have complained about or wanted to complain about. Ask them to say what the problem was and what they did or could have done to complain.

1 Spend a few minutes discussing the two questions together. It might be interesting to compare smokers' and non-smokers' opinions.

2 Give students sufficient time to carefully read the instructions and all the input material. Then answer the five questions designed to help them determine content and style.

3a Students select what is and isn't relevant for the letter.

3b They then need to decide how to organise the points into paragraphs.

4a Some of the phrases are more formal and so are suitable for the letter.

4b Here, students select a suitable opening, considering the purpose and style of the letter.

4c They can choose to use these expressions or reject them if they wish.

5/6 Time how long it takes students to write the composition and check it systematically, ensuring that they write it once carefully from the plan and don't waste time writing it then copying it out again. Ask them what they were checking for and if they followed the checklist on page 188.

Key

2 1 A part-time worker in a restaurant; a ban on smoking has been proposed in public places.
2 Letter: to complain to the City Council about a proposed smoking ban in public places
3 Leaflet: council explaining ban to local businesses (with your reactions); letter: to a fellow student saying what happened in your country and your plan to write a letter
4 Leaflet: reason for ban; timing/extent of ban; consequences of flouting; letter: details of the experience
5 Aim of letter: to get council to change its mind; style: formal, diplomatic but firm, not too forceful (to persuade the reader)

4a **1** drop **2** take **3** introduce **4** damage **5** round **6** majority **7** wider **8** oppose **9** suffer **10** down **11** effect **12** right

4b 1 correct level of polite formality 2 too informal

5 Suggested answer

I am writing to express my great concern at your proposed smoking ban. I fear that such a move would damage businesses, and many workers would lose their jobs.

At present, I am working for Café Noir, a well-known restaurant popular with visitors, many of whom have told me that if such a ban were introduced, they would take their custom elsewhere.

On top of that, I think the amount of notice you have given people is unreasonable. If you are going to make such a huge change, there needs to be much wider public debate, and if a majority of people agree, they should be given enough time to get used to the idea.

To support my argument, I would like to tell you what happened in my country when a ban was introduced. Firstly, it was almost impossible to enforce. There simply weren't enough police to go round checking every public place. Secondly, business suffered enormously, with many people objecting that they should be given the right to smoke if they wished. In the end, the government had to back down and instead tried to persuade restaurants and cafés to provide non-smoking areas.

This approach strikes me as much more realistic, and I urge you to reconsider before it is too late.

(214 words)

Module 7: Review p.118

1 **1** D **2** C **3** A **4** B **5** C **6** D **7** A **8** C **9** D **10** C **11** D **12** B

2 **1** If we club/clubbed together, we could get her a great leaving present.
2 I got/became hooked on TV soap operas when I was ill in bed recently.
3 He spends money (just) for the sake of it.
4 I apologise for not ringing you.
5 She's not short of money.
6 There's no point (in) pushing him too hard.
7 I never tempt fate by saying I'm going to win in case I lose!
8 Her speech had/made a powerful/great/huge/profound impact on me.
9 What drives him is the desire for success.
10 There's no chance of me ever giving / that I'll ever give up eating chocolates!

3 **1** complimentary; practice **2** procedure; entered **3** realise; priceless **4** Unbelievably; lose **5** two-year; swimming **6** Unknown; idle **7** heels; receipt **8** There; unprofitable **9** beautiful; behaviour **10** truly; ineffective

4 **1** as **2** were/was **3** could **4** would **5** were **6** only **7** had **8** sooner/rather **9** were **10** what **11** wish **12** had

Module 8 Making life better?

This module includes topics such as the effects of television, modernisation and change, the 24-hour society, and advertising before moving on to aspects of law and order, including fighting crime and crime prevention.

Lead-in p.119

One way to begin would be to start with books closed and brainstorm different ways or places to advertise. Alternatively, ask students to talk in groups about a favourite advert from television, print or elsewhere, and one that really annoys them.

Then look at the photos and discuss the questions. If students say that they are not influenced by advertising, ask them why they think companies spend so much money on it. The photos show: an advert for a snack painted on the side of a bus (top); a plane towing a banner advertising a lottery (middle); the neon advertising panels in Piccadilly Circus, London (bottom left); and a magazine advert for a mobile phone (bottom right).

8A A changing society

Reading p.120

With books closed, ask students to spend a few minutes talking about television. Ask them how much they watch and the type of programmes that they most enjoy watching. Then (to link to but contrast with the text), ask them whether they have always had television at home and how their lives would change if they suddenly started to live without it.

1 Get students to look at the picture, title and subheading of the text, asking them not to read further at this stage. Ask them what, if anything, they know about Bhutan, and what effect television might have had on such a remote country.

Background

Bhutan, which lies at the eastern end of the Himalayas, with Tibet to the north and India to the south, is certainly a little-known country; current estimates of its population vary between 700,000 and two million. It has one of the lowest urban populations in the world, with only 30,000 living in the main city Thimpu, said to be the only capital in the world without traffic lights. Seventy-five per cent of the people are Buddhist, with 25% Hindu of Nepali origin. An Indian protectorate, it has not always been the peaceful paradise it is made out to be; the monarchy was only established in 1907 after 200 years of fighting and feuding. There are currently 100,000 refugees living in Nepal following ethnic unrest. The government welcomes tourists, but only on organised tours and charges them $200 a day.

2 Give students a suitable time limit (two to three minutes) to skim the text to get a general understanding of its scope and style and to look for any mention of the effects television has had.

3 Elicit a suitable strategy and the type of clues students should be looking out for to help match the paragraphs (grammatical links, such as adverbials, pronouns, articles, and other reference or linking words, and lexical links such as repetition or parallel phrases). Give students 10–15 minutes to complete the task, reminding them to finish by quickly reading through the complete text to make sure it all fits together grammatically, logically and cohesively.

4 Allow students enough time to scan the text for references to life before and after television so they have plenty of ideas for the discussion. Focus on/elicit useful vocabulary during this discussion – make sure students mark relevant parts of text (e.g. crime wave vs. only social vice … *was over-indulgence in rice wine*, etc). Discuss how the suddenness of the introduction of TV was a major factor, compared with the country's previous isolation from the world, forced on it by its leaders.

Key

3 1 D – the first sentence (*The explanation …)* answers the question before the gap; *the change* refers back to the sudden crime wave. In line 16, after the gap, *Only a few years later* refers back to lifting the ban in 1999 in paragraph D; it

refers back to the new cable television service.
2 B – *In the real Bhutan* contrasts with the mention of the fictional Shangri-La; *his* in line 28 refers back to *the current Dragon King's father*
3 F – *It* (= *outlook*, line 39); *wasn't completely spiritual either* contrasts with *not quite so materialistic. Later that same year (*line 40) refers back to the mention of *In 1998 …*
4 A – Three clear links between the first sentence of paragraph A and the previous sentence: *Beneficial or not* links to *will be good*; *it is certainly omnipresent* refers back to *television*; and *to this crowded country* refers back to *for our country*. Also *the impact of foreign influences such as this* (line 58) refers back to *a violent Australian programme.*
5 E – *pupils* refers back to *children*; *this onslaught* refers back to the negative impact mentioned in the previous paragraph; *They* (line 68) and *their parents* (line 69) refer back to the children mentioned in the paragraph.
6 C – *this new materialism* refers back to the examples of consumerism mentioned in the previous paragraph.

Vocabulary p.122

1a The first part focuses on verb phrases which are idiomatic expressions or strong verb collocations. As the verb phrases are all from the text, students could first match the halves, underlining the key part, before finding them in the text to check and to see them in context. Check students understand the meaning of them by asking them to paraphrase them (e.g. *bowed to popular pressure* = finally agreed to do what the people wanted him to do).

1b The passage uses paraphrases, which students need to replace with forms of the expressions in Exercise 1a. Follow up by finding out how many of the class enjoy playing computer games. The passage refers to the Sony Playstation and the Microsoft Xbox.

2a Students should first skim the text to get an idea of what it is (a political party's policy leaflet) before completing it with verbs taken from the text, using a dictionary to help if needed. Again, it would be useful to compare their use here with their use in the original text to develop understanding of their meaning and connotation.

2b This could be just a quick discussion or extended by getting students, possibly in groups, to write their own manifestos.

3a This section aims to review aspects of word formation, looking at the relationship between adjectives and nouns. When students have completed the table, either from their knowledge of the words, from a dictionary or by finding them in the text, point out that sometimes the adjective is formed from the noun (*idyll – idyllic*) and sometime vice versa (*content – contentment*).

3b The important point for students to note is that in some words (e.g. *content – contentment*), there is no shift in stress, but in others there is, and some suffixes follow generalised stress patterns.

3c To complete this text, students will need either a noun or an adjective in each gap.

Background

The are many styles of yoga, such as Hatha (traditional), Sivananda and Ashtanga (or power yoga), all variations of the basic philosophy which dates back thousands of years. Yoga became popular in the western world in the 1960s and remains so. B.K.S. Iyengar has been teaching his style, which emphasises precision in each posture, for over 60 years.

Key

1a **1** e (*bow to (popular) pressure*) subheading
2 d (*play a (crucial) role*) line 46
3 g (*capture the market*) line 61
4 a (*think nothing of -ing*) paragraph C
5 h (*can't be bothered to*) paragraph C
6 c (*lift a ban on*) paragraphs D/G
7 b (*reel off a list*) paragraph E
8 f (*stand accused of -ing*) line 16

1b **1** captured the market **2** reel off a list of **3** think nothing of playing **4** can't be bothered **5** play a crucial role **6** to stand accused of **7** to bow to popular pressure **8** lift the ban

2a **1** initiate **2** consume **3** transform **4** maximise **5** erect **6** inhibits **7** relent **8** confide in

3a/b **1 materialistic**, materialism
2 influential, **influences**
3 preoccupied, preoccupation
4 content, **contentment**
5 popular, **popularity**
6 controllable, control
7 spiritual(ly), spirituality
8 idyllic, **idyll**
9 beneficial, benefit
Adjectives (and nouns, e.g. *cynic, fanatic*) ending in *-ic* are stressed on the syllable before (e.g. *idyllic, materialistic*), as are nouns ending in *-ity* (e.g. *spirituality, popularity*) and *-tion* (e.g. *preoccupation*).

3c **1** materialistic **2** beneficial **3** influential **4** popularity **5** preoccupied **6** spiritual **7** contentment **8** control **9** idyllic

▶ **Student's Resource Book, page 83**

Listening 1 p.123

1 The lead-in is to create interest in the topic of living in a 24-hour society. Students could also be asked if they have shops that are open 24 hours where they live and if so, whether they use them late at night, and whether any of them use Internet banking services out of hours.

2 Give students 30 seconds to look at the questions before playing the recording twice, with only 30 seconds between. Give them a chance at the end to compare and justify their answers, but remind them that, in the exam, picking the right letter is enough. Remind them also that if they are unsure, they should guess. Compare strategies. One way is to attempt both parts the first time they listen and then check/confirm the answers the second time. The alternative is to answer the first task the first time they listen and the second task the second time. The strategy will determine how much they read at first.

3 Ask them to pick up on any interesting points the speakers made, encouraging them to support their arguments with reasons of their own.

Key

2 **Task One**
1 A – *sending their kids to history classes … find myself marking exercise books …*
2 E – *many employers like myself …I need people to help me get that urgent deal sorted out*
3 H – *pets and zoo animals don't take any notice of surgery hours*
4 C – *you have to be if you work on newspapers … My job involves both writing and sub-editing*
5 F – *people are always wanting lifts to and from airports at ridiculous times … added extra to the sum on the meter*
Task Two
1 D – *I can't see people sending their kids … in the middle of the night*
2 H – *there's a danger that some bosses might take advantage*
3 A – *I'm not really too bothered one way or the other … it's six of one and half a dozen of the other.*
4 B – *and of course more cars at night would mean not so many during the day, which would have to be an advantage for everyone*
5 F – *it's just not good for you – it plays havoc with your body clock … there are risks.*

Language development 1 p.124

This section focuses on an aspect of grammar that may be new to some students, as it is not usually studied below this level. It relates to earlier writing sections, such as 'Making your writing more interesting' on page 62, by looking at ways certain parts of a sentence can be made more emphatic.

The two ways highlighted here are by using a number of fixed negative expressions at the start of the sentence and by changing the word order.

1a The idea of the first part is just to recognise the emphasis and identify how it is achieved. Get students to look at the three pairs of sentences and then answer the four questions before consulting the Grammar reference on page 183.

1b Students need to make the necessary transformations. Do the first one together as an example, eliciting the changes required: highlight the change of word order, the inversion of *we* and *will* and the change from *not … any* to *no*. Point out that, in some sentences, they will need to add other appropriate auxiliary or modal verbs. Other expressions that can be used in this way include *seldom, scarcely, only by, not until, nowhere, so.*

2 Start by asking students if they use an Internet banking service and whether it is efficient or not, then ask them to skim the passage and ask what the writer thinks. The exercise is designed to give further practice of these structures, but it would be worth pointing out that the type of words that they need in the gaps are similar to those required in Paper 3 Part 2, the open cloze.

3 This section focuses on the method and purpose of changing the word order of a sentence to give more emphasis to a certain part. Put a sentence using such a structure on the board with and without the emphasis (e.g. *I looked everywhere for my glasses. Finally, I put my hand in my pocket and they were there / there they were.*) and elicit the differences. Point out that the second is more emphatic and *there* links more closely to *my pocket.*

3a The idea here is to see how the emphasis is achieved with a variety of structures. By changing them back into a less emphatic form, students should be able to see what is emphasised and how the emphasis was achieved.

3b Students first identify the part to emphasise and then, using the patterns in Exercise 3a, rewrite the sentences.

Extra!
If there is time, extend the exercise by asking students to complete some sentences in a way that is true for them. Suggestions include:
Under no circumstances would I ...
At no time have I ...
Not for months/years have I ...
Seldom do I ...
So ... am I that ...
Try as I might, I ...
I ..., as do/does ...
... I may be, but ... I'm not!

Photocopiable activity

Activity 8A can be used here. It is a type of board game that students play in groups, using key words to complete sentences to revise the language of this section and the vocabulary from page 122.

Key

1a **1** 1 b 2 b 3 b
2 By starting each sentence with an emphatic construction.
3 Subject and verb are inverted after an emphatic construction (as in a question form).
4 Auxiliary verb *did* with verbs in the present or past simple.

1b **1** Under no circumstances will we enter into correspondence with competitors.
2 On no account will the judges discuss their decision.
3 Not only is this slogan witty, it is also original.
4 Not since winning the school poetry prize have I been so excited.
5 Rarely do you get the opportunity to visit such a remote place.
6 Hardly had we arrived, when we were besieged by reporters.
7 Only now are they beginning to realise what a mistake they made.
8 Not a word was said (by him)/did he say all evening.
9 No sooner had I dropped off to sleep, than there was a knock at the door.
10 No way will I go / am I going to the party wearing that!

2 **1** only **2** but **3** did **4** had **5** when **6** sooner **7** than **8** was **9** Never **10** has

3a **1** 'He's going to resign.' '**I find that** hard to believe.'
2 We arrived at the base of the mountain. Then **the long trek to the summit began**.
3 **I spent hours** thinking of a slogan for the competition!
4 **It may be difficult**, but **it isn't impossible**.
5 The restaurant serves excellent food. **Their starters are (the) best of all**.
6 **The response to our offer has been so great** that the deadline has been extended.
7 The King is very worried about the situation, **and his government is, too**.
8 **However hard we try**, we will never surpass their achievements.

3b **1** **So tense** was the competition that tempers flared.
2 **Try** as they might, they weren't able to overtake the leaders.
3 Shakespeare wrote many plays, but *Hamlet* is his best-known work. / The best-known of Shakespeare's many plays is *Hamlet*.
4 We stayed in a hotel in the old part of town. **Opposite** it / the hotel was a statue of the city's founder.
5 So **good** is his work that he deserves the Nobel prize for literature.
6 I'm anxious for news, **as are** the other team members.
7 **Weeks** it took us to finish the project.
8 'It's a beautiful place and very cheap.' '**Beautiful** it may be, but **cheap** it isn't.'

Use of English 1 p.125

1a Compare the meanings of the five nouns given. Get students to work out which of the words fit each of the sentences and establish that *grasp* is the only one that fits all three. Compare the meaning in each sentence: understanding, ability to achieve something, way of holding.

1b Give students a few minutes to do the exercise and then compare their answers in groups. Point out that if they can't find a word that fits all three sentences but can find one that fits two, they should have a guess as it might have a use that they are unfamiliar with.

2 The task analysis looks at the way the various meanings of a word, although different, may be connected. For example *serve* in question 4 has various meanings related to doing something for others. *Gentle* in question 5 has three meanings that all suggest something not tough.

Students might be able to think of other meanings of some of the words. For example, *serve*: to present a legal document to someone, to start a point in games such as tennis.

Key

1a grasp

1b **1** flat : lacking interest, a note slightly lower than it should be, level with the ground
2 lay: past of lie, transitive verb – to put something down, to prepare a table before a meal
3 term: an expression, a period of time, the division of an academic year
4 served: to give someone food, to spend time doing useful work/military work, to provide people with something useful
5 gentle: kind and careful, not extreme or strong, not steep

▶ **Student's Resource Book, page 84**

Writing 1 p.126

This section focuses on using certain types of phrases to indicate attitude to facts or opinions that are important in certain writing styles such as reports or proposals.

Start by asking students about their attitudes to technology, and whether they are technophiles or technophobes. Ask them how knowledgeable or well informed they feel about recent developments.

1a Students compare extracts A and B and answer the two questions that are designed to draw their attention to the use of attitude phrases and the effect they have on style.

1b Have students look at the box and the functional categories that these phrases can be used for. They should then read the other extracts C and D and, from the context, identify the functions of the phrases in italics. Then look at the Writing reference on page 203, which lists these and other examples.

2 Here, students should first decide what function the phrase has, and from that decide which one is more appropriate. For example, in pair 1, the appropriate phrase should generalise, not give an opinion about the first point.

3 There could be different ways to do this. Either students could choose which topic to interview others about, or the class could be divided into three groups, with each group being given one topic to prepare and ask questions on before producing a report.

Key

1a **1** Generally speaking
2 What was particularly noticeable was that

1b Generalising: *generally speaking*
Opinion: *understandably*
Commenting: *as we shall see, surprisingly, evidently*
Emphasising: *indeed*
Admitting: *admittedly*

2 **1** Generally speaking **2** For example **3** In fact **4** Not only that **5** Judging by what they said

3b **Suggested answer**

REPORT ON ADVANTANGES AND DISADVANTAGES OF TECHNOLOGY

Introduction

The aim of this report is to present the views of a number of people interviewed on the benefits or otherwise of technology in three key areas.

1 Education

Generally speaking, the people we spoke to believed that bringing the latest technology into education was a good thing, although, arguably, in some areas, like the use of calculators in maths lessons, some felt it had made students more lazy. In the main, though, our interviewees appreciated the fact that there are so many more sources of information that are readily available nowadays.

2 Shopping

Surprisingly, perhaps, when it came to shopping, people were less convinced of the benefits. Of course, they liked the ease with which they can find and buy things on the Internet, but most regretted that it encouraged people to spend too much time indoors behind their computers instead of meeting people in town. They noted that, as a result, many town centres were dying.

3 Housework

On the other hand, people were very positive about technology helping them perform mundane chores in the house. Some said they couldn't imagine life without their washing machine or microwave and, clearly, many of them would welcome all jobs being taken over by robots.

Conclusion

All things considered, people were positive about benefits of technology, only expressing minor doubts in relation to education and more serious doubts about the effect of changing shopping habits.
(242 words)

▶ **Student's Resource Book, page 86**

8B Law and order

The second half of the module focuses on the issues of crime, crime prevention, catching criminals and dealing with crime.

Listening 2 p.127

One way to begin would be, with books closed, to elicit names for different crimes, for example, but going round the class, with students being knocked out if they can't think of another one quickly, until there is one winner left. Alternatively, introduce the topic by drawing a stick woman on the board with the name Laura Norder underneath and ask the class if she is more likely to be a criminal or a policewoman. Clearly they will have no idea until they start saying the name aloud and hear the link with the title of the section, *Law and order*. The serious part is to point out various aspects of pronunciation in the phrase; the linking of the first two words, the intrusion of the /r/ sound between the two vowel sounds ('lɔːrən), and the elision of the *d* in the unstressed *and* (a feature common in such binomial expressions: *fish'n'chips, black'n'white*).

1 Students discuss the type of crimes/offences more commonly associated with young people. It will be harder for them to answer the second question, but might hopefully raise some of the points mentioned in the listening.

2 Elicit what candidates are required to do and a suitable strategy. Refer to pages 170 and 171 if necessary. Give students 30 seconds to skim the introduction and the questions only. Ask one or two concept questions. E.g. Who are Martin, Mary and Glen?

Give them a little longer to read the answers before playing the recording twice.

3 Discuss the questions in small groups, asking if they think such a scheme would work in their countries.

Key

2 1 A – *certain youngsters becoming potential future offenders … everyone pulling together to stop this happening*
2 D – *of far more significance in my view is how well a child relates to his parents and siblings*
3 B – *I felt quite glad in a way because it meant that something was finally going to be done*
4 C – *I came across a leaflet from the library*
5 D – *I was encouraged to give him loads of attention … rather than just being negative all the time*
6 C – *he will now sometimes give me a hug and show me he does actually care*

Speaking p.128

This section covers different ways of fighting crime. Start by asking students about their attitude to the police, if they do a good job and whether or not the police are generally respected in their country (assuming there are no police officers in the class!).

1a Get students to look at the photos and identify the connection (they all show different aspects of police work). Then ask students to spend a few minutes looking at the expressions in the box that all relate to policing, dividing them into those they know and those they don't. Next, by asking classmates or using a dictionary, get them to check those expressions they don't know. Finally, in groups, ask them to describe the pictures using as many of the expressions as they can.

1b Ideas that might come up here include: surveillance, undercover work, stop and search, questioning witnesses, community work, attending car accidents, etc.

1c Again, make sure students understand all the points, perhaps by going through them one by one quickly with the whole class, eliciting the meaning or examples, before they decide which they would support.

1d The results of the British poll are on page 207. Although it doesn't give the age range of the people polled, it shows that the police are generally held in high regard, with a huge percentage wanting to help them by giving them information, more resources and through greater co-operation.

1e Other suggestions might include the death penalty, for which there is no evidence of success, and a popular policy in USA known as 'three strikes and you're out' (from the baseball term), meaning (here) that anyone convicted for a third minor offence automatically receives a long (20-year) prison sentence.

2a Ask students to skim the article and say which of the points in Exercise 1c it relates to (reporting information to the police). Then, either by working in groups or with dictionaries, ask them to complete the passage by choosing the correct words, thinking also about what the others mean or how they would be used.

2b Ask students if they have a television show (like Crimewatch in the UK) where the police ask the public for help in solving specific difficult crimes and finding wanted people and viewers call in with leads.

3 Again, make sure students are familiar with all the vocabulary before they attempt the discussion (e.g. *libel* is a written or printed untrue statement that gives a bad impression, whereas *slander* is spoken false statement).

Background

Anti-social behaviour orders (ASBOs) are a new approach in the UK to tackling anti-social activity such as: vandalism, graffiti, harassment, causing disturbances, being noisy late at night, threats of violence, drug abuse. An ASBO can be taken out, through the courts, to prohibit the person/people causing the trouble from certain acts. They are community-based and are designed to protect the public rather than punish the perpetrator. They can include fixed-penalty fines, banning people from certain areas, prohibiting people from congregating with other named individuals, confiscation of sound systems, and forcing drug addicts to undergo treatment.

Photocopiable activity

Activity 8B can be used any time after this to provide further practice of the language of crime. Students work in groups to decide suitable punishments for a number of different offences.

4 The focus here is on the collaborative task (Part 3) and the following three-way discussion (Part 4).

4a Students listen first to the task instructions. Ask how many things they have to do and what they are. Also ask how long they have to do it (four minutes).

4b/c Students get a chance to do the task and report back to the class.

5a Remind students that the questions they are asked in Part 4 follow and extend the theme of Part 3. The five questions here are examples of how that might happen. Stress that students should not actually answer them at this stage, just think about the content of their answers.

5b As students listen, they should mark the questions that the examiner asks and comment on the students' performances.

6a Play the recording again. Students listen for the language used. Remind students of the use of past tenses to talk about an unreal present, referring back to Language development 2 in Module 7 if necessary.

6b Encourage students to add comments saying why they agree or disagree.

6c Students follow the patterns in Exercise 6a to rewrite the sentences, but warn them to consider whether each one is expressing a real or unreal concept.

7a/b Now students should have the ideas and the language to do the task themselves in an exam format, and then compare their ideas with others in the class.

Key

1a **Suggested answers**
A the police have arrested somebody.
B the police officer is carrying out desk work.
C The police are carrying out crowd-control duties. They have erected crowd barriers.
D The police officer is carrying out protection duties. He is protecting Prince William.
E The police officer is using a speed gun. He is trying to catch speeding motorists.
F The police have sealed off the area while they attend the scene of the crime.

1b **Suggested answers**
Crime prevention, community liaison, investigate/solve crimes, keep the peace, respond to emergency calls, help victims of crime

2a **1 general** public **2** drug-**dealing** **3** scheme
4 anonymity **5** written **statement**
6 come forward **7** reprisals **8** entitles
9 take **up** (= accept) **10** set **up** (= established)
11 arrest of **12** recovered
13 charged with (compare *accused of, arraigned for*)

4a **1** Discuss how challenging each aspect is.
2 Decide which **two** photos would be most suitable for a police recruitment brochure.

5b **1** Questions 1 and 5
2 Their answers respond to the questions and expand them. However, the female candidate interrupts the male candidate and dominates the discussion. She also wanders off the subject, starting to talk about car accidents.
3 *Well, that's a difficult one.*

6a **1** were waiting **2** would realise
3 put more effort **4** didn't show **5** was watching

6c **1** It's not as if we need more prisons. Crime isn't any higher than it used to be.
2 It's time they found an alternative to prison.
3 I'd rather the courts passed more Community Service Orders.
4 I wish the government would ban violent computer games.

▶ **Student's Resource Book, page 88**

Use of English 2 p.131

1 Students may have heard of Butch Cassidy from the movie (see Exercise 4).

2a Students skim text to get a general overview of his life.

2b Elicit the type of words that students will need to complete this type of open cloze (generally structural words such as articles, pronouns, auxiliary verbs, prepositions, etc.).

3 The task analysis is both to focus students attention on the type of words used and to link to the following Language development section on comparatives.

4 Extend the discussion by asking if they enjoy such films and whether they think it is right for films to romanticise criminal activities.

Key

2b **1** unlike **2** whom **3** from **4** of **5** himself **6** one **7** into **8** than **9** were **10** for **11** over **12** most **13** it **14** why **15** their

3 **1** pronouns: 2, 5, 13, 15
prepositions: 1, 3, 4, 7, 10
comparative structures: 8, 11, 12
2 *one of the most prolific bank and train robbers*

Language development 2 p.132

Start by asking students which they think are the most dangerous cities in the world.

1a Students should first skim the text and say if New York is a dangerous city. Then they should go through it more carefully, correcting the mistakes. They all relate to the form and use of comparative structures.

Background

The photo shows a female bank clerk in New York in 1922 being shown how to use pistol. This was due to the rising level of crime in the city at that time.

1b The idea here is to look at more advanced and more specific comparative structures, those that don't just compare, but say by how much. The exercise uses a simple adjective (*safe/safer/safest*) throughout so students can focus on the more complex structures.

Look at the example and do one more together before leaving students to match the halves. Afterwards, pick up on some of the expressions and ask students to compare them in terms of meaning, register and impact (e.g. *it's not as safe / It's nowhere near as safe*).

1c The opportunity here is for students to use the structures about a topic that they have some knowledge of. Younger students may have to make informed guesses.

2a The section focuses on other structures that are used to make comparisons. Students should choose the correct option, then highlight the form used. Spend some time focusing on the structures by giving more examples and by referring to the Grammar reference on page 184.

2b Using the structure in Exercise 2a, students complete the sentences.

3 The two texts here review language from Exercises 1 and 2. Note also the link back to the open cloze in English in Use in the type of words that are missing.

Extra!
Finish by asking if any of the students have ever done a self-defence course, if they think it would be necessary and what they would want to learn.

Photocopiable activity

Activity 8C can be used any time after this section. It is a pairwork activity to review the language of comparatives and superlatives. Students find missing words to complete some expressions.

Key

1a (1) New York was once notorious as one of the world's most dangerous **cities**. (2) Nowadays, however, it is far **safer**. (3) There are considerably **fewer** crimes, and people are not as ~~much~~ afraid to walk around the streets. (4) The crime rate is now much the same **as** in other comparable cities and nothing like it was. (5) A key reason for the improvements was the introduction of new laws that were **a** great deal stricter. (6) Penalties are now by far **the toughest** the city has ever seen. (7) There are some people who say life is not as much fun these days and that they feel **more** restricted. (8) (correct)

1b far/considerably less safe / safer
nowhere near as safe
nothing like as safe
by far the safest
just about the safest
one of the safest
somewhat/slightly less safe / safer
(not) nearly as safe (as)
a great deal safer
much the same (as)

2a 1 *than – I'd sooner A than B*
2 *prefer – I'd prefer to do A (to B)*
3 *a terrible – such a* + adjective + noun (compare *so* + adjective)
4 *as; like – as* = he really was, a fact; *like* = not identical, but similar to
5 *more strange – more A than B* (the *-er* form is not used, even with short adjectives, as it is not 'how strange', but more about being strange rather than suspicious)
6 *The more; the less* – a causal relationship: the more that A happens results in an increase in B.
7 *more and more* – to emphasise a constantly changing situation (with short adjective use *-er and -er*, e.g. *it is getting tougher and tougher*)
8 *as – as A as B*

2b **Suggested answers**
1 The more people read about crime, the more they become afraid. / People become increasingly afraid the more they read about crime.
2 I am not so much happy as relieved that he has been caught.
3 He's not such a reliable witness as Jim.
4 Crimes using weapons are happening more and more. / There are more and more crimes using weapons.
5 Some people are too afraid of crime to go out.
6 I'd sooner have more police officers on the street than more CCTV cameras.
7 The new regulations are too complicated for me to understand.

3 A 1 much 2 least 3 the 4 the 5 too 6 such
B 1 As 2 more 3 more 4 equally 5 as 6 nowhere 7 as 8 deal

▶ **Student's Resource Book, page 89**

Writing 2 p.133

1 Ask students to look at the photo and say what they think it shows. Having established what a hall of residence is, give students a moment to discuss the two questions.

Background

Halls of residence (known to some students as 'dormitories') are large blocks of single or shared rooms, sometimes with shared kitchen facilities and sometimes with meals in a canteen. They can be either on the main university/college campus or away from it.

2 Students read the task, highlighting the key points. The hope is that, at this stage, they will be in the habit of asking the right questions to help plan their answer.

3a/b Elicit as much as possible from the students, as they should be able to do this themselves by now.

4a The box gives examples of the formal language and structures that would be useful in such a report. The aim is for students to notice which part of a report each would be used in.

4b Remind students that, in the exam, they would need to imagine their own points to include, but here they are given quotes that need to be turned into more formal suggestions. Point out that part of their planning process would include noting down suggested actions.

5/6 Get students to write the report in class so that it is done without further consultation and within a restricted time period. They could then check their own work or each other's.

Key

2 Who is writing to whom about what? – You, as Student Representative, to the School Administrator about security issues.
What is the purpose of the report? – To explain the current situation and suggest improvements.
What style will it be written in? – A concise, factual and fairly formal style.
What points have you been asked to cover? – 1 Describe the current situation; 2 Give examples of security issues/concerns; 3 Outline suggestions for improvement.
What will make the reader think it's a well-written report? – If it is well presented and laid out, and presents the points in a clear and concise way with suggestions that relate to the problems.

3a Possible headings: Introduction, The general situation, How the information was obtained, Security worries, Recommendations

4a **Suggested answers**
This report outlines the general security situation on the summer school.
In recent days, several/a number of security problems have arisen.
I conducted a formal survey by questionnaire and spoke to students informally.
According to some students, a lot of valuable items have gone missing of late.
In light of the above, we believe the front door should be locked during the day.

4b Suggested answers

1 The front door should be locked during the day.
2 There should be an unarmed security guard to patrol the premises.
3 In the long term, security cameras and security lighting should be installed.
4 The college authorities should consider introducing individual safes in the main office for students to store their valuables.
5 A system of identity cards could be introduced which the security guard would check.

5 Suggested answer

Introduction
This report outlines:
1 the general security situation at the summer school;
2 how the information was obtained.
It also gives instances of specific security worries and makes recommendations for improvement.

The general situation
In recent days, several security problems have arisen, and in general, many students feel the college is not a safe place to live and work in.

How the information was obtained
I conducted a formal survey by questionnaire and spoke to students informally. Many asked me, in my role as Student Representative, to prepare a brief and urgent report for the college authorities and to convey their specific suggestions.

Security worries
According to some students, a number of valuable items have gone missing recently. During the daytime, the front doors of the residence are unlocked, and unwelcome outsiders, pretending to be students, have come in off the streets and wandered around the premises. Of course, it's quite possible that the thefts have been committed by other students, but we feel this is unlikely.

Recommendations
In light of the above, we believe:
1 the front door should be locked during the day;
2 there should be an unarmed security guard patrolling the premises at all times challenging any suspicious-looking people.
In the long-term, we suggest the college authorities should consider:
1 installing security cameras and security lighting;
2 introducing a system of identity cards which the security guard could check;
3 providing every student with a safe for valuables, to be kept in the main office.
(250 words)

► **Student's Resource Book, page 91**

Module 8: Review p.134

1 1 C **2** A **3** D **4** A **5** D **6** A **7** C **8** B **9** C **10** C

2 **1** The company lifted the ban on people working nights.
2 Only one witness has come forward.
3 The newspapers seem (totally) preoccupied with young people.
4 He was the victim of a robbery last night.
5 We need to reduce congestion on our roads.
6 He didn't realise he had committed an offence (= *a crime*) /caused an offence (= *upset someone*).
7 It's time we initiated a public debate on terrorism.
8 Suddenly there was a crime wave. / There was a sudden crime wave.

3 **1** I'd sooner ~~to~~ have more police officers on the beat **than** more CCTV cameras.
2 Little **did** I realise that there were **so** many serious crimes in **such a** small place.
3 The hackers' attacks are nowhere near **as** intense **as** last week's.
4 No way **will I** work **as** a community officer in my spare time!
5 **Generally** speaking, **the** less we know, **the** better **it is**.
6 He acted more **like** a criminal **than** a police officer!
7 On no account **should you** feel too scared **to** report the crime.
8 Not once **did** anyone **break** into the house!

4 **1** I'm not so much desperate as disheartened at the slow progress we're making.
2 Not since the first PC has there been so much interest in a piece of new technology.
3 It isn't such a simple device as I thought.
4 To get one anywhere is almost impossible.
5 It's somewhat quieter / less noisy than the other one.
6 Hardly had it appeared on the market when a major flaw came to light.
7 The bug was too complicated to fix immediately.
8 The more I think about the problem, the more worried I am/get/become about it.

Exam practice 4 TRB p.197

Paper 1: Reading
1 B **2** G **3** C **4** E **5** F **6** D

Paper 3: Use of English
Part 3
1 reliant **2** invigorated **3** irritable **4** politely **5** uneasy **6** dependent **7** excessively **8** unable **9** consumption **10** eventually

Part 4
1 give **2** dealt **3** making **4** fix **5** dull

Paper 2: Writing
Suggested answer

Report
The Mulvane Outdoor Pursuits Centre
To: Niall Sanchez, Principal
From: Lucia Delaney
This report has been compiled from the feedback received from students who took part in the Spring Break programme. Although in general their reactions were positive, a few areas have been identified by the participants in which there is room for improvement in the facilities and level of service on offer.
Range of activities
The standard of activities and teaching was generally felt to be good, but complaints were registered concerning the fact that there was not a daily choice – that is, not all activities, or categories of activity, were available every day. Watersports are given as an example of this.
Quality of facilities and tuition
Equipment and tuition were considered first-rate by all, but students criticised the accommodation: they expected more space and more privacy. For example, each bathroom served five students. This was felt to be unacceptable at a sports centre where washing facilities are of the utmost importance.
Suggestions for improvements
I would suggest requesting greater availability of sports and improved accommodation. The staff at the centre seemed keen to receive feedback from visitors, and I am sure that if this report is forwarded to them, they will take the points raised into consideration when planning future programmes for us.

Paper 4: Listening
1 A **2** C **3** A **4** B **5** D **6** B

Module 9 Communication

This module covers much more than speech, with activities on animal communication, visual signals and body language, image and fashion statements, and hype.

Lead-in p.135

Start, with books closed, by asking students to come up with non-verbal ways in which we communicate, consciously or otherwise, with those around us. Elicit the following: posture, body language, facial expression, gesture (hands, winking, etc.), dress (what you wear and the condition it is in), make-up and general appearance, and lifestyle. Most of these come up later in the module.

Then ask students how much they can tell about someone from those factors and if they are reliable indicators of personality or mood.

Finally, look at the photos on page 135 and discuss the questions together. Do students agree what is being communicated? If so, how do they all know that? Is understanding non-verbal communication learnt or instinctive?

Background

Fomkin and Rodman in *An Introduction to Language* (1983) describe communication as 'a system for creating meaning' and claim that up to 90% of the meaning of a message is transmitted non-verbally. The importance of non-verbal communication has also been shown in other studies. For example, when verbal and non-verbal signals contradict each other, it has been shown that people are more likely to trust the non-verbal.

Key

Suggested answers

The monkeys are grooming or nit-picking, an important social ritual suggesting friendship, companionship and trust.

Through their clothes and the way they are standing and looking, the youths are communicating a sense of solidarity and shared identity, and the desire to be the same.

The stag could be communicating dominance of female deer and/or a threat to others.

The man in the couple is communicating affection, but also, from the way he is holding his girlfriend, possession, dominance and perhaps jealousy or fear of losing her. The girl exhibits signs of love and submission, but also, from the way she is instinctively protecting her neck, a lack of trust.

9A Something to say

Reading p.136

Start with books closed and ask the class if animals have language. Ask for examples or reasons why they don't. Establish a difference between language and communication, and elicit ways in which animals can communicate.

1 Students share how much they know about animal communication, following on from the introductory discussion. Encourage them to give examples of how different animals can communicate.

2 When students have looked at the list of seven animals, ask them if they know what they all are and if it matters. For the sake of doing the task in the exam, it wouldn't matter if they had no idea what a *lemur* or a *stink bug* were; it is understanding what is said about them that is important. (In this case, there are some pictures, which the exam text will not have.)

3 Students should now do the task under as close to exam conditions as possible.

4 Remind students that, in the exam, they would have about 20–25 minutes to do this task, including the time to transfer their answers to the answer sheet.

5 Having gone through the answers to the exam questions, get students to find examples of rephrasing (e.g. question 1 *feel a part of the social group* / *bond them to the family or tribal unit*). Then give students a few minutes to look back through the text to answer question 1 here. Follow it up with a discussion of questions 2 and 3.

Background

Human language is said to differ from animal communication in a number of ways. Chomksy said that it is the creative aspect that sets human communication apart. Three aspects of this are:

1 Displacement: animals are limited to talking about here and now. Humans can talk about the past and the future, about what is not here and about imaginary situations.
2 Productivity: we can create new words for new situations.
3 Duality: animals tend to have one signal for one meaning, whereas human language is on two levels. There are few single sound meanings. Sounds combine to form words so /d-ɒ-g/ is very different from /g-ɒ-d/. In English, only 44 sounds can create over 1 million words.

Key

3 1 G – line 102 *grooming … to bond them to the family or tribal unit*
2 A – line 14 *rhythm of five scratchy pulses … sent out by the female … and a reply from the male; pure-toned pulses …*
3 B – line 47 *If this doesn't scare away a predator …*
4 E – line 76 *maintain their group's territory with scent-marking*
5 F – line 90 *look to the starling for signs of danger*
6 B – line 53 *where dense vegetation prevents them from keeping an eye on one another*
7 A – line 20 *the female can tap out a different rhythm to make him go away*
8 H – line 116 *can describe individual people in detail … whether they are carrying a gun*
9 E – line 83 *the smellier tail wins and the overwhelmed contender backs off*
10 B – line 62 *can identify each other's rumbles at a range of over a mile*
11 and **12** in either order
F – line 92 *a sign of a predator approaching on the ground … while a clear whistle … will make them look to the sky*
H – line 111 *they have evolved a special alarm call for each one*
13 and **14** in either order
A – line 12 *conducted … along a network of branches*
D – line 69 *by marking trees*
15 B – line 60 *Elephants recognise the calls of about 100 other herd members*

5 1 warning of danger, finding a mate, keeping in contact with group, parent locating baby, marking territory against rivals, bonding

Vocabulary p.138

1a Continuing the idea of the importance of paraphrasing, this exercise looks at alternative ways of expressing ideas in the text and also at collocation.

Students search the text for words or phrases that have been rephrased or explained. Look at the example together and how it paraphrases the words in line 5. Students should note how you *learn new things* ***about*** *something* but *get insights* ***into*** *something*.

1b Here, students have the opportunity to select words that they find useful, interesting or just hard to remember. Emphasise the importance of personalising their vocabulary books, and adding to them constantly. Point out that pages such as this are designed for a wide range of students and that some of the words they may know already and others may not seem of value to them. The point of the exercises is to show patterns of vocabulary for them to build on individually.

2a The text uses many words to refer to animal sounds. Two ways to remember them are to associate them with the animal that makes them or to try to break the sound down into parts, as in the exercise. Remind students that a *starling* is a type of bird (the one that vervet monkeys look to for signs of danger). Note that all the words can be used as both verbs and nouns.

2b There could be some discussion here if students disagree (e.g. to get attention, do people whistle or hiss?). The point is to process the vocabulary and to extend its range seeing, for example, that a *roar* is not just an animal sound but could be applied to an engine, the wind or a crowd of people.

2c This exercise shows that the sounds can also be used to add detailed description to how people speak, commenting on their mood or emotion.

3a This section focuses on the use of idioms to add expression to language. Some will translate directly into other languages, others will seem very strange. The ones featured here are all reasonably common, all include an animal and are connected to communication in some way. First, go through the list of animals checking that students know what they are (elicit, in particular, that a donkey is smaller than a horse but has bigger ears, a goose (plural *geese*) is bigger than a duck, a rat is quite different from a mouse). It might be more fun to do the activity in small groups if there is argument about which animal fits where. Point out that they can use a certain amount of logic (e.g. you are unlikely to let a donkey out of a bag, as you would have a job putting it there in the first place), but also that it is the meaning of the whole phrase that is important, so for question 1, it is not a question of which animal smells the most.

3b Students put the idioms into a context. When they have finished, ask them if they know of any other animal idioms. They could use them to make another exercise similar to Exercise 3a or 3b.

Key

1a **1** infinitesimally small (vibrations) (line 12) **2** comes in on (the conversation) (line 19) **3** crack (a code) (line 27) **4** pervasive (form(s) of communication) (line 28) **5** convey (messages) (line 36) **6** emits (a loud trumpeting sound) (line 46) **7** scare away (a predator) (line 47) **8** account for (the telepathic way) (line 58) **9** traces (line 68) **10** plaintive (call) (line 78) **11** intimidate (rivals) (line 79) **12** pungent (substance) (line 80) **13** responds in kind (line 82) **14** backs off (line 84) **15** look to (the starling) (line 90) **16** take to the trees (line 94) **17** exhibit (very sophisticated behaviour) (line 107) **18** evolved (a special alarm call) (line 112) **19** down to (their size) (line 117)

2a **1** c loud, high, long, musical
2 e loud, low, long
3 c quiet, very low, long
4 g loud, high, short, could be musical
5 d quiet, high, short
6 a quiet, long
7 f quiet, very low, long
8 b loud, low, short

2b **1** roar **2** rumble **3** whistle/roar (with laughter)/hiss (in a pantomime) **4** whistle/roar **5** squeak **6** hiss **7** whistle/roar (of the crowd) **8** trumpet **9** squeak/croak **10** hiss/growl/roar **11** whistle/hiss

2c **1** growled (quietly) / roared (loudly) **2** trumpeting **3** hissed (to show disapproval) **4** squeak(ed) **5** croaked **6** whistled **7** roared **8** rumbled

3a **1** rat – c (dishonesty)
2 cat – f (think of the cat leaping free)
3 horse – b (possibly from horse-racing tips)
4 donkey – e (known for endurance)
5 rabbit – d (probably from rhyming slang: *rabbit and pork = talk*)
6 goose – a
7 parrot – g (learns words without knowing the meaning)

3b **1** (straight) from the horse's mouth **2** talk the hind legs off a donkey **3** let the cat out of the bag **4** wouldn't say boo to a goose **5** smell a rat **6** rabbits on **7** parrot fashion

▶ **Student's Resource Book, page 94**

Listening 1 p.139

The discussion is about behaviour at job interviews. One way to start would be to briefly look back at the texts on page 18, which gave advice on attending interviews.

1 Students start by sharing their ideas on the best way to behave at job interviews.

2 Do not allow students too long to look at the questions before playing the recording through twice under exam conditions. When checking the answers, ask if students can remember any of the phrases used that helped them choose the correct option, while stressing that it isn't important if they can't.

3a Students compare what they said initially with what they heard.

3b If necessary, play the recording once more for students to make a note of do's and don'ts.

Photocopiable activity

Activity 9A can be used at this point. It is a version of bingo, with students matching descriptions of different gestures to pictures of them.

Key

2 **1** A – *It's an alarming thought, and I'm not 100% convinced*
2 B – *if you present a false impression of yourself … you're hardly going to feel comfortable*
3 A – *The thing to remember … is that it's talking about subconscious impressions … it seems that they are influenced, … even if that's not what they notice at the time*
4 D – *You want to look as if you've made an effort*
5 B – *Candidates who project an image of vitality and energy come across as more capable*
6 C – *but don't get hung up on them*

3b **Suggested answers**

Dos
Give a good first impression.
Send out the right message.
Greet people appropriately.
Be yourself.
Find out what people wear to work there and dress accordingly.
Try to fit in.
Sit up straight.
Speak clearly.
Look people in the eye.
Smile.
Monitor your body language.
Don'ts
Don't rely on a smart suit and a firm handshake.
Don't wear clothes you wouldn't normally wear.
Don't pretend to be someone you're not.
Don't fold your arms, cross you legs or look at the floor.
Don't get hung up on body language.

Language development p.140

The assumption here is that students are familiar with the basic aspects of reported speech (tense shift, change of pronouns and time references). It focuses on more advanced aspects, such as the use of reporting verbs and impersonal report structures.

Start by putting a quote such as *Kathy said, 'Yes, I lied at the interview'* on the board and ask students to put it into reported speech. Elicit the forms:

Kathy said that yes, she had lied in the interview.
Kathy admitted that she had lied at the interview.
Kathy admitted lying at the interview.

Compare the forms, establishing that, where possible, it is better to use a reporting verb and that they follow a number of different patterns. Some verbs, such as *admit*, can be used in two ways, others in more patterns and others in only one.

1 In the first exercise, all of the options fit in terms of meaning, but one does not fit grammatically. Students need to identify which is the odd one out and rewrite the sentence using it. Do the first one together as an example.

2a Here, students have to first identify and summarise the function of the quotations by matching them to a verb before deciding what pattern it follows.

2b In pairs, students ask each other about their television-watching habits. Elicit additional questions they could ask each other (e.g. *Have you ever been upset about missing a programme?, What type of programme do you think there is too much/not enough of?)*

2c Students report what their partner told them using the reporting verbs in Exercises 1 and 2a.

3a Look at the example together. Get students to say how the same idea could be expressed in other ways (e.g. *Some people say that dolphins are highly intelligent*). Establish that the impersonal form is more formal and more suitable in certain writing styles (e.g. report writing). Point out that the impersonal structure is a passive construction. (*They say dolphins … / Dolphins are said …).* Highlight the other common report structures.

2b Students use the structures in Exercise 2a to make the ideas more formal/impersonal. Establish which would be more suitable in a report on animal communication.

2c Students discuss the points, referring back to the reading text on page 137 to support their arguments if necessary.

Photocopiable activity

Activity 9B can be used after this section to review these structures and the vocabulary on page 138. It is a pairwork activity in which students complete gapped sentences with reporting verbs and names of animals to complete idioms.

Key

1 **1** A **2** C **3** D **4** C **5** C **6** B **7** A
8 D (*objected to our using* is considered more formal and old fashioned)
Suggested rewrites
1 She complained to him about the problem.
2 Mike confessed to having eaten all the ice cream.
3 The lawyer advised me that I should contact him immediately.
4 Jackie requested me to write the letter for her.
5 Peter advised me to email you.
6 A number of people noticed how easy she was to talk to.
7 The report suggests people spend more time together as a family.
8 They didn't allow us to use the phone.

2a 1 Emily *apologised* for forgetting to record the programme.
2 Mike *announced* that the programme he (had) made would be on TV the next day.
3 Kevin *advised* me/us to get rid of the TV.
4 Claudio *blamed* Laura for breaking the TV.
5 Stella *reminded* me/us to switch the TV off before I/we went to bed.
6 Richard *regretted* staying up late to watch the film.
7 Doug *admitted* (that he had been) watching a lot of TV recently.
8 Susanna *explained* that the reason (why) she had the TV on was that she didn't like being on her own.

3a 1 *It is claimed* that gorillas are as intelligent as humans.
2 *It is known* that Penny Patterson taught a gorilla, Koko, to communicate.
3 *Koko is reputed* to have acquired 645 words.
4 *It's been suggested* that Koko understands grammar.
5 *It is hoped* that more research can be done in future.

3b 1 It is said that Koko has an IQ of 85–95.
Koko has been said to have an IQ of 85–95.
2 It has been reported that she can make logical sentences.
Koko has been reported to make logical sentences.
3 It has been hinted that Koko's trainers imagined she is cleverer than she is.
4 It has been argued that Koko knows only words and not grammar.
5 It has been suggested that the word order she uses is either memorised or random.
6 Human language is believed to be outside of the capacities of other species.
It is believed that human language is outside the capacities of other species.
7 It is (generally) accepted that human language is a unique phenomenon without significant similarity in the animal world.

▶ **Student's Resource Book, pages 95–96**

Use of English 1 p.141

1a Have a quick review of good exam technique and then leave students to do the task in as close to exam conditions as possible.

1b After a suitable time limit (10 minutes) stop the students and get them to compare and justify their answers before checking them with the whole class.

2 Use the task analysis to highlight any vocabulary or structures the students are unfamiliar with.

Key

1b 1 Tina *prevented/stopped/banned me from going* into the house.
2 There *has been a noticeable change in* public opinion in recent years.
3 We usually *agree with each other* about most things.
4 I was *always (getting) under my mother's feet* when I was a child.
5 She *is reputed to have made* a lot of money in recent years.
6 I managed to get good grades *despite not being a very* good student.
7 Philip *used to come across as* a very nervous man.
8 Whether or not I *get the job depends on how* I do in my exams.

2 1 Example, 1, 5

Writing 1 p.142

Most of the focus in the writing sections has been on the overall result looking at: task achievement, planning, text organisation, style and cohesion. However, writing in Paper 2 is also assessed on grammatical accuracy. This section concentrates on accuracy at sentence level, including spelling and punctuation and common grammatical errors such as word order, subject verb agreement, missing subjects, run-ons etc.

1 Students first work individually then in small groups to both identify the type of errors and then correct them.

1a Do the first one together as an example.

1b When students have completed this, have a look at the notes on run-ons in the Writing reference on page 204 together.

1c The mistakes here are spelling or with confusing words.

Extra!
Look at the writing strategy notes at the top of the page. They link to the checklist on page 188. The point to stress is that it is no use just reading through a composition once looking for mistakes; they probably won't notice many of them. Stress that they need to be looking for specific mistakes. The best approach is to look through the work a number of times, looking for a particular type of mistake each time. For example, first check all the spellings. Then check that all the subjects and verbs agree, then stop at each noun and check if it requires an article, and if so, which, etc. The secret is for each person to focus on the mistakes that he/she is most likely to make.

2 Students plan a composition about the difficulties that they have with English. This should help them to focus on their weak points.

Key

1a 1 word order: *a new series of wildlife programmes starts on TV*
2 run-on: … *subtle. However,* …
3 fragment: … *language, because* …
4 run-on: … *money, as the pay* …
5 fragment: *This is a great opportunity* … / … *courts, providing a great opportunity* …
6 run-on and wrong word order: … *experience; their willingness to learn is much more important.*
7 punctuation: … *parents, learning from* … / … *parents and learn* …

1b 1 subject–verb agreement: *There* ***are*** *a lot of things* …
2 subject–verb agreement: … ***is*** *my favourite* ***meal.***
3 verb form: *have never* ***heard*** *of = present perfect*
4 wrong verb form after *Why not:* ***take*** *up*
5 wrong conditional tense form: *If I* ***hadn't thrown*** *away … could have* ***got*** …
6 wrong relative pronoun: *All* ***that*** *concerns* …
7 inversion, missing subject, subject–verb agreement: … ***not only does*** *the Swan Hotel* ***have*** … *but* ***it*** *also* ***takes*** *care of* …

1c 1 beleive > believe; rise> increase
2 presantation > presentation; farther > further; inform > information; (cooperation > co-operation;) wish > hope
3 sugested > suggested; accepted > agreed; exept from > except for; as I am > as I was; say > express; practice > practise; bump > rush

2 **Suggested answer**

LEARNING ENGLISH – JOY OR MISERY?

How can a language with so many phrasal verbs ever be easy? That's what a friend said to me once, and she'd been trying to learn English for years!

But for me, it hasn't been as hard as I thought it was going to be. One of the good things is that there are so many English words used everywhere these days that even in my beginners' class not all the words were unfamiliar. Also, I'd been listening to English-language pop songs and watching English-language films for years, which meant I had learned a lot of the language without thinking about it.

The one thing I still find very difficult, though, is the link between spelling and pronunciation. How is it that words like row, lead and refuse can have different pronunciations and have different meanings! Ridiculous! Not surprisingly, my teacher says I still need to work on my spelling.

Although I've learnt a lot in my classes, I think what really helped me improve was spending a few weeks in London last summer, working in a café. I really had to understand what everyone was saying or I would get the customer's order wrong, and of course I had to learn to speak politely in English.

Overall, I have really enjoyed leaning English. My teacher is very young and he makes the classes very interesting, encouraging me all the time to speak in English and have fun using the language. But in the end, there's nothing like being in a native-speaker environment to make you learn quickly!

(263 words)

9B Making a statement

Listening 2 p.143

1 Start with books closed and ask students what they think is the most universal item of clothing. Having elicited *the T-shirt*, ask them how many they have. Then look at the questions in the book together.

2 This is a Part 2 task, which students last did in Module 6. Give students 30 seconds to read the notes before playing the recording twice.

3 It would be useful to let students compare their answers before they are given the correct ones. This is to help them focus on the clues that signal the parts containing the answer.

4 There is plenty of scope for discussion here. In question 1, ask what they think of famous designers putting their name to a plain white T-shirt.

In question 2, examples of garments originally intended for other uses would be denim jeans (and cowboy boots) originally designed for cowboys, trainers designed for athletes but now worn by older people and for general leisure wear, and baseball caps, which are seen everywhere.

In question 3 there are plenty of examples of clothes that say you fit in, such as wearing a plain dark suit and a tie to work in a bank, or fashions such as punk, gothic or grunge that say you want to be different (but not from your peers!).

Key

2 **1** French **2** mass production **3** sleeves **4** sportsmen **5** 'skin-tight' **6** rebellion **7** advertising **8** political

Speaking p.144

Here students get a chance to do a complete Paper 5 speaking test. Although it is clearly removed from exam conditions it should help them to get a better sense of how the various parts all fit together.

1a Start by doing the quiz to remind them of the important points. One way to do this would be with books closed as a lead-in. Read out the questions, with the class either shouting out the answers or noting them down individually.

1b Spend a few minutes reading through the information on pages 171–172 before students start.

2 Divide the class into groups, preferably fours but minimum threes. If students know who their partner will be in the real exam and they are in the same class, they should partner up now. Make sure students are familiar with the roles and responsibilities of the interlocutor (controls the test, gives instructions, asks the questions and keeps an eye on the time) and the assessor (says 'hello' and listens to the candidates, noting strengths and weaknesses).

2a/b The interlocutor asks the first pair some of the introductory questions on page 207, involving both candidates equally and keeping the conversation going for three minutes. They swap over roles and use some of the remaining questions.

3 The instructions are on page 207 and students use the pictures of rooms on page 145. When the assessor has given feedback on the first pair, let them change over and give the second pair a turn.

4 This time let the second pair go first.

5 Keep the discussion going for the final part before going back and repeating parts 3 and 4 for the second pair.

6 Give students a chance to assess themselves and each other and discuss ways of improving their performances.

Extra!
If there is time, pick up on some of the expressions students used to describe the rooms or when discussing news media.

Key

1a 1 Fifteen minutes (for two candidates)
2 Two (three if there are an odd number on the day)
3 Two: an interlocutor and an assessor
4 Four: Part 1, three minutes; Part 2, four minutes; Parts 3 and 4, eight minutes.
5 Part 1: answer interlocutor's questions, ask each other questions
Part 2: talk about a set of photos
Part 3: discuss a social issue with each other using prompts
Part 4: answer general questions related to the issue in Part 3
6 Your ability to: compare, contrast and speculate; discuss, evaluate and select.
7 In general: listen carefully to the instructions/questions and do what they ask, show interest in your partner, keep speaking but don't dominate, don't say 'I don't know', don't worry if your partner seems much better or worse than you, it will not affect your mark.
In Part 1: show interest and respond appropriately.
In Part 2: keep talking and answer both parts of the instruction
In Parts 3/4: be sensitive to turn taking, offer opinions, agree or disagree giving reasons, reach a conclusion.

Use of English 2 p.147

1 Ask students to read the title of the text, look at the poster and try to guess the connection between the two.

2a Students should skim read the text quickly to check their answer to Question 1.

2b Give students 10 minutes to do the task, reminding them that that in the exam they would have to skim, do the task and transfer their answers all within that time. Remind students that they may have to make two or more changes to alter the word class/type and that adjectives may need to made negative and nouns plural. At least one word will need a prefix.

Remind students that some words have more than one form in the same word class (e.g. two nouns or adjectives). Where there are two nouns, one is often uncountable and the other countable. Adjectives are formed from both *-ing* and *-ed* participles. Some adjectives (attributive) can only be used before a noun (e.g. *live, elder*), others (predicative) can only be used with a verb (many beginning with *a* e.g. *alive, asleep*) and not before a noun. (You can't say an *alive/asleep/alone animal.*)

4 Merchandising related to films is now very popular and sometimes it is hard to know if the products on sale or given away in cereals and fast-food restaurants are there to promote the film or if the film is there to promote the merchandise! A nice example of the link between the two was in *Jurassic Park*, which featured a gift shop actually stocked with Jurassic Park merchandise. Years ago, hype was created in other ways, for example by widely advertising a film, then releasing it on only a few screens, so long queues built up, creating the image of a must-see movie.

Photocopiable activity

Activity 9C can be used any time after this section, as it reviews vocabulary in the entire module. It is a pairwork activity in which students complete a crossword by identifying the wrong words in sentences and replacing them with the correct words.

Key

2b 1 relations: plural noun after adjective, collocates with *public*
2 reality: noun after preposition *with*
3 significantly: adverb before adjective *inferior*
4 Originally: adverbial discourse marker
5 appearance: noun after possessive *his*
6 unforgettable: adj to describe *character* contrast with *silly but* ..
7 atmospheric: adjective between adverb *darkly* and noun *series*
8 substantial: adjective between a and noun *sum*
9 marketing: noun (uncountable) after preposition *on*
10 excessive: adjective after verb *be*

3 markets – where you sell the products; marketing – how you promote the product (un)forgettable – something that can(not) be forgotten; forgetful (no negative possible) – used for a person who frequently forgets things

Writing 2 p.148

1 Look at the lead-in questions together. Find out if any students have ever been to an outdoor centre and what they did there.

2 Give students a few minutes to read the instructions and then read through the input material carefully. Then they should look at the five focus questions.

3a Students go back through the input, highlighting key parts and organising the points into related areas.

3b Students look at the headings and decide which points would be covered within another and therefore would be subheadings of others.

3c Students use the headings as a basic paragraph plan.

4a Look at each of the structures in the boxes, eliciting some sentence completions to highlight the meaning and the form.

4b Note that *it would be a good idea* is followed by a *to*-infinitive, whereas *suggest* and *propose* are followed by *-ing*.

5/6 Students write, edit and correct their work. Remind them of the editing procedure on page 142.

Key

1 **Suggested answer**
Water sports: rowing, canoeing, sailing, kayaking; climbing; orienteering; archery; etc.
Attract customers by: having a wide range of activities, good atmosphere, low prices, excellent publicity, etc.

2 1 You are the Senior Administrator in the Bookings Department; you need to write a proposal suggesting how to increase sales, mainly through raising the profile of the centre.
2 To make suggestions (for the Board of Directors) about ways of increasing sales through raising the centre profile.
3 Memo: no increase in bookings, need more money for promotion and new facilities, need to raise profile of centre
Feedback: positive, but centre doesn't publicise itself well
4 Like a report, there will be a clear layout, probably with headings/subheadings. Also, there will be some analysis and suggestions. With a report, there is more emphasis on the analysis (probably with some concluding recommendations); with a proposal, there is more emphasis on a set of suggestions (possibly with some analysis).
5 Consistently formal or neutral.

3a Main focus: recommendations

3b 1 Introduction
2 Suggestions and recommendations
A Publicity material
The website
B Facilities
C Pricing policy
D Special events
3 Conclusion

4a **Suggested answers**
It is clear from customer feedback that too few people know about our centre.
The last year's results have been disappointing for the centre.
With regard to the website, the general view seems to be that it could be made more user-friendly.
To raise the profile of the centre, we should display attractive posters in prominent locations.
The aim of this proposal is to suggest ways of getting more people into the centre.
Perhaps the most effective way of increasing bookings would be to offer special deals.
If the centre is to attract more customers, it is vital that we overhaul the website.
A programme of visits to local schools could be organised.

4b **Suggested answers**
1 We recommend that a specialist web-design company should redesign the website.
2 It would be a good idea to invite journalists to visit the centre.
3 We suggest contacting Tourist Information Centres and asking them to display posters of the centre.
4 We urge the board to increase the promotional budget for next year.

5 **Suggested answer**

PROPOSALS FOR INCREASING CENTRE ATTENDANCE

Introduction

The main aim of this proposal is to make suggestions for improving sales. Having looked at customer feedback I have considered whether it is necessary to extend our facilities and looked at different ways of increasing our public profile.

Suggestions and recommendations

Facilities
It is clear from feedback that our customers think our facilities are already of a high standard, sufficient in number and do not need to be extended.

Publicity material
With regard to our website, the general view seems to be that it is inefficient. Search engines frequently fail to locate it, and it is poorly designed and very slow making it difficult for customers to find their way around.
We propose getting a specialist web-design company to overhaul the site.

Special events
The centre has not been featured in the local media for some time, and we are therefore currently underexposed. It would be a good idea to invite local journalists to the centre on a regular basis.
Perhaps the most effective way of increasing visitor numbers would be to offer discounted rates for school parties at quiet times.

Conclusion

If the centre is to increase sales, it is vital that we introduce some, if not all, of the above measures.
(210 words)

▶ **Student's Resource Book, page 104**

Module 9: Review p.150

1 **1** A **2** C **3** D **4** B **5** A **6** B **7** A **8** D **9** C **10** B

2 **1** His arguments were (very) persuasive.
2 There was gradual acceptance of the need for extra publicity.
3 Getting them to speak clearly is a seemingly impossible task.
4 She wore a very/highly fashionable dress.
5 He was uncharacteristically quiet today.
6 I'd like a job somewhere in the local area/this locality / this area.
7 Her nervousness showed in her voice.
8 The previous owners had left the place in a mess.

3 **1** I reminded Tom to bring his laptop.
2 He didn't want to comment on the report.
3 It was suggested that the advertisers should be sacked / that they should sack the advertisers.
4 We were all urged to work out what the message meant. / We urged them all to work out what the message meant.
5 She complained about the high cost of visiting Europe.
6 She refused to have anything more to do with him.

4 **It** has been known for some time that learning languages can stimulate intellectual development in young **children. Now** research in Canada has suggested that **speaking** more than one language also helps us to stay mentally alert in old age. Indeed, researchers there have noticed ~~**about**~~ how much more quick-thinking older bilinguals are than non-bilinguals. They claim that not only **are bilinguals** more mentally efficient at all the ages they tested, but their memories decline less rapidly in old age. They admit that, as yet, they **have managed** to find no evidence to show that learning a language below bilingual level **makes** a difference to adults. However, they hope **to do** further research in this area shortly and have promised they will publish the results at the earliest opportunity. All the same, they remind us **of/about** the growing body of evidence suggesting that *any* intellectual activities may have a beneficial effect on the health of our brain.
1 *There* > *It* (line 1) *It has been known …*
2 run-on sentence (line 3)
3 *to speak* > *speaking* (line 4)
4 no preposition after *notice* (line 6)
5 subject and verb inverted after *not only* (lines 8–9)
6 present perfect tense after present tense reporting verb (line 11)
7 present tense after present-tense reporting verb (line 13)
8 *to*-infinitive after *hope* (line 14)
9 vocabulary: *do research* (line 14)
10 wrong preposition with *remind* (line 17)

Module 10
The world of entertainment

This module embraces a wide range of leisure activities such as comedy, shopping, opera, television, sport and music. It includes features on enjoying them both as a spectator and performer.

Lead-in p.151

With books closed, brainstorm different forms of entertainment. Ask what is more suitable for children, teenagers, young adults, middle-aged people and older people. Ask for forms of entertainment that need a lot of money and those suitable for people with very little, or for active and inactive people. Then open books, look at the photos and discuss the questions together. The photos show people dancing in a nightclub (left), a teenager shopping for sports goods (top right) and a group of friends watching something exciting on TV (bottom right).

10A You have to laugh

Reading p.152

With books closed, elicit different types of humour or comedy, such as slapstick, mime, satire (these three are used in the lead-in to the text), stand-up, story-telling, puns, double acts, situation comedies, double-entendre, visual wit, etc.

1 Get students to look just at the photo and the headings and answer the two questions. Point out that the title *Fears of a clown* is an adaptation of the phrase *tears of a clown*, the suggestion that inside every funny person there is unhappiness. Elicit the meaning of the phrases *rubber-faced* and *fixing a plug* (electrical).

2 Give students two minutes to skim the article to explain the three points in the subheading

3 Elicit the strategy for multiple-choice questions to remind students, and give them 12–15 minutes to do the task.

4 The discussion allows students to give a personal opinion of types of comedy.

Key

2 Rubber-faced – his ability to pull funny faces in his visual comedy
The burden of comedy – the pressure to perform, the fear of failure
The joy of fixing a plug – his love of electronics and engineering

3 1 A – line 15 *unless it was perhaps the desire to break out and rebel*
2 D – line 22 *his comic persona exists in a parallel world dominated by his lifelong passion for cars and machinery*
3 B – line 34 *since adolescent self-consiousness set in at the age of 11*
4 A – line 38 *he needs an audience and the formality of staging or a camera before he can be somebody else*
5 B – line 46 *I constantly believe that there is a better performance just out of reach*
6 B – line 69 *he must represent to adults from many nations the child within them*
7 C – line 80 *It's myself and the audience out there who I'm interested in*

Vocabulary p.154

This section looks again at paraphrasing and also at collocation, noun phrases and the meaning of certain prefixes.

1a This exercise is similar to the one in Module 9. Students find fixed expressions, phrasal verbs, idioms and other phrases used in the text.

1b Elicit the type of words needed in the collocations (in 1–5, the adjectives are missing in the collocations with nouns; in 6–8, the verbs are missing in collocations with nouns.)

1c Students should note down any other individual words or phrases they find interesting or that would be useful to them.

2a Start with the two phrases from the headings of the reading text, *the fears of a clown* and *the burden of comedy*. Draw students' attention to the way the two nouns are linked with *of* and point out that this is a common pattern, but there are other prepositions that can be used to link two nouns.

Students should then try adding the missing prepositions to the seven phrases before finding them in the text to check. Stress that in theses cases, the preposition is determined by the preceding noun, for example *a passion* ***for*** *something*, *the key* ***to*** *something*. Check students know the meaning of all the expressions (e.g. *a knack* (always singular) is a natural skill or ability).

2b Tell students to read only the first line of the text. Ask them if it is about artists of the Impressionist movement, such as Monet and Renoir (it isn't). Tell them to use each of the seven expressions from Exercise 2a once each to complete the passage.

2c Students ask a partner the questions and report back.

3a Ask students to think carefully about the meaning of the prefixes as used in these examples. Point out that some of them (see key) have more than one meaning.

3b Students could use logic and instinct to decide which prefixes can be used with each word before checking in a dictionary.

3c Students should find two or more examples of words using each prefix. Check which of the different uses of the prefixes their suggestions come under. Examples include: *well-informed, well-read, well-off* (NB some words such as *well-built, well-educated* have aspects of both quantity and quality); *off-piste* (skiing), *off-side* (football); *self-absorbed, self-critical, self-esteem* (noun*), self-depreciating*; *low-pressure, low-spirited, low-lying.*

3d Students complete the sentences using words from Exercise 3b

3e The idea here, as well as giving further opportunity to process the vocabulary, is to raise the issue of connotation, with students hopefully noting that *self-assured, self-confident* and *self-respecting* are generally positive, whereas *self-satisfied* and *self-important* are generally negative.

Key

1a **1** never uttered a word (line 6) **2** stumbled across (line 9) **3** forging (line 11) **4** underpin (line 12) **5** all-consuming (line 26) **6** trademark (line 35) **7** nondescript (line 36) **8** contrived (line 42) **9** debilitating (line 47) **10** shudders (line 51) **11** vicious (line 52) **12** is adamant that (line 60) **13** phenomenal (line 65) **14** relish (line 80)

1b **1** phenomenal **2** vicious **3** nondescript **4** contrived **5** all-consuming **6** forge **7** utter **8** shudder

2a **1** for **2** for **3** of **4** to **5** of **6** to **7** of

2b **1** knack for **2** target of **3** prospect of **4** attention to **5** key to **6** passion for **7** risk of

2c **1** for **2** of **3** for

3a **1** *well* – (quantitative) a large amount or to a great degree
Compare with other use of *well-* (qualitative) (e.g. *well-acted, well-written*) to mean 'pleasing' or 'successful'.
2 *off* – not happening or located in usual place.
Compare with other use of *off-* (*off-balance, off-peak*) which means 'not the case'.
self – how people feel about themselves.
Compare with other uses of *self-* (*self-made, self-defence*) in which *self-* means actions done to or by oneself, or *self-cleaning, self-locking* in which *self-* means 'automatic'.
3 *low* – not high or complex.

3b well – established, paid, travelled, worn
off – duty, guard, line, road, track
self – assured, confident, important, respecting, satisfied
low – level, paid, profile, risk, tech

3d **1** well-established **2** self-important **3** off-track **4** self-respecting **5** low-paid **6** off-guard **7** low-profile

▶ **Student's Resource Book, page 105**

Listening 1 p.155

1 To link to the previous section pick up on the noun phrase in question 2 (*a sense of humour*). Give students a few minutes to discuss the importance of a sense of humour. Ask them what type of things they laugh at. Raise the issue of why people laugh when they are nervous or embarrassed.

2 Having recapped the strategy for dealing with multiple-choice questions, give students a moment to read the text before playing the recording twice. Check the answers, highlighting how the points are paraphrased between the recording and the written options.

3 Humour can vary greatly from place to place but some comedy programmes (such as *Friends* or *The Simpsons*) have wide appeal. If there is time, get students to tell a joke to the class. Joke telling can be very hard in another language but some jokes, especially those that disparage certain sections of a community, are often the same the world over. Have a joke ready and be prepared to set an example by telling the first joke.

Key

3 **1** C – *people are always quoting bits from it in the office the morning after and I felt a bit left out*
2 A – *it runs the risk of getting a bit repetitive*
3 B – *Also unique here is the way you're keen not to be 'over the top'.*
4 B – *desire the British have not to take themselves too seriously*
5 C – *nobody had the courtesy to let me know.*
6 C – *word of mouth recommendation … you're certainly not going to get this from me!*

Language development 1 p.156

This section looks at the use of participle and infinitive clauses. The assumption is that, by now, students will be able to use relative clauses with confidence, so this is a natural progression to more complex clauses.

1a Look at the examples from the text, and at the Grammar reference on page 186. Draw students' attention to the way these clauses act as an efficient way to combine information, reducing the length of sentences by omitting redundant subjects and auxiliary verbs.

1b It is important to be able to rewrite the structures using finite verbs to fully understand their structure and use,

2 Allow students to refer to the Grammar reference as they try to incorporate participle clauses. Encourage them to think about whether the part in italics is expressing time, contrast or reason/result. Check any difficult items of vocabulary (e.g. *a household name* = widely known, *sitcom* = an abbreviation of *situational comedy*).

Do the first one together as an example. Get students to think how the information could be expressed using a relative clause: *John, who had appeared in a popular sitcom, became a household name*. If this was reduced, it would become *John, having appeared in a popular sitcom, became a household name*. Finally, note the similarity to the participle clause that puts the clause at the front: *Having appeared in a sitcom, John ...*

3a Point out to the students that the idea is to express the same information, but in a more succinct way. Note that in question 1, the clause *Having seen the film already* expresses not only reason, but also an element of time. The distinctions in the Grammar reference overlap to some extent.

3b Look together at the use of *to*-infinitive clauses in the Grammar reference (purpose, consequence, result and condition) before students attempt the exercise.

4 The idea here is both to practise the structures and to show how a piece of writing can be improved by using them to reduce redundancy and express information more succinctly. When students have finished the task, ask them which is better and why.

Photocopiable activity

Activity 10A can be used here. It is a pairwork activity in which students unjumble and then match sentence halves to complete sentences that practise participle and infinitive clauses.

Background

The pantomime *Dick Whittington* is a popular show for families, mainly performed around Christmas time. Dick is a poor boy who goes to live in London with his cat, intending to make his fortune. There are several versions of the story, most involving Dick going on a sea voyage in which the ship is plagued by rats (lead by the evil King Rat) until his faithful cat kills them all. After his adventures, he returns to London to a hero's welcome, is elected Mayor of London and marries the girl he loves, Alice Fitzwarren.

Key

1a/b **1** time clause – *Less than three years **after he (had) left** university, ...*
2 contrast clause – ***Despite the fact that he is*** *acknowledged ...*
3 *-ing* clause to replace *and* + co-ordinate clause – *works hard to extract maximum leverage from his talents **and (he) pays** incredible attention to detail* – and an *-ed* clause to express reason/result *attention to detail, **because he is terrified** of the risk*
4 contrast/passive – *Although **he is impressed** by*

2 **1** Having appeared in a popular sitcom, John became a household name.
2 Glancing at the TV page, I saw that my favourite comedy was on later.
3 Being very witty, Sam is a great performer.
4 Having spent five years working in the theatre, Sarah has lots of stories to tell.
5 Bored with the normal TV channels, I decided to get cable TV.
6 Generally speaking, few young people like opera.

3a **1** Having already seen the film, I didn't go to the cinema with the others.
2 Amazed by the special effects, I went to see the film three times.
3 Thinking I would get a better view, I sat in the front row.
4 Having had a lot of trouble getting tickets, I had hoped the concert would be better.
5 Finally Beyoncé walked off stage, blowing kisses to the audience as she went.
6 Knowing how much my sister likes *Swan Lake*, I've bought her tickets at the Mariinsky Theatre.

3b 1 To see her perform live, you'd think she's been doing it for years. (condition)
2 My grandmother saw *Casablanca* enough times to know all the lines off by heart. (result/consequence)
3 I rushed home to watch *Friends* (purpose), only to discover that it was a repeat. (consequence)
4 Phil has directed enough plays to know what he's talking about. (result)
5 I like opera, but to understand the story, I have to read the programme notes. (purpose)
6 The show is not for everyone. To put it another way, it's pretty unusual. (purpose)

4 One day, **feeling generous**, I volunteered to take my five-year-old niece Amy to a pantomime.
Thinking *Dick Whittington* would be one she'd enjoy, I booked two tickets.
Planning the outing well would ensure that / Planned well, all would go smoothly and we would have a great time.
Having stopped a few times on the way to the theatre, we arrived just as the show was starting.
We found our seats, **only to discover** that we were sitting behind a very tall family.
Not being able to see, Amy had to sit on my lap for the full three hours!
At first, she was spellbound. **To see her face**, you'd think she was really having a great time.
But **frightened by King Rat**, Amy started howling.
To distract her, I offered to buy her an ice cream.
Creeping through the darkness, we found somewhere **to buy one**, then returned to our seats. **Having apparently got over her fear**, she sat back in my lap and enjoyed the rest of the show.
At the end she stood **clapping** until the curtain came down for the final time.
I don't know how the actors felt at the end of the evening, but **after seeing / having seen the show** with Amy, I was exhausted! Nevertheless, **all things considered,** it was a great success.

▶ **Student's Resource Book, page 106**

Use of English 1 p.157

1a Students have had plenty of practice of Gapped Sentences by now so it shouldn't take long to recap the task strategy.

1b Give students a suitable time limit (5–8 minutes) to do the task and then let them check the answers using a dictionary.

2b If students cannot think of other uses of the words, they can use the dictionaries again to find them.

Key

1 1 casual – relaxed; temporary, not regular; not formal
2 joined – went to meet; connected; began to take part in an activity others are doing
3 low – small level; quiet/deep sound; unhappy/depressed
4 pulling – (phrasal verb) pull apart – separate/examine closely; move your body; change expression
5 poor – not good; lacking money; unlucky, showing sympathy

2 *Casual* can also mean – not serious (casual relationship); not planned (a casual remark)
Join can also mean – become a member (join a club); do something together (join me in a glass of wine)
Low can also mean – not much left (we are running low on petrol); dim (low lighting); small amount of heat (set the boiler on low)
Pull has many meanings such as – move something towards yourself (e.g. sign on a door); tow (pull a trailer); injure (pull a muscle)
• *Low* (adjective) can also be an adverb (flying low) and a noun (highs and lows)
• *Pull* (verb) can also be a noun (give a rope a sharp pull)

Writing 1 p.158

Read through the writing strategy notes at the top of the page. As mentioned in the previous module, much of the writing strand has focused on important factors such as task achievement, planning, organisation and cohesion. This section again focuses on control at sentence level and reviews some of the structures covered earlier in the course.

1a Spend some time looking at the sentence patterns on page 204, comparing the types of clauses and how they are linked.

1b This could be done individually or in pairs, with students pooling their knowledge and ideas.

2a Students should look at the task and highlight key parts.

2b Students combine the sentences in the same way as in Exercise 1a.

3a Here, students have the choice of finishing the sample answer or starting their own. While following the scope of the task, their focus should be on using well-constructed sentences.

Key

1b Suggested answers

1 Although I love various forms of entertainment, like the cinema and the theatre, being someone who relishes a challenge, I also like to learn something new in my spare time. (*or* Although I love … , I am also someone who … , which is why …)

2 After working hard all day, Mark needs to relax in the evenings and he does so/this by playing computer games, which is unusual for someone who enjoys the company of others so much.

3 Our ancestors, not being slaves to the work ethic, had more leisure time than we do, in spite of the fact that they had fewer labour-saving devices. (*or* Because our ancestors were not … , they had … in spite of having …)

4 Even though some people think board games are very old-fashioned, they are still very popular because they are a great way for families to have fun together. (*or* Despite the fact that some people … *or* Although board games may be considered old fashioned by some people, they are still …)

5 I like relaxing in a hot bath in the evenings because it helps me to get a good night's sleep, which is important, as I need to be wide-awake and ready for work the next morning. (*or* Relaxing in a hot bath in the evenings helps me to get …)

2b Suggested answer

The problem is that these days many people lead very busy lives, as a result of which, they don't have so much time on their hands, particularly when they start work. Having had a long hard day at the office, the last thing they feel like doing is making much of an effort to go out. Many hard-working people don't go to the theatre or cinema, preferring to slump in the armchair in the evening and watch TV.

3a Suggested answer

These days people can enjoy themselves doing things they couldn't do 150 years ago, such as watching TV or going to the cinema. However, most people lead very busy lives. They get up early in the morning and go to their jobs, where they work till late, and have little time for fun. That's why, instead of going out to the theatre or the cinema, many hard-working people prefer to slump in their armchair in the evening and watch TV.

In the old days, people didn't work quite so hard to earn money and, because they had more leisure time available, were able to spend more time chatting to friends and playing games. What's more, if they went out to the theatre or the opera, they made it into much more of a big occasion, dressing up and taking time over it.

In my view, as time is precious in today's fast-moving world, the best way of having fun is to learn to enjoy the simple things in life, like reading a book on the train, or cooking a meal for a friend on their birthday. Being quite a shy individual, I also like to spend an hour or two sitting on a park bench in the warm sunshine watching the world go by and feeding the birds.

Believe it or not, once you realise you don't have to spend lots of time and money going out to shows, having foreign holidays or spending the weekend at Disneyland, life becomes a lot more enjoyable.

▶ **Student's Resource Book, page 108**

10B Taking part

Listening 2 p.159

Start, with books closed, by asking students if they can sing. Ask them when and where they like singing and what type of music or songs they like. Ask if they have ever considered singing (or another aspect of music) as a career.

1 Get students to read the task rubric in Exercise 2 and discuss the two questions in small groups. Ask also if they have ever heard of such an idea (they may have heard of similar ideas in the world of pop music, such as *Pop Idol* or *The X Factor*) and if they would ever enter such a competition.

2 Remind students that the answers follow the same order as the recording and that they should listen for the interviewer's questions as cues for the section with each answer.

3 Continue the discussion about students' own musical abilities, including instruments they play, which will link into the following section.

Key

2

1 D – *our primary aim was to bring out the best in non-professional singers; open up avenues for them*

2 B – *people expect opera singers to lead exotic lives, whereas mine is anything but – you know, I'm just a supermarket cashier*

3 C – *all this time I've harboured a nagging feeling that maybe I blew my big chance of a career in music*

4 C – *hasn't got what it takes in terms of determination, energy and stamina*

5 A – *They were, of course, much less used to the speed and intensity … longer to adapt what they had prepared*
6 A – *Mind you, I did feel a bit defenceless at times*

Speaking p.159

As in earlier speaking sections, there is a lot of attention here on developing vocabulary; this time, it is all music related.

Photocopiable activity

Activity 10B can be used at any time here, either to introduce the topic of musical styles or to review the vocabulary. Students work in small groups to identify musical genres from descriptions of them.

Background

Ravi Shankar, born 1920, has been a bridge between Indian and Western music, developing a distinctive style of sitar. He worked with the famous violinist Yehudi Menuhin, and the Beatles' George Harrison, from whom the quote comes, became his pupil. In the 1960s, he played at major music festivals, including the legendary Woodstock festival.

1a The box contains 15 musical styles/genres. The exercise could be done as a light-hearted competition with students working in pairs, racing to be the first to come up with a piece of music or song and a singer/composer for 12 of them.

1b First check students are familiar with the categories asking concept questions such as *Do you hit it or blow it? What is it made of?* Students then try to match the instruments to the genres above.

Background

A bodhran, pronounced /'baurən/, is an Irish frame drum, like a tambourine, that is hit with stick.
Maracas are hollow containers, traditionally gourds, filled with seeds, beads, etc. with handles, played by shaking, usually in pairs.
A marimba is African in origin and is like a xylophone, but with wooden bars that are hit with a stick.

2 Here, students work individually as a preparation for the following task.

3 Using their ideas in Exercise 2, students compare music tastes.

4a In this passage, students are looking to discriminate between the words in each pair, either from their meaning or from their usage.

Extra!
Ask them to explain when or how the other word in each pair could be used.

4b This task could be spoken or an additional piece of writing. For students that haven't been to any live concerts, ask them to talk about other types of live performance, such as theatre or comedy.

5a/b This is designed to be a further illustration of how a Part 3 task develops into a Part 4 discussion, giving examples of the type of questions asked. Give students four minutes for the first part, then let them feed back to the class.

6 In pairs, they develop the topic, discussing the questions.

Key

1b **Strings:** electric guitar, acoustic guitar, harp, banjo, cello, violin, double bass, sitar
Woodwind: saxophone, clarinet, flute, panpipes
Brass: trumpet, French horn
Percussion: drums, marimba, bodhran, maracas
Keyboard: piano, organ, synthesiser

4a 1 performing – as musicians; artists and designers show their work
2 live – there in front of them; *living* is the opposite of *dead*
3 venue – the place/building where concerts take place; *location* = *position*
4 holds – maximum capacity; *contains* = number in it now
5 off – *pull something off* = *succeed*; *pull through* = *survive*
6 meant – *caused*; *resulted in* = result
7 despite – + noun/gerund; *although* + clause
8 recording – music; *filming TV/cinema*
9 comprising – + noun; *consisting* ***of*** + noun
10 what – *What* + (adjective) + noun; *How* + adjective
11 but – *couldn't help but do something*

▶ **Student's Resource Book, page 110**

Use of English 2 p.162

1 Use the picture from the film *The Sound of Music* to create awareness and generate interest in the topic.

2 Give students 10 minutes to both read the text for gist and answer the questions.

3 Discuss why people go along and what musicals students would like to sing along to!

4 Follow up by getting students to produce sentences with some of the alternative words to help show the differences in both meaning and use.

Key

2b **1** D – introduce = to bring a new experience to a place
2 B – collocates with *latest*
3 A – new level suggests higher, raising the standard
4 B – suggests *many*
5 C – collocates with *recommended*
6 A – people are dressing up for other people
7 C – phrasal verb *warm up* = to excite
8 D – collocates with *prizes*
9 B – phrasal verb = relax in a seat
10 B – idiomatic expression – top of their voices = very loud
11 A – extended
12 C – only one that forms a superlative with *best* (the others collocate with *most*)

Language development 2 p.163

This section looks at dependent prepositions following nouns, adjectives and verbs, including some that cause confusion, and also at what can follow the preposition.

1a Students match the halves, either by recognising the sentence or by identifying the pattern required. Do the first one with them.

1b Students use the sentences in Exercise 1a to complete the rules, then look at the Grammar reference on page 186.

2 As well as correcting the mistakes, students should notice what has caused them.

3 These verbs can all be used with two prepositions, depending on the situation. Students should find the correct preposition and identify the difference.

4 Some adjectives can be followed by either a *to*-infinitive or a preposition + *-ing*. Note that in the example, *keen to learn* and *keen on learning* have the same meaning. Students need to choose the correct prepositions in order to transform the sentences.

5 Students should choose the correct option in each pair, checking in a dictionary for those they are not sure about. In Paper 3 Part 3, if they notice a preposition is required, they need to think carefully about which one.

Start by giving students the sentences 1 *Paper is made ___ wood.* 2 *The table is made ___ wood* and ask students what the missing prepositions are and what the difference between them is (1 *made from* – the original material is physically/chemically changed and is unrecognisable; 2 *made of* – original material is still recognisably there). Note also *made in China, made to order, made for me, made by my father*.

Key

1a **1** d – *dream of* (verb + preposition)
2 f – *object to* (verb + preposition)
3 a – *interest in* (noun + preposition)
4 b – *discuss something with* (verb + object + preposition)
5 h – *strange about* (adjective + preposition)
6 e – *depend on someone for support* (verb + preposition + object + preposition)
7 g – *(no) objection to* (noun + preposition)
8 c – *discourage someone from* (verb + object + preposition)

1b **A** noun; verb
B verb
C possessive (e.g. *their* ***daughter's*** *taking up the sport*)
D the same (e.g. *object to/objection to*)

2 **1** Someone presented ~~to~~ the singer with a big bouquet of flowers. (verb with two objects, so two forms possible, compare *presented a big bouquet to the singer*).
2 The critic aimed most of his comments **at** the writer. (*aim something at someone*)
3 First we had a discussion about the venue, then we discussed ~~about~~ the dates. (*a discussion* (noun) + *about*; *discuss* (verb) without preposition)
4 The success of the comedy saved **the theatre from disaster**. (verb + object + preposition)
5 The producer blamed ~~on~~ the press for the lack of ticket sales. (two objects: *blame something on someone; blame someone for something*)
6 When his jokes failed, he resorted to **shouting** insults at the audience. (*to* as a preposition followed by *-ing*)
7 The reason that I don't like music festivals is that I'm unaccustomed to **sitting** in muddy fields all day. (as question 6)
8 What's the point **in/of buying** a ticket if you can't see anything? (*the point of/in* + *-ing*)

3 **1** with (someone); about (something)
2 to (someone); for (something)

3 about (something); to (someone)
4 on (someone); for (something – the topic)
5 with (someone); about/over (something – the topic)

4 1 We were annoyed **at finding** someone else sitting in our seats.
2 I'd like a new CD, but I'm nervous **about asking** again.
3 My parents advised me **against going** to drama school.
4 They insisted **on everyone buying** a ticket.
5 The doorman suspected **me of buying** the tickets on the black market.
6 I'm sorry **for/about losing** your CD.

5 1 **a** heard about (be told news)
b heard of (know exists)
2 **a** shouted; to (difficulty hearing)
b shout at (in anger)
3 **a** anxious for (strong feeling of want)
b anxious about (worried)
4 **a** cares about (think it is important)
b care for (in negative to not like)
5 **a** laughed about (something that involves you)
b laugh at (something that doesn't involve you)

▶ **Student's Resource Book, page 111**

Writing 2 p.164

1 Students start by discussing their attitude to reviews. Do they read reviews before or after they have seen something? Can they think of examples when they have seen (or not seen) something just because of a review they read?

2 Here, they are given a choice of reviews to write, whereas in the exam they would not. However, in the exam, the review would be an optional question; here it is not! Ask students to read both tasks and to decide which one they would rather answer.

3a Students should underline key points in the task they have chosen and select which two programmes or CDs they are going to review and prepare some notes about them related to the task.

3b/c The two paragraph plans show alternative approaches for comparing and/or contrasting two things. The first is a 'point-by-point' approach, i.e. compare and contrast A and B in relation to one point, then compare and contrast them in relation to another point. The second is a 'one-at-a-time' approach, i.e. look at all the points or pluses and negatives for A, then do the same for the B. Both approaches finish with a summary. Students could use either.

4a The list of adjectives could be used to describe CDs or TV programmes or both. Apart from *recorded* (which depends on the adverb used with it, e.g. *beautifully/badly*), they all have either a positive or negative connotation. Students decide which and note the words that collocate with them to make useful phrases.

4b Students could look back at Exercise 1b in Language development 2 of Module 8 (page 132) for further phrases for comparing and contrasting. Some of the phrases for balancing an opinion or summarising came up in the section on attitude phrases in Writing 1 of Module 8 (page 126).

5/6 Students write and check the work as they have done in previous modules, but this time using a structured approach, working through the checklist and looking at both the sentence level and the piece as a whole.

Key

4a **Positive:** *memorable, original, lively, popular, riveting, stylish, entertaining, moving, sophisticated, hilarious* (unless it is unintentionally funny), *different, beautifully recorded*
Negative: *predictable, boring, unconvincing, over the top, inaudible, overrated, flat*

5 Suggested answers

TASK ONE

If you like to watch ordinary people make fools of themselves, then these two programmes could be for you.

In *Big Brother*, ten contestants are locked into a house, with no contact with the outside world. Every week, the TV audience votes one of them out, and the final person left in the house wins the prize. During the week, the producers give the contestants tasks to make the show interesting. Of course, how interesting the programme is depends on the personalities and how they all get on. Some people have found the show riveting, but personally I found it excruciatingly boring.

Survivor is considerably more interesting and, at its best, is absolutely hilarious. Whereas *Big Brother* stretches over an interminable nine weeks, *Survivor* lasts only 39 days and has 16 contestants, thrown together in a remote part of the world. The basic idea is that they have to compete against each other and survive a number of physical and psychological challenges, then vote each other out of the game one by one. Of course, hidden cameras pick up everything, and there are some terrific shots, not only of the contestants making fools of themselves, but of exotic wildlife, too. However, despite the many amusing moments, I have to admit I found the programme a bit flat at times, depending on which personality was being focused on.

If I had to choose one programme, though, there's no doubt which one I'd choose. Quite simply, *Survivor* is fun. *Big Brother* is a bore!

(252 words)

TASK TWO

You might think that Diana Krall and Eminem could have nothing in common, but you'd be wrong.

In *The Girl in the Other Room*, Diana Krall, the marvellously talented jazz pianist with the wonderful singing voice, explores her musical depths in a number of moving interpretations of modern standards. In *The Eminem Show*, the notorious rapper, whose controversial lyrics have shocked the world, produces tracks that are raw with thundering rhythms. While Krall's music takes us back to an earlier era and appeals to the mature young and a middle-aged audience who like their music safe, Eminem seems to threaten society as we know it and is the standard-bearer for angry, dispossessed youth. And yet both have a kind of sincerity and honesty in what they do, as they try to explore the possibilities of their art. In this album, Krall is experimenting with different formats and exposing her audience to new material, in an attempt to create something that is truly original. These beautifully recorded songs are a mixture of her own and adaptations of such singers as Joni Mitchell and Bonnie Raitt. Each song is wonderful, and Krall's voice and style fits in perfectly with each song she sings. Although they are quite sad, they are no way depressing. Eminen, too, likes his material downbeat, as he explores his life, loves and failures with surprising insight and honesty. In every song there is genius. So two very different artists, but what they have in common is creativity and integrity. They are two great talents.

(256 words)

▶ **Student's Resource Book, page 113**

Photocopiable activity

Activity 10C is designed to be used at the end of the course, as it reviews some of the target language from all modules. Students work in groups to answer questions on cards, each of which covers an aspect of the course.

Module 10: Review p.166

1 **1** B **2** C **3** A **4** C **5** B **6** A **7** D **8** C **9** A **10** C **11** D **12** B

2 **1** As soon as he came on stage, I burst out laughing.
2 I stumbled across one of her old records in a second-hand shop yesterday.
3 I have no objection(s) to you/your going to the show.
4 It's a well-established/long-established tradition.
5 Many amateur stage productions formed/provided the basis for his later success in movies.
6 He has an all-consuming passion for music.
7 Do you think there's any prospect of his performance ever improving / that his performance will ever improve?
8 He's off duty / not on duty tonight.

3 **1** with; about **2** Having **3** to
4 Frightened/Disturbed/Deafened **5** finding
6 against buying/getting **7** at
8 watching/seeing

4 **Having** had my car stolen the day before and still **feeling** completely devastated, I decided to dance the night away in my favourite club. Annoyed **at** finding it full when I got there, I had to resort to **arguing** with one of the security guards to let me in. Even then, he insisted **on** me waiting until some people had left. **To understand** my frustration, you need to realise that, **having been** a regular there for many years, I love the place and don't much care **for** being treated as an ordinary clubber. Anyway, having complained **to** the manager about being left out in the cold, I was let in and, **losing** myself in the music and the dance, I forgot my troubles in an instant.

Exam practice 5 TRB p.203

Paper 1: Reading
1 E **2** C **3** A **4** B **5** D **6** C **7** A **8** E **9** C **10** A/C **11** C/A **12** D **13** E **14** A/B **15** B/A
Paper 3: Use of English
Part 1
1 D **2** C **3** D **4** B **5** B **6** A **7** C **8** A **9** A **10** C **11** C **12** A
Part 2
1 for **2** such **3** Instead **4** addition **5** which **6** with **7** Unlike **8** more **9** on **10** a **11** over **12** Having **13** other **14** When/Once **15** there
Part 5
1 is famous for **2** prevented me from entering **3** no circumstances should this door **4** I was forever **5** not so/as easy to use as **6** not only lost his job **7** wishes that she had not got **8** was unaware of the change
Paper 2: Writing
Suggested answers
1

I am sure all our readers have heard about the concert held in the remains of the medieval castle up on the hill behind the town last Saturday night. I went because my boyfriend persuaded me to go, although I was not at all keen. I've never been a fan of classical music, and this promised to be the usual boring group of white-haired, unsmiling violinists whose only aim seems to be to send the audience to sleep.
Well, I was wrong.
The first thing that took me my surprise was the players – all young and fashionably dressed. The orchestra was conducted by a woman, which really impressed me, and they played jazz! After the initial shock, I was taken over by the rhythm and intensity of the music, and couldn't keep my feet still. I could see that other people felt the same and soon the whole audience was moving to the sound of the beat.
Then a young people's choir sang, and I was amazed to see how enthusiastic they were, and how they conveyed their joy of singing to everyone present. At the end, they sang some well-known songs – and we all joined in!
I don't know whether it was the atmosphere created by the ruined castle under a moonlit sky, or the enthusiasm conveyed by the musicians. But other musical events I have been to seem boring in comparison with this one.
And the best thing? I've been inspired to join the choir. So who knows – you might see me up on the stage next time!

2

Dear Sir/Madam,
Thank you for your letter concerning the customer feedback form I filled in. I am happy to expand on the points I found unsatisfactory and hope my comments will enable you to make improvements.
The first point you mention, speed of service, speaks for itself. If you define your outlets as 'fast-food restaurants', then not only the food but also the service must be 'fast'. On my visit, I had to wait 20 minutes for my meal, when the restaurant was only half full.
During that time, I complained twice (and this brings me to the second point), and was told off for not waiting my turn as there were others in front of me! The attitude of the staff was, I'm sorry to say, that of people bored with their jobs for whom customers are a nuisance. I nearly got up to leave, and almost wish I had.
When my meal finally arrived, I was pretty hungry, so you can imagine my disappointment when I discovered that it was almost inedible! I ordered your 'cheese surprise'. The only 'surprise' was the grit in the lettuce (obviously unwashed) and the rubbery cheese. I had the impression that my meal had been cooked at least ten minutes before I got it.
It seems clear to me that the underlying problem at that particular restaurant is the quality of the service. As your chain of restaurants has, on the whole, a solid reputation, I would suggest you devote some attention to staff training.
Yours faithfully

3

FILMS ETC.

AIMS OF THE CLUB

Films etc. is an entirely new concept in entertainment. We hope to bring together people who want to make film-watching a total experience. What do we mean by that? We don't only want to give you a good time, but we hope to tell you something more about the actual making of the films we show. Our aim, of course, is to help you improve your English; we feel, however, that watching a film, although enjoyable and instructive, is often a one-way experience. So we aim to invite actors and film directors (amateurs of course – our budget would not run to Jude Law or Anthony Minghella!) to explain what goes on behind the scenes.

TIMES

We hope to have twice-weekly meetings, at 8 p.m. Occasionally, on Sundays, we may have special performances of longer, epic movies.

RANGE OF FILMS AND ASSOCIATED ACTIVITIES

We hope to cater for all tastes, so the films we choose will cover all genres, and come from different historical periods. One thing will characterise them all, however – they will all be quality films (in our opinion, anyway). The fact that we shall have guest actors and film directors coming along will give you a chance to practise your English in the discussion session afterwards. There will also be refreshments!

ADVANTAGES OF BECOMING A MEMBER

Membership costs €40 a year, and entitles you to unlimited viewing of films. Besides this, you will be given a card which allows you discounts at many cinemas in the town.
Come along! Enrolments starting October 1st.
Secretary's office – ask for Stella.

4

REVIEW OF TWO ENGLISH-LANGUAGE WEBSITES

Have you ever looked for a website to help you learn English? I decided to check out two of the most popular ones: English Online and PC English.

Ease of navigation

To make an assessment of how easy each site was to navigate, I decided to look for two specific areas in both sites – grammar exercises and short stories for children. English Online is definitely better in this aspect – I was able to find exactly what I wanted in quite a short time. PC English was more complicated, and at times the instructions were not clear, and I had to go back to the homepage more than once.

Range of information available

English Online scores higher on this point, too, with exceptionally good grammatical explanations and interesting texts. There are also listening exercises for all levels. On the other hand, PC English has a section on pronunciation which the other doesn't have, and students can speak into the microphone and record their pronunciation.

Special features

English Online provides the possibility to contact tutors for special problems, but PC English doesn't have this facility, although it advertises several local centres where students can seek help.
In general, I would say that English Online is better for all the reasons listed above, but also because it caters for all age groups and many different interests; PC English, by comparison, doesn't have an extensive kids' section, and although it does have great appeal for teenagers, might not be so appealing to adults.

Paper 4: Listening

1 G **2** H **3** D **4** C **5** F **6** B **7** D **8** G **9** E **10** A

UNIVERSITY *of* CAMBRIDGE
ESOL Examinations

Do not write in this box

Candidate Name
If not already printed, write name in CAPITALS and complete the Candidate No. grid (in pencil).

Candidate Signature

SAMPLE

Examination Title

Centre

Supervisor:
If the candidate is ABSENT or has WITHDRAWN shade here

Centre No.

Candidate No.

0	0	0	0
1	1	1	1
2	2	2	2
3	3	3	3
4	4	4	4
5	5	5	5
6	6	6	6
7	7	7	7
8	8	8	8
9	9	9	9

Examination Details

Candidate Answer Sheet

Instructions

Use a PENCIL (B or HB).

Mark ONE letter for each question.

For example, if you think B is the right answer to the question, mark your answer sheet like this:

0	A B C D E F G H

Rub out any answer you wish to change using an eraser.

1	A B C D E F G H	21	A B C D E F G H
2	A B C D E F G H	22	A B C D E F G H
3	A B C D E F G H	23	A B C D E F G H
4	A B C D E F G H	24	A B C D E F G H
5	A B C D E F G H	25	A B C D E F G H
6	A B C D E F G H	26	A B C D E F G H
7	A B C D E F G H	27	A B C D E F G H
8	A B C D E F G H	28	A B C D E F G H
9	A B C D E F G H	29	A B C D E F G H
10	A B C D E F G H	30	A B C D E F G H
11	A B C D E F G H	31	A B C D E F G H
12	A B C D E F G H	32	A B C D E F G H
13	A B C D E F G H	33	A B C D E F G H
14	A B C D E F G H	34	A B C D E F G H
15	A B C D E F G H	35	A B C D E F G H
16	A B C D E F G H	36	A B C D E F G H
17	A B C D E F G H	37	A B C D E F G H
18	A B C D E F G H	38	A B C D E F G H
19	A B C D E F G H	39	A B C D E F G H
20	A B C D E F G H	40	A B C D E F G H

A-H 40 CAS

denote Print Limited 0121 520 5100

DP594/300

UNIVERSITY *of* CAMBRIDGE
ESOL Examinations

Do not write in this box

Candidate Name
If not already printed, write name in CAPITALS and complete the Candidate No. grid (in pencil).

Candidate Signature

Examination Title

Centre

Centre No.

Candidate No.

Examination Details

0	0	0	0
1	1	1	1
2	2	2	2
3	3	3	3
4	4	4	4
5	5	5	5
6	6	6	6
7	7	7	7
8	8	8	8
9	9	9	9

Supervisor:
If the candidate is ABSENT or has WITHDRAWN shade here

Instructions
Use a PENCIL (B or HB).
Rub out any answer you wish to change.

Part 1: Mark ONE letter for each question.
For example, if you think B is the right answer to the question, mark your answer sheet like this: 0 A B C D

Parts 2, 3, 4 and 5: Write your answer clearly in CAPITAL LETTERS.
For Parts 2, 3 and 4, write one letter in each box. 0 EXAMPLE

Candidate Answer Sheet

Part 1

1	A	B	C	D
2	A	B	C	D
3	A	B	C	D
4	A	B	C	D
5	A	B	C	D
6	A	B	C	D
7	A	B	C	D
8	A	B	C	D
9	A	B	C	D
10	A	B	C	D
11	A	B	C	D
12	A	B	C	D

Part 2

		Do not write below here
13		13 1 0 u
14		14 1 0 u
15		15 1 0 u
16		16 1 0 u
17		17 1 0 u
18		18 1 0 u
19		19 1 0 u
20		20 1 0 u
21		21 1 0 u
22		22 1 0 u
23		23 1 0 u
24		24 1 0 u
25		25 1 0 u
26		26 1 0 u
27		27 1 0 u

Continues over →

CAE UoE DP597/301

Part 3

		Do not write below here
28		28 1 0 u
29		29 1 0 u
30		30 1 0 u
31		31 1 0 u
32		32 1 0 u
33		33 1 0 u
34		34 1 0 u
35		35 1 0 u
36		36 1 0 u
37		37 1 0 u

SAMPLE

Part 4

		Do not write below here
38		38 1 0 u
39		39 1 0 u
40		40 1 0 u
41		41 1 0 u
42		42 1 0 u

Part 5

		Do not write below here
43		43 2 1 0 u
44		44 2 1 0 u
45		45 2 1 0 u
46		46 2 1 0 u
47		47 2 1 0 u
48		48 2 1 0 u
49		49 2 1 0 u
50		50 2 1 0 u

denote 0121 520 5100

UNIVERSITY of CAMBRIDGE
ESOL Examinations

Do not write in this box

Candidate Name
If not already printed, write name in CAPITALS and complete the Candidate No. grid (in pencil).

Candidate Signature

Examination Title

Centre

Supervisor:
If the candidate is ABSENT or has WITHDRAWN shade here

Centre No.

Candidate No.

Examination Details

0	0	0	0
1	1	1	1
2	2	2	2
3	3	3	3
4	4	4	4
5	5	5	5
6	6	6	6
7	7	7	7
8	8	8	8
9	9	9	9

Test version: A B C D E F J K L M N   Special arrangements: S H

Candidate Answer Sheet

Instructions

Use a PENCIL (B or HB).
Rub out any answer you wish to change using an eraser.

Parts 1, 3 and 4:
Mark ONE letter for each question.

For example, if you think **B** is the right answer to the question, mark your answer sheet like this:

0 A B C

Part 2:
Write your answer clearly in CAPITAL LETTERS.

Write one letter or number in each box.
If the answer has more than one word, leave one box empty between words.

For example:

0 N U M B E R _ 1 2

Turn this sheet over to start.

CAE L   DP600/304

Part 1

1	A	B	C
2	A	B	C
3	A	B	C
4	A	B	C
5	A	B	C
6	A	B	C

Part 2 (Remember to write in CAPITAL LETTERS or numbers)

		Do not write below here
7		7 1 0 u
8		8 1 0 u
9		9 1 0 u
10		10 1 0 u
11		11 1 0 u
12		12 1 0 u
13		13 1 0 u
14		14 1 0 u

Part 3

15	A	B	C	D
16	A	B	C	D
17	A	B	C	D
18	A	B	C	D
19	A	B	C	D
20	A	B	C	D

Part 4

21	A	B	C	D	E	F	G	H
22	A	B	C	D	E	F	G	H
23	A	B	C	D	E	F	G	H
24	A	B	C	D	E	F	G	H
25	A	B	C	D	E	F	G	H
26	A	B	C	D	E	F	G	H
27	A	B	C	D	E	F	G	H
28	A	B	C	D	E	F	G	H
29	A	B	C	D	E	F	G	H
30	A	B	C	D	E	F	G	H

denote Print Limited 0121 520 5100

Photocopiable activities: teacher's notes

Pre-course: CAE exam quiz

• Use at the start of the course, before Module 1.

Aim	To raise awareness of the various aspects of the CAE exam and to answer some common questions
Time	15–20 minutes
Activity type	Pairwork/Groupwork. Students find out how much they know about the exam by doing a quiz.
Preparation	Make one copy of the quiz (page 137) per student.

Procedure

1 Ask students what *CAE* stands for. Is it:
A Cambridge Advanced Expert
B Cambridge Advanced Exam
C Certificate in Advanced English
(Answer: C) Tell the students that they are going to do a quick quiz to learn more about the exam.

2 Give out a copy of the quiz to each student and set a time limit (five minutes) to complete it. Students should first have a go on their own and then compare with a partner or in groups.

3 Refer students to the Exam overview on page 6 of the Coursebook and get them to check their answers.

4 Discuss answers with the class and answer any other questions about the exam that the students have.

Follow-up

Show students where they can find the Exam reference in the Coursebook (page 168) and explain that they can find more detailed information about the exam there.

1

Paper 1	c) Reading	iv) 1 hour 15 minutes.
Paper 2	d Writing	v) 1 hour 30 minutes
Paper 3	e) Use of English	i) 1 hour
Paper 4	a) Listening	iii) Approximately 40 minutes
Paper 5	b) Speaking	ii) Either approximately 15 or 23 minutes (15 minutes for two candidates, 23 minutes for three candidates)

2 The Speaking test is done first. The Listening paper is done second. Papers 1, 2 and 3 are then done on the same day.

3 There are four parts in the Reading paper, with a total of 34 multiple-matching, multiple-choice and gapped-text questions.

4 Both are false. No dictionaries are allowed in any part of the exam.

5 False. No marks are removed for wrong answers, or for not answering a question.

6 There are two parts. You must answer Part 1, and answer one of the five questions from Part 2.

7 You wouldn't be expected to write a poem, a story or a postcard.

8 Part 1: 180–220 words; Part 2: 220–260 words. This is the number of words that would be adequate to answer each question effectively.

9 vocabulary, grammar, spelling, punctuation, word-building, register, cohesion

10 Five parts: 50 questions/items.

11 a) False: there are four parts.
b) True
c) True
d) False: it consists of two sections which have to be done at the same time.

12 a) three
b) four (These are the other candidate(s), the interlocutor – the person who speaks to you – and the assessor – the person who decides how you perform. At the end of the test, the interlocutor and assessor discuss your performance together.)
c) four
d) grammar
e) vocabulary
f) management (your ability to communicate clearly, effectively and with coherence and cohesion)
g) communication (how well you work with the other people in the room)
h) pronunciation

13 About 60% to get a Grade C. If you miss a paper, or do very badly on one of the papers, you will fail the exam, even if you do exceptionally well in the other papers.

1A Different sentences, same mistakes

• Use this activity after Language Development 1 (CB p. 13).

Aim To identify and correct common structural mistakes
Time 15 minutes
Activity type Pairwork. Students identify and correct the same mistakes in two different lists of sentences without actually seeing each other's sentences.
Preparation Make one copy of Activity 1A (page 138) for each pair of students. Cut each sheet into three sections along the dotted lines.

Procedure

1 Divide the class into pairs. Ideally, each student should sit facing his/her partner.

2 Give students a copy of the Student 1 or Student 2 section. They should not show these to each other. The Answers section should be placed so that they can both see it.

3 Explain that both students in each pair have a different set of sentences, but each set contains the same grammatical mistakes For example, the mistake in sentence 1 on Student 1's list matches the mistake in sentence L on Student 2's list.

4 They should work together to identify, match and correct the same mistakes in their sentences, but they should do this verbally without looking at each other's paper. When they have identified which mistakes in Student 1's list match those in Student 2's list, they write the appropriate letter from Student 2's list on their answer paper.

5 Allow them about 15 minutes for this, then ask the whole class to discuss their answers, explaining why the sentences are wrong and how they should be corrected.

Variations

This activity also works well with groups of four (two students taking the Student 1 sentences and two students taking the Student 2 sentences).

Alternatively, you could do this as a whole-class activity from the start, with half the class taking the Student 1 sentences and half the class taking the Student 2 sentences. As they find, match and correct the mistakes, you should write these on the board.

Follow-up

Working in pairs or small groups, students write five sentences of their own and deliberately include a mistake in each one. These are then passed to another pair/group for correction.

Encourage students to self-correct their mistakes (this is especially useful while and after writing an essay). Some teachers also encourage peer-correction in the classroom.

1 L **2** P **3** N **4** E **5** I **6** O **7** K **8** D **9** A **10** M **11** G **12** B **13** J **14** F **15** H **16** C

These are the corrected sentences. Students might be able to identify some other possible answers.

1 This film **is** the worst film I **have ever seen**.
2 Robert **has lived** in Edinburgh for more than ten years.
3 I **have been** far better off financially since I **started** working here in April.
4 By the time we **arrived** at the station, the train **had left**.
5 I **have played / have been playing** football for this team for two years, and before that I **played** tennis for my school.
6 If he continues to spend money at this rate, he **will have spent** everything before his holiday **ends / has ended**.
7 I usually work in London, but at the moment **I'm spending** some time working part-time for a company in Dumfries.
8 Normally I **enjoy** going to parties, but now I'd like to enjoy my own company for a while.
9 Last Saturday, I **saw** a really great film at the cinema in the town centre.
10 The film **had already started** by the time we got to the cinema.
11 We **were playing** golf when it started raining.
12 I'm going on holiday tonight, so this time tomorrow **I'll be lying** on a beach in the Bahamas!
13 They're bound **to** be late this evening; they always are.
14 He was on the point **of** leaving the house when suddenly the phone rang.
15 Here you are at last. We were **about** to leave without you.
16 Thanks for being so understanding. I thought you would **be** really angry with me.

A We had a great time last summer: we **visited** our family in Australia.
B Good luck in your exam tomorrow. **I'll be thinking** of you then.
C Everyone said she would **be** happy when we told her the news, but she was really upset.
D She usually **visits** her parents at the weekend, although this weekend she is seeing her friends for a change.
E When the teacher finally **arrived** in the classroom, she realised that half her students **had left**.

F They were so angry that they were on the verge **of** shouting each other.
G While I **was watching** television, I suddenly heard a strange scratching noise at the window.
H Thanks for calling me. In fact, I was just **about** to phone you myself.
I I **have worked / have been working** here since the beginning of the year, and prior to that I **worked** in an office in Canterbury.
J Everyone says he's sure **to** pass his exam tomorrow. He's never failed one yet.
K Normally he works hard, but for the time being he**'s taking** things easy because of his poor health.
L They **are/were** the rudest people I **have** ever met.
M When I got to Bob's house, I discovered he **had already left** for the airport.
N We **have known** each other since we **met** at a party last year.
O I'm taking too much time off work, so if I'm not careful I **will have used up** all my holiday leave by the time summer **has arrived / arrives.**
P Janine **has worked** at Pembury and Co. since she got married.

1B In other words ...

• Use this activity after Writing 1 Exercise 2 (CB p. 14).

Aim	To review alternative ways of expressing ideas (especially informal to formal)
Time	15 minutes
Exam focus	English in Use: Paper 3 Part 5. (Also useful for formal letters, reports, etc. in Paper 2 Part 1.)
Activity type	Pairwork or small-group activity. Students complete a crossword by thinking of alternative ways of expressing the same idea based on words and expressions in the clues.
Preparation	Make one copy of Activity 1B (page 139) per two or three students.

Procedure

1 Divide the class into pairs or groups of three. Give each pair/group a copy of the crossword and clues.

2 Explain that each clue contains two sentences. The words and expressions in bold in the first sentences can be expressed in another (usually more formal) way in the second sentence. These all appeared in the Writing 1 section of their coursebook.

3 Working in their pairs or groups, they should decide what these alternative words/expressions are, and write the answers in their crossword grid (an example – 1 Down – has been done for them).

4 Allow them about ten or fifteen minutes to do this, then review their answers.

Variations

If students do this in pairs, you could make this activity more communicative by separating the across and down clues, and giving each student one or the other of these clues. They then have to read their sentences to each other (without looking at their partner's clues) and complete the crossword.

If students do this in groups of three, do the same as above, except give one of the students the crossword grid, which he/she does not show to the other two.

Follow-up

When you review the answers with students, write them on the board. Students should then return the crossword to you (or put it somewhere they cannot see it), look at the words/expressions on the board and try to remember the original words/expressions in sentence 1 of their clues – which they should then use to write their own sentences.

Down
1 presented **2** rigorous **3** reputable **5** exhausted **6** tuition **7** advantages **10** publicised **11** suitable
Across
4 organised **8** irritated **9** attend **12** involved **13** outstanding **14** beginners **15** completion

1C Job qualities

• Use this activity after Speaking Exercise 1d (CB p. 16).

Aim	To review adjectives of character and personality
Time	20 minutes
Activity type	Groupwork. Students identify adjectives from a series of brief descriptions, then use these to complete a grid and reveal a 'hidden' new word.
Preparation	Make one copy of Activity 1C (page 140) per group of three or four students. Cut into cards along the dotted lines. Make sure you keep Part 1 and Part 2 separate.

Procedure

1 Divide the class into groups of three or four. Give each group a copy of the adjectives in Part 1.

2 Explain to / remind students that these adjectives are all useful to describe personal qualities that might be

useful in certain jobs and occupations. Working in their groups, they should discuss the types of jobs that might require an employer with these qualities.

3 Ask them to return the list to you, then give them a copy of the Part 2 cards (including the numbered grid). Explain that each sentence on the cards can be completed with one of the adjectives from the list they have just discussed. These adjectives can then be written in the numbered grid (students will have to work out which order to put them in; there will be some trial-and-error which should promote some discussion).

4 If they put the words in the correct order in their grid, they will reveal another 'hidden' word in the shaded vertical strip. The first group to reveal this word is the winner.

Variation

With smaller classes, or in classrooms where students have room to move around, make just one copy of the cards (preferably enlarge them when you photocopy them) and put these on the walls around the classroom. Students then walk around the classroom in their groups reading the sentences and 'collecting' the answers, before writing them in their grid.

Follow-up

Ask students to brainstorm other 'job quality' adjectives. Write these on the board (try to get one adjective for every two students in the class). Then ask each pair of students to choose one of the adjectives (remove these from the board as they are chosen) and to write a sentence using that word. They then give their completed sentence to you. You read each sentence out loud to the class, without saying who wrote it, and without reading out the adjective in the sentence. The other students should then try to decide what the missing adjective is.

The cards are in the correct order on page 140. The grid should be completed as follows:
1 patient **2** sensible **3** friendly **4** tactful **5** assertive **6** resilient **7** energetic **8** persistent **9** tolerant **10** gregarious **11** decisive
The following word appears in the shaded vertical strip: *industrious*

2A Larry the lion

• Use this activity after Reading Exercise 4 (CB p. 25).

Aim	To identify words and expressions that give cohesion to a piece of written narrative
Time	15–20 minutes
Activity type	Groupwork jigsaw reading. Students put a piece of written narrative into the correct order. If done correctly, this reveals a 'punchline' to a joke.
Exam focus	Reading: Paper 1 Part 2
Preparation	Make one copy of Activity 2A (page 141) per group of three or four students. Cut into cards along the dotted lines.

Procedure

1 Divide the class into groups of three or four. Give each group a set of cards.

2 Explain to students that they should arrange the cards so that they tell a joke. They should do this by identifying linking devices (e.g. pronouns, synonyms, parallel expressions, etc.). The first one has been done for them.

3 Also explain that the joke is missing its punchline (the funny part at the end of the joke). When they have arranged their cards, they should look at the underlined letters on each card. These letters form words (one word per card). The first one has been done as an example (= *There's*). They should write these words on the 'Part 2: The punchline' card.

4 If they have arranged their cards correctly, they will reveal the punchline to the joke.

Variation

For a longer but more communicative version, do this as a whole-class activity, giving one or two cards to each student (if you have more than 13 students, some will have to share a card). They take it in turns to read out their card, then work together to decide which order the cards should go in.

Follow-on

In small groups, students write their own joke (including the punchline), then put the sentences or paragraphs in the wrong order before giving their joke to another group, who should rearrange it.

The text is in the correct order in the book. The punchline is: 'There's no need to get angry just because you don't know the answer'.

2B Relative-clause noughts and crosses

• Use this activity after Language development 1 Exercise 4 (CB p. 29).

Aim	To review relative clauses, expressions using relative pronouns and reduced relatives
Time	About 20 minutes
Activity type	Competitive group activity. Students play a grammar version of noughts and crosses (tic-tac-toe).
Preparation	Make one copy of Activity 2B (pages 142–143) per three students. Cut into sections along the dotted lines.

Procedure

1 Begin by playing a quick game of noughts and crosses with the class. Do this on the board.

2 Divide the class into groups of three. Give two students a copy of the Game 1 grid. Give the third student a copy of the Game 1 Referee answers. He/she should not show this to the other two students in his/her group.

3 Explain that the first two students in each group are going to play a game of noughts and crosses on their grid. They do this by choosing a square on the grid and completing it with a relative pronoun or expression using a relative pronoun. The number of gaps in each sentence shows the number of words that they should use to complete the sentence. They should be careful, however, as there may be some sentences that do not require a relative pronoun.

4 The first two students in each group play the game, and the third student (the referee) tells them if they are right or wrong. If they are wrong, the sentence remains 'open' and they can try it again on their next go.

5 The first student to get three squares in a row (horizontally, vertically or diagonally) wins the game.

6 Steps 3–5 above are then repeated with Games 2 and 3, with each of the other students taking the referee's role for Games 2 and 3 (so that each student gets a chance at being the referee).

Variation

With smaller classes, do this as a whole-class activity. Divide the class into two groups. Draw a large noughts and crosses grid on the board. Give each group a copy of Game 1 (make sure that everyone in the group can see it). The groups then take it in turns to complete the sentences. You write the appropriate answers in the grid on the board. Repeat this for Games 2 and 3. The group that wins two of the three games is the winner.

Follow-on

Working in small groups, students devise a general-knowledge 'quiz' using relative clauses, reduced relatives, etc. They write some factual statements about famous places, people, etc., then remove a key word from each of the sentences and give them to another group to complete (e.g. '________, the venue for the 2004 Olympics, is one of the world's most highly populated cities'.)

The answers are on the referee's cards on page 143.

2C Trading words

• Use this activity after Language development 2 Exercise 4 (CB p. 35).

Aim	To review vocabulary from the Vocabulary and Language development 2 sections in Module 2
Time	15 minutes
Activity type	Pairwork. Students replace incorrect or inappropriate words in sentences by 'trading' words with their partner.
Preparation	Make one copy of Activity 2C (page 144) per two students. Cut each sheet into two sections, Student A and Student B.

Procedure

1 Divide the class into pairs. Give one student in each pair a copy of the Student A sentences, and the other a copy of the Student B sentences. They should not show these to each other.

2 Explain that each of their sentences contains a word which is incorrect or inappropriate. They will need to identify and change each incorrect word. Their partner has the words that they need. Often, these are words which have similar meanings but which are used in a slightly different way (e.g. depending on their meaning, their form or how they collocate with other words in the sentence).

3 Working together, but without looking at each other's sentence's, they should 'trade' words. An example has been given to them (Student A's *well-paid* in sentence 1 can be traded for Student B's *lucrative* in sentence O so that both sentences become correct). In some cases, they will need to change the form of the words.

4 The first pair in the class to correctly trade all their words is the winner. Alternatively, set a time limit of ten or 15 minutes, and then see who has correctly traded the most words.

Follow-on

Many words and expressions in this activity are synonyms. Encourage students to develop a vocabulary 'bank' of similar words (i.e. words that have similar meanings and/or are often confused), giving a definition and sample sentence for each word or expression. You might also like to introduce students to the Longman Language Activator, which groups related words under different themes, key words, etc.

The following 'trades' should be made:

1	lucrative	**O**	well paid
2	digit	**N**	number
3	see	**M**	visualise
4	private	**D**	intimate
5	tricked	**F**	fooling
6	deceptive	**H**	deceptively
7	cheat	**B**	trickster
8	press	**I**	tapped
9	stroking	**J**	patted
10	scratch	**K**	rubbed
11	have	**L**	has
12	every	**C**	each
13	little	**G**	few
14	neither	**A**	both
15	a	**E**	the

3A Meet the celebrities

• Use this activity after Vocabulary Exercise 3 (CB p. 42).

Aims	To review lexis from the Vocabulary section To review aspects of text cohesion for reading and writing purposes
Time	20–25 minutes
Activity type	Groupwork. Students match different sentences from four different texts, and complete the texts with missing words.
Preparation	Make one copy of Activity 3A (pages 145–146) per four students. Cut along the lines to make eight cards (seven sentences plus the missing words). You can also cut the cards into individual sentences; this makes it easier for the students to see each sentence and will promote more interaction, but if you do this, you will need to put the number of the sentence on each one.

Procedure

1 Divide the class into groups of four. Give each group a set of cards.

2 Explain that seven of the cards contain four extracts from newspaper articles about interviews with famous people. Each card contains one sentence from the article.

3 Each article also contains six missing words, which can be found in the word list on their eighth card.

4 Each student in the group begins by choosing one of the celebrity interviews on their 'First sentence' card. He/she then has to find the follow-on sentences on the other sentence cards. He/she should explain to the others in the group why he/she thinks the sentences follow on from each other. If the others think a mistake has been made, they should explain why.

5 When the students have matched all the sentences to form four articles, they should work together to fill in the gaps with words from their word card.

6 The first group in the class to match all the sentences to form the articles, then fill in the gaps with the missing words, is the winner.

Variation

For a shorter version of the activity, divide the class into pairs, and ask them to form just two of the articles.

Follow-on

Working individually, students imagine that they have just interviewed a celebrity and write an article.

1 The popular press describe him as a suave, sophisticated demi-god with charm, personality and intelligence.
When I met him, however, I began to wonder whether any of these journalists had actually met him, or were just playing a joke on their reading public.
What struck me the most was his complete lack of any of these qualities, with social skills coming right at the bottom of the list.
Although he is in his late twenties, I felt like I was interviewing an adolescent schoolboy.
He couldn't sit still, couldn't concentrate for more than two minutes at a time, answered my questions in monosyllabic grunts, and not once made eye contact with me.
There was also an air of barely suppressed anger about him, and it seemed that, at the slightest inconvenience or provocation, he would blow his top like a volcano.
I guess a more sympathetic person would feel sorry for him, but I couldn't see beyond the spoilt child that he really is, and although I have looked hard for his 'appeal factor', I am still thoroughly perplexed at the blind devotion he receives from his fans.

2 He's made a big-screen living playing a vast array of violent roles, whether as a gangster, a corrupt policeman or a sadistic army sergeant.
So to meet him in the flesh really came as a pleasant surprise.
Here is a polite, intelligent young man who, unlike others, hasn't let fame go to his head.
Not for him the intolerable rudeness and arrogance that celebrity can bring.
And despite his reputation for avoiding public scrutiny (which presumably means journalists like myself probing into his private life), he seemed quite happy to sit and answer all my questions.
True, he has strong religious convictions and political beliefs, but he doesn't ram them down your throat, and he spoke objectively about the things he feels passionate about.
It all made a refreshing change from interviewing the usual self-centred celebrities who fill our screens, newspapers and magazines, and our interview was the most enjoyable two hours I've spent in ages.

3 The public have an insatiable appetite for news of their favourite celebrities, and I must admit that I was quite excited to get the chance to interview her.
Here, after all, was the most glamorous female star to have appeared on our cinema screens for years.
But what we see on film and what we get in reality are often two different things, and this was no exception.
Not for me the tall, slim, sharp-dressed starlet with a twinkle in her eyes, a toothpaste smile and charming manners.
When I was shown into her dressing room, I got a rather short, slightly overweight, bad-tempered teenage girl in scruffy casual clothes who scowled at me all the way through the interview, refused to answer half my questions and couldn't pay attention for more than a few minutes.
It certainly gave me a different perspective on fame.
I guess rudeness just goes hand in hand with those like her who achieve celebrity status!

4 Thanks to her role as Peggy on television's *Alimony Street*, hers is a household name.
And much like the TV character she plays, she exudes an air of innocence and helplessness that has found her a place in the general public's heart and which has taken her so far in such a short space of time.
However, initial appearances can be deceptive, and I soon learnt what people mean when they say that we shouldn't take others at face value.
After only ten minutes in her presence, I began to suspect that under the childlike veneer she employs so well on the screen there lurks a mature and determined woman who doesn't suffer fools gladly, enjoys having power over others and never, ever takes 'no' for an answer.
So it didn't come as too much of a shock to me when I found myself on her wrong side by asking about her current boyfriend.
Without warning, she dropped the sweet, innocent act and told me, in no uncertain terms, to 'change the subject right now or get out'.
Here, then, is a private wolf in public sheep's clothing.

3B Moral dilemmas

• Use this activity after Speaking Exercise 6 (CB p. 49).

Aim	To practise discourse markers in spoken English and other general speaking skills, including: practising speaking together, encouraging others to speak, getting a point across and paraphrasing
Time	20–25 minutes +
Activity type	Whole-class discussion. Students discuss hypothetical situations involving moral dilemmas.
Exam focus	Paper 5 general
Preparation	Make one copy of Activity 3B (page 147) for the whole class. Cut into cards.

Procedure

1 Hand out the cards at random to ten students, but tell them not to look at them yet. They should not show their cards to the other students in the class.

2 Explain that the cards contain situations which involve some kind of moral dilemma. Give an example of a moral dilemma (e.g. You find a wallet containing a lot of money lying in the street. The wallet also contains the name and address of the owner. You recognise the owner as being a notoriously mean local millionaire. You are currently doing some work for a local charity which is desperately short of money. Do you return the wallet (and its contents) to the millionaire, or do you give the money to the charity for which you are working?).

3 The students with the cards should look at their dilemmas and decide what they would do in their situations. They should write a brief answer, but they should not use any of the words in bold on their cards.

4 While they are doing this, the students without cards should work in pairs or small groups to

predict/discuss what their classmates' moral dilemmas might be (this step can be omitted if you have ten or fewer students in the class).

5 The students with the cards then take it in turns to read out their written answers. The other students in the class (including those with cards) should try to decide/find out what the dilemma is. They can ask the speaker questions, but the speaker can only answer 'yes' or 'no'.

6 After each dilemma has been identified, hold an open-class discussion on what the others in the class would do in a similar situation. When students say what they would do, they should try to use the discourse markers introduced in Exercise 4a on page 49 of the Coursebook. Encourage everyone in the class to speak, and encourage students to encourage one another to speak.

Note: This activity doesn't need to be done on the same day. You can carry it over to the next lesson, or spread it over a few lessons.

Follow-on

If there are more than ten students in the class, ask those who didn't have cards to read out the dilemmas they predicted in Step 4 of the above procedure. Students can then each choose one of these and write a brief description of what they would do, giving their reasons.

3C Modals plus bingo

• Use this activity after Language development 2 (CB p. 50).

Aim	To review modals and other expressions used to express necessity, permission, advice, ability, possibility, probability and deduction
Time	20–25 minutes
Activity type	Pair- or groupwork (depending on class size). Students identify suitable responses to spoken sentences in a variation of the traditional 'bingo' game.
Preparation	Make one copy of Activity 3C (pages 148–149) for the whole class. Cut into six 'bingo' cards along the dotted lines.

Procedure

1 Divide the class into six teams of roughly the same size. Give each team one of the bingo cards. Make sure that everyone in the team can see the sentences on their card.

2 Explain that the sentences on their cards are responses to some other sentences that you are going to read out to them. Working in their groups, they should look at each response on their card and decide/predict what might be said in order to prompt those responses. (You could give them an example: Response = *You shouldn't have taken it without asking her.* Possible prompt = *I borrowed Isabel's bicycle when she was out.*)

3 Warn students that one of the sentences on their card is grammatically or structurally wrong or inappropriate. They should identify this sentence, as they will not hear a prompt sentence for it.

4 When students have discussed possible prompts for their responses and identified the incorrect sentence, begin the game by doing the following:
   - Choose a sentence at random from the list below and read it out aloud twice, clearly and at normal speed.
   - Allow the teams a few moments to see whether they have a suitable response on their card. If they think they have a response, they should put a line through it.
   - Choose another sentence at random and read it out.

5 Continue doing this until one team has matched all of their responses (with the exception of their incorrect sentence) to your prompts, at which point they shout 'Bingo!'

6 Check their answers. If their responses match your prompts, they win the game.

Variation

You could ask one or two of the students to read out the prompt sentences.

Follow-up

Ask students to return their cards to you (or put them somewhere they can't see them). Choose five or six of the sentences below – it doesn't matter if you repeat some that were used in the game – and read them out. Working individually, in pairs or in small groups, students write suitable responses using modals or other expressions from Language development 2 on page 50.

Read these out in random order. Delete them after you read them to make sure that you don't repeat any.
The numbers in brackets refer to the correct responses/answers, e.g. 5/3 = Team 5, sentence 3.
I'm surprised Elizabeth didn't join us last night. (5/3)
The exam begins at nine, doesn't it? (4/6)
I've invited the Jenson's for dinner tonight. (6/5)
Did Ben take my camera? (3/4)
Tom and Maria aren't coming tonight. (1/3)

I borrowed Jack's gold pen yesterday, but then I lost it. (1/1)
Our flight leaves in less than three hours. (2/3)
Are afternoon classes compulsory? (5/4)
I don't think we've got enough money left. (2/1)
Are you going on the trip this afternoon? (2/6)
Today's test has been cancelled. (2/4)
I cleaned your room for you. (2/2)
How much will it cost? (3/5)
I haven't heard a word from Lucy all day. (6/1)
There's an important exam coming up next week. (1/5)
Were you too busy to join us yesterday? (5/2)
I don't think I'll be able to finish my essay by tomorrow. (1/6)
Is Samantha coming to dinner tonight? (3/6)
When does the course begin? Monday? (3/1)
Has Jane arrived yet? (4/3)
How many people are coming to your party? (4/2)
Carl came late again. (4/1)
Is your teacher here yet? (4/4)
I'm afraid I won't be able to come to your class this morning. (5/6)
Eddie went for a walk in the snow in only shorts and a T-shirt. (3/2)
Have you got a light? (1/4)
I knocked on his door for ages, but he didn't answer. (6/2)
How much have you managed to put aside? (5/5)
Did you miss your flight? (6/3)
Why did you ask Jenny to your party? She's mad! (6/4)

The following sentences are incorrect or inappropriate:

1/2 This should be: *It's advisable **to** leave early if you don't want to be late.*

2/5 This should be: *You **had better**/ You'**d better** book early to avoid disappointment.* (*You better* ... is sometimes used in non-standard spoken English)

3/3 This should be: *You are **under no** obligation to attend the classes* (or *You are not under **any** obligation to attend the classes*).

4/5 This should be: *He could **have** telephoned yesterday to let me know* ...

5/1 This should be: *We felt **obliged** to give him plenty of warning.*

6/6 This should be: *I'm afraid smoking is **forbidden** in the building.*

4A Proverbs

• Use this activity after either Reading (CB p. 56) or Vocabulary (CB p. 58).

Aim	To match proverbs to short paragraphs which summarise the main meaning of each proverb
Time	15–20 minutes
Activity type	Pairwork. Students compete to match sentences with paragraphs on cards.
Exam focus	Reading: Paper 1 Part 4
Preparation	Make one copy of Activity 4A on pages 150–151 for each pair. Cut the paragraphs A–T into cards along the dotted lines.

Procedure

1 Write the word *proverb* on the board and ask the class if they know what this word means. Give them the example *Blood is thicker than water* and explain or elicit through a sample sentence that this is a proverb which means that family relationships are more important than any other kind. (Sample sentence: *I know that what my brother is doing is wrong and everybody else is right, but I'm still going to support him. After all, blood is thicker than water.*)

2 Divide the class into pairs, and give each pair a copy of the sentences 1–20 and a set of cards. The cards should be placed face down on the table.

3 Explain that the sentences are all examples of English proverbs. Working in pairs, students should look at them and decide what they might mean.

4 Students then take it in turns to pick a card, read the sentence or paragraph on it, and match it with one of the proverbs. The proverbs are a summary of the main meaning of the sentence or paragraph. They write the appropriate letter to the right of the proverbs.

5 After about ten minutes, ask them to stop and review their answers. The student in each pair who made the most correct matches is the winner.

Variations

Instead of giving each pair a set of cards, just make one copy for yourself. Read the text on the cards out aloud to the class. Working in pairs, they take it in turns to match what you have said to the appropriate proverb. (Note that this then becomes a listening rather than a reading activity.)

Alternatively, make a copy of just one set of cards and put them up on the walls around the classroom. Students walk around in their pairs matching their proverbs with the cards.

Follow-up

Ask students to work in small groups and think of some proverbs in their own language. They should then try to explain what they mean in English (either verbally or on paper).

Alternatively, they could work in small groups to look at the proverbs 1–20 in the activity and decide whether there is an equivalent in their own language(s).

1 O **2** R **3** Q **4** S **5** H **6** B **7** K **8** E **9** J **10** P **11** D **12** A **13** T **14** M **15** G **16** L **17** C **18** I **19** F **20** N

4B Word formation

• Use this activity after Language development 1 (CB p. 60).

Aim	To review word formation (prefixes, suffixes, spelling changes, etc.)
Time	15 minutes
Activity	Groupwork. Students identify and correct wrong word forms on a playing board.
Exam focus	Use of English: Paper 3 Part 3
Preparation	Make one copy of the playing board (Activity 4B on page 152) for every three students. You will also need three counters (students can use three different coins) and one die per three students.

Procedure

1 Divide the class into groups of three and give each group a copy of the playing board, a die and three counters (if students are not providing their own).

2 Ask students to quickly look at the sentences on their board and to tell you if they can see anything wrong with them. They should identify that there is one word in each sentence which is incorrect (i.e. the form of the word is wrong).

3 Taking it in turns, each student rolls the die and moves their counter along the board (beginning from the 'start' box and following the arrows/numbers). Each time they land on a box, they should identify and correct the wrong word in the sentence. Sometimes the word will need a prefix, sometimes a suffix, and sometimes it may need both. In some cases, a spelling change may also be necessary. It is important that they look at the whole sentence carefully, as this will give them the context.

4 They write their answers down on a separate piece of paper, then initial the box on the board. If they land on a box that has already been initialled, they should move their counter to the next 'free' box.

5 Let them play for about 15 minutes, then tell them to stop and check their answers with you. The student in each group with the most correct answers is the winner.

Variation

Put students into groups of four. One of them has a copy of the answers (below) and is the 'referee'. The other three play the game as above. If they get an answer wrong, the referee tells them that they are wrong, but does not give them the correct answer. Instead, the student has to move back along the playing board to his/her previous box. When one student reaches the 'Finish' box, the total number of correct answers is added up to find the winner.

Follow-up

Working individually, students choose one of the 'wrong' words from the playing board and write a word family for it (using a dictionary to help them). They then give this to another student, who should try to use each word form from the word family in sentences of his/her own.

1 influence = influential **2** energy = energetic **3** diverse = diversity **4** believe = unbelievable **5** persist = persistence **6** history = historical **7** Participate = Participation **8** depend = dependable **9** understand = misunderstood **10** Jealous = Jealousy **11** pay = underpaid **12** fiction = non-fiction **13** cruel = cruelty **14** deep = depth **15** accurate = accuracy **16** large = enlarge **17** fortune = unfortunately **18** confront = confrontation **19** strong = strengthening **20** reverse = irreversible **21** prove = proof **22** broad = breadth **23** please = pleasure **24** tolerate = intolerant **25** appear = disappearance **26** rich = enrich **27** responsible = responsibilities **28** develop = underdeveloped (*or* undeveloped) **29** draw = overdrawn **30** populate = over-populated **31** conform = non-conformist **32** education = co-educational **33** danger = endangered **34** date = predates **35** work = co-workers **36** cook = under-cooked (*not* uncooked) **37** aggression = aggressive **38** general = generalise (*or* generalize)

4C First to 85

• Use this activity after Speaking (CB p. 64).

Aim	To review useful vocabulary for talking about family relationships
Time	15 minutes
Activity type	Pairwork. Students compete against each other to complete gapped sentences with appropriate words.
Preparation	Make one copy of Activity 4C on page 153 per two students. Cut into two sections, Student 1 + Student 2.

Procedure

1 Divide the class into pairs and ask each pair to check with each other that they understand the target vocabulary on page 64 of their Coursebook.

2 Ask them to close their books and work in their pairs to brainstorm as many of the expressions as they can remember. They don't need to write these down, but it will help them in the next part of the activity if they do.

3 Give each pair a copy of the Student 1/Student 2 sentences.

4 Explain that each sentence on their list can be completed with one word from page 64. The words they need to complete each sentence are in alphabetical order. If they add all the letters of all the answers on their sheet together, they should come up with a total of 85 letters.

5 Working individually, they try to complete the sentences. The first student in each pair to come up with words with a total of 85 letters is the winner.

Variations

With smaller classes, do this as a whole-class activity. Divide the class into two teams. One of the teams works together to complete the 'Student A' sentences, and the other team completes the 'Student B' sentences. The first team to get to 85 letters is the winner.

If students are struggling, you could give them the first letter of each missing word.

Follow-on

Exercise 2c on page 64 provides a useful follow-on exercise. Students could also work in their pairs to use the target words from their sentences in a short story or article based on an imaginary family (or perhaps a 'fictional' family from a television programme, book or film, e.g. the Osbornes, the Simpsons, etc.).

Student 1
1 caring **2** dependent **3** each **4** enthralled **5** extended **6** inseparable **7** knit **8** look **9** runs **10** responsible **11** siblings **12** strong
Student 2
1 close **2** devoted **3** disappoint **4** envious **5** expectations **6** furious **7** hit **8** protective **9** resentful **10** sheep **11** start **12** takes

5A Which company?

• You can use this activity at any stage in the module, but you might like to use it after Listening 1 (CB p. 75), as it develops the theme of big-business ethics and the impact of big companies on local environments and people.

Aim	To develop the theme of big-business ethics and global problems
Time	30–45 minutes
Activity type	Whole-class simulation. Students work together to negotiate and make decisions.
Preparation	Make one copy per student of the newspaper article *Trouble in Paradise* on page 154. Make one copy of the company information sheet on page 155. Cut into cards. Make five or six copies of the 'Notes for Babarrie Investment Committee' on page 156.

Procedure

1 Give each student a copy of the newspaper article *Trouble in Paradise* and ask them to quickly read through it (allow them about three minutes). Then tell them to turn the paper over and ask them what they can remember.

2 Explain that four major international companies have expressed an interest in setting up businesses on the island of Babarrie. These companies will bring both advantages and disadvantages to the island.

3 Divide the class into two groups. Divide one of these groups into four smaller groups. Explain that these smaller groups are the companies that want to set up their business on the island. Give each group one of the company information cards and ask them to read the information.

Explain that the other, large group is the Babarrie Investment Committee. They are going to have to decide which of the companies is allowed to set up shop on their island. Give this group the copies of the Babarrie Investment Committee notes and ask them to read through them.

4 While the Babarrie Investment Committee are reading their notes, explain to the four 'companies' that the Committee have discovered some facts about the companies that might jeopardise their chances of establishing their business on the island. They should try to predict what these facts might be.

5 Begin the simulation by asking one of the companies to explain to the Committee the benefits that their company can bring to the island. When they have done this, the Committee should challenge them by pointing out the negative factors that are/might be involved (using the facts on their sheet). They can then negotiate on a settlement (e.g. the company can offer to increase salaries, improve working conditions, etc.).

6 Repeat step 5 for the other three companies.

7 When all four companies have argued their point and negotiations have been made, the Committee should choose one of them. This company will be allowed to set up business on the island and is therefore the 'winner'.

Variation

Instead of giving your students the newspaper article, read it out to them (or ask one of the students to read it to the rest of the class) while they make notes.

You can make this activity shorter by reducing the number of companies to just two or three.

5B Modifying adjectives

• Use this activity after Language development 1 (CB p. 77).

Aims	To review/practise modifiers with gradable/ungradable adjectives
Time	15 minutes
Activity type	Pairwork. Students complete modifiers and match them with adjectives hidden in a wordsearch grid.
Preparation	Make one copy of Activity 5B on page 157 for each pair.

Procedure

1 Divide the class into pairs. Give each pair a copy of the activity.

2 Explain that each sentence has two words missing. The first is a modifying word which they have already met in Language development 1. The first and last letters of each modifier are given to them. The second word is an adjective, which they will find hidden in the grid at the bottom of the sheet. Most of the adjectives appear in Language development 1, but some of them may be new. They should use each adjective once only.

3 Working in their pairs, they should try to complete each sentence with the modifier and an appropriate adjective from the grid.

4 Allow them about 12–15 minutes for this, then check their answers.

Follow-on

Working in their pairs, students choose five pairs of modifiers/adjectives, and write a sentence for each. They then remove the modifier/adjective from each sentence, then pass them to another pair to complete.

1 painfully slow **2** totally obvious **3** extremely important **4** deeply divided **5** seriously rich **6** absolutely furious **7** completely opposed **8** heavily dependent **9** perfectly beneficial **10** highly desirable **11** terribly worried **12** bitterly disappointed **13** widely recognised **14** virtually identical **15** relatively cheap **16** fairly similar **17** rather alarmed

5C Word steps

• Use this activity after Language development 3 (CB p. 83).

Aims	To review/recycle target vocabulary and language from the Vocabulary section and Language developments 1 and 2
Time	12–15 minutes
Activity type	Individual + groupwork. Students complete sentences with appropriate words.
Preparation	Make one copy of Activity 5C on page 158 per student.

Procedure

1 Divide the class into groups of three, Give each student a copy of the activity.

2 Tell each student in each group to choose one of the Student 1, 2 or 3 sentence sets and fill in the set number at the top of the grid. They should not choose the same set as another student in their group. Explain that the sentences all contain one missing word which has appeared in the Vocabulary or Language development exercises in Module 5.

3 Working individually, they should look at their sentences and write the missing words in the table at the top of the activity. Each answer is one letter longer than the previous answer.

4 After three minutes, say 'Change'. The students pass their paper to another student in their group, who should check their answers, and try to complete any that haven't been done.

5 After another three minutes, say 'Change' again. Step 4 above is then repeated so that each student in the group has checked/tried to complete each of the sentence groups.

6 After another three minutes, say 'Stop'. Students should then work together in their groups for another two minutes to brainstorm possible answers that they haven't been able to find when working on their own. At the end of this time, stop the activity. The group who have completed the most number of tables is the winner.

Variation

For a faster version of this activity, divide the class into three groups and allocate each group a different set of sentences. Students then work together to try to complete the sentences. The first group to do this is the winner.

Follow-on

Divide the class into three groups. In these groups, students write their own sentences using the target words and expressions from either the Student 1, 2 or 3 set. They then read out their completed sentences, minus one of the key words, to the rest of the class, who should try to decide which word is missing (they should turn their paper over when listening so that they cannot see the answers).

Student 1
1 on **2** But **3** long **4** peace **5** unless **6** promote **7** quantity **8** implement **9** livelihood
Student 2
1 in **2** bit **3** from **4** drain **5** nibble **6** outlive **7** Provided **8** lifestyle **9** epitomises
Student 3
1 to **2** not **3** life **4** right **5** living **6** imagine **7** somewhat/terribly/slightly **8** virtually **9** absolutely

6A Health quiz

• You can use this activity at any stage in Module 6A, but you might find it particularly useful at the beginning, as it reviews a lot of vocabulary related to health and fitness, which your students might find useful as they work through the module.

Aims	To review key vocabulary used to talk about health and fitness
Time	15 minutes
Activity type	Pairwork. Students do a quiz about health and fitness.
Preparation	Make one copy of Activity 6A on page 159 for each pair.

Procedure

1 Divide the class into pairs. Ask them to work together to brainstorm words and other expressions they know that are related to health and fitness. They could divide these into three groups: nouns, verbs (including phrasal verbs) and adjectives.

2 Give each pair a copy of the quiz and let them work through the questions. Allow them about 10–12 minutes for this.

3 Discuss the answers with the whole class.

Variation

After step 1 above, put two pairs of students together and give them the quiz. They look through the quiz, then take it in turns to choose questions to give the other pair. The winning pair is the pair in each group who correctly answered the most questions.

Follow-on

Working in their pairs, students use the words and expressions from the quiz in a short written text about keeping fit and healthy.

1 b **2** c **3** c **4** a **5** a) a vegetarian; b) a vegan **6** Steaming is the most healthy; frying is the least healthy. **7** False: Fat cannot be turned into muscle, and muscle cannot be turned into fat. **8** False: They contain protein. **9** b **10** sedentary **11** vitamins; minerals; protein; calories **12** swimming **13** This is an idiomatic expression for a very sedentary person who sits around watching television, eating fast food, etc. **14** Junk **15** You go to a health club to get fit; you go to a health centre to see a doctor or nurse about a health problem. **16** False: Your body produces it naturally because you need it, but too much cholesterol is bad for you. **17** All of them. **18** expectancy **19** b **20** a

6B Cohesion connections

• Use this activity after Writing 1 (CB p. 94).

Aims	To review words and expressions used to provide cohesion in written texts
Time	15 minutes
Activity type	Groupwork. Students match two sentences together and complete the second sentence with a word or part of an expression that provides cohesion to the sentences.
Preparation	Make one copy of Activity 6B on page 160 per four students, and cut into cards.

Procedure

1 Divide the class into groups of four and give each group a set of cards. Ask them to find the card that say 'First sentence of first pair'.

2 Explain that the sentence on this card can be followed by another 'follow-on' sentence on another of the cards. They will find this 'follow-on' sentence on the left-hand side of the card. On the right-hand side of the card, they will see another sentence which also has another 'follow-on' sentence on yet another card and so on. Note that the sentence pairs are not connected with one another (i.e. they don't tell a story or form a complete single text).

3 The aim of the activity is to match all the sentences with their follow-on sentences, then complete the gap in the second sentence with a suitable 'cohesion' word. They should not use the same word more than once. This can only be done if the sentence pairs are matched correctly.

4 Let students do the activity for about 15 minutes, then tell them to stop. The winning group is the group who managed to correctly match the most sentence pairs and correctly complete the most gaps.

Follow-on

In their groups, students write their own sentence pairs, then remove one of the 'cohesion' words and give their sentences to another group to complete.

The sentence pairs are in the correct order on the activities page. The missing words are in bold below:
I have my own reasons why I refused **it**, but I don't intend to explain why
His intentions were good, but I don't think it's right to be protected from difficult or unpleasant situations.
If I ever went **there**, however, I would probably spend my time hanging out on a beach somewhere.
If **so**, perhaps you should consider something a little less dramatic – like changing your diet, for example.
One of **these** is to take regular exercise or take up a sport or other physical hobby.
I love it, **which** is why I never seem to save up enough money to buy myself any luxuries.
Another is golf, although not everyone can do this unless they have easy access to a golf course.
Neither has my wife, which probably explains why we spend so much time sitting in front of the TV and moaning about the weather.
Since **then**, there have been some remarkable technological developments and improvements.
To us, doing **this** is completely normal, but it would have been almost inconceivable to our grandparents.
Such a move was risky, of course, but I gave the matter a lot of thought and careful consideration.
What's **more**, it's been proven that it can be terribly dangerous for you
I'm not sure if it's **their** upbringing that's to blame, or just the fact that they are plain miserable and bad-tempered.
Secondly, I'm sceptical of claims made by certain food manufacturers that their products can reduce cholesterol.
Whatever the reasons, a lot of companies are making a lot of money out of people.

6C English language quiz

• Use this activity after Language development 2 (CB p. 98).

Aims	To review target language from the Vocabulary section and Language developments 1 and 2.
Time	20–25 minutes
Activity type	Groupwork. Students identify correct and incorrect sentences in a 'language quiz' activity.
Preparation	Make one copy of Activity 6C on pages 161 and 162 per three students.

Procedure

1 Divide the class into groups of three. Give each group a copy of the activity.

2 Explain that some of the sentences on the sheet are correct, and some of them are wrong (for example, there might be a missing word, a wrong word, a wrong word form or some other aspect that makes it incorrect). Also explain that they have 100 points. They can risk these points, based on how certain they are that the sentence is right or wrong. They should write the number of points risked in the first column of the table ('Points risked').

3 Allow students about 15 minutes to decide whether the sentences are right or wrong, and to write their points risked on their table.

4 At the end of the allocated time, students should pass their paper to another group.

5 Read out the answers (but don't correct the wrong sentences) while the groups check one another's answers and add or remove the risked points accordingly. They write the total in the box at the bottom of the table before returning the paper to the original group.

6 The winning group is the group which ends up with the most points.

Follow-on

Working in their original groups, students correct the mistakes in the wrong sentences. Check as a class.

They can then write some of their own sentences, using the target language in Vocabulary and Language developments 1 and 2. Each sentence should contain a mistake. They then pass these to another group to correct.

1 Wrong (... *I'm confident* ***that*** ...)
2 Right
3 Wrong (... *their eating habits* ***overnight****, especially while* ...)
4 Right
5 Right
6 Wrong (*Governments are* ***pumping*** *a huge amount* ...)
7 Wrong (... *banks are offering* ***interest-free*** *loans* ...)
8 Right
9 Wrong (... *for two hours* ***that*** *I remembered* ...)
10 Right
11 Wrong (*All I* ***have ever*** *wanted to do* ...)
12 Right
13 Wrong (... *I've been considering* ***spending*** *some time* ...)
14 Right
15 Wrong (***I've tried applying*** *for* ...)

7A The wrong word

• Use this activity after Language development 1 (CB p. 108).

Aim To look at some other words that are easily confused
Time 15 minutes
Activity type Pairwork. Students identify and correct the wrong words in sentences.
Preparation Make one copy of Activity 7A on page 163 for each pair.

Procedure

1 Divide the class into pairs, and give each pair a copy of the activity.

2 Ask them to look at the example, which contains the word *effect*. This is a wrong word in the context of the sentence. The correct word is *affect*. They will see this word, written backwards, in the box at the top of the page.

3 Explain that all the other sentences contain a word (or words) which are wrong in the context of the sentence. They should identify this word, then decide on a correct word to replace it. They will find the answers, written backwards, in the box (but not in the same order as the sentences).

4 Allow students about 12–15 minutes to identify and correct the wrong words, then check their answers.

Follow-on

Students can then use the wrong words in sentences of their own.

Note that the sentences in this activity have been taken or adapted from the *Longman Dictionary of Contemporary English*. If your students have this dictionary, they can use it to help them check that their answers to the activity are correct.

1 process→procession **2** continual→continuous **3** priceless→worthless **4** chances→possibilities **5** respectable→respectful **6** rise→raise **7** practice→practise (*although* practice *is also a verb in American English*) **8** lie→lay **9** intolerable→intolerant **10** however→moreover **11** harm→damage **12** controlled→checked **13** permission→permitted **14** looking at→watching **15** while→during **16** work→job **17** criticisms→objections **18** inconsiderable→inconsiderate **19** conscious→conscientious **20** appreciable→appreciative **21** shortly→briefly **22** damages→injuries **23** inventing→discovering **24** prevent→avoid **25** besides→beside **26** presumption→assumption **27** activity→action **28** advice→advise

7B What has just happened?

• Use this activity after Speaking (CB p. 113).

Aims To review vocabulary of emotions
To practise describing a situation and its implications
To extend Exercise 7 in the Coursebook
To practise using paraphrasing techniques, etc. to say things in another way
Time 15–20 minutes
Activity type Pairwork. Students have to identify what their partner is talking about.
Exam focus Speaking: Paper 5 Parts 2 and 4
Preparation Make one copy of Activity 7B on page 164 for each pair. Cut into two sections, Student 1 + Student 2.

Procedure

1 Divide the class into pairs. Give each student a Student 1 or Student 2 card and ask them to read through the situations. They must not show their situations to each other.

2 Student 1 begins by talking about the first situation on his/her card. He/she should try to talk for one minute and explain how the people feel and how their lives will change as a result of their situation. The words in bold must not be used while talking.

3 At the end of one minute, he/she should stop talking. Student 2 has to decide what has just happened. It is then Student 2's turn to talk about the first situation on his/her paper. Steps 2 and 3 are repeated until all the situations have been described.

Variation

With smaller classes (up to 14 students), make this into a whole-class activity. Make just one copy of the activity. Invite students to come up one at a time to the front of the class, and show them one of the situations. They then talk to the rest of the class for one minute. At the end of the minute, the class try to decide together what has just happened.

7C What if?

• Use this activity after Language development 2 (CB p. 115).

Aim	To review expressions with hypothetical meanings
Time	15–20 minutes
Activity type	Groupwork. Students match three parts of different sentences together.
Preparation	Make one copy of Activity 7C on page 165 for each pair.

Procedure

1 Divide the class into groups of four. Give each group two copies of the activity. Explain that they will be working in teams of two to play a game against the other two in their group.

2 Explain that, at the top of the activity, they will see 18 sentences (nine for Team 1 and nine for Team 2). Each sentence has a follow-on sentence in the boxes underneath. These follow-on sentences have been split into three sections.

3 Working in their pairs, the two teams should take it in turns to choose a sentence from their list, then find the follow-on sentence in the boxes. The two teams should time each other, allowing a maximum of one minute for each sentence.

4 Stop the activity after about 15 minutes. The winning team in each group is the pair who made the most correct matches.

Follow-on

Working in the same groups, students write their own follow-on sentences for those at the top of the activity (using expressions with hypothetical meanings).

Team 1
I wish I could get better grades in my exams.
I wish you wouldn't smoke in here all the time.
If only they knew how difficult it can be at times.
He really wishes he had gone to an English school in Oxford instead.
If only she had studied harder and paid more attention in her class.
It's high time you started making more of an effort and coming on time.
I'd prefer it if they didn't play their Coldplay albums until three in the morning.
You talk to me as if I were a child.
Suppose he did just that one day and walked out on you?

Team 2
But what if it doesn't work in practice?
I'd sooner stick my head in a food processor!
If only I had stayed at the Carlington instead.
You looked as if you knew what you were doing.
However, I'd rather run the risk of getting it wrong than doing nothing at all.
If only he could be more motivated and industrious.
Supposing you broke down in the middle of nowhere and needed help?
It's not as if I was risking life and limb by trying.
It's high time you got rid of it and bought a new one.

8A First to the right word

• Use this activity after Language development 1 (CB p. 124).

Aim	To review language and structures from Vocabulary and Language development 1
Time	15–20 minutes
Activity type	Groupwork. Students compete to find the correct words to complete sentences.
Preparation	Make one copy of Activity 8A on page 166 per three or four students. You will also need some dice (one per group) and counters (one per student; they can use coins or make their own).

Procedure

1 Divide the class into groups of three or four. Give each group a copy of the activity. They should make sure that they can all see it.

2 Explain that there are 25 sentences. Each sentence has a missing word. The word can be found on the playing grid. There are also several words on the grid

that do not fit in any of the sentences.

3 The students place their counters in the 'Start' space at the top of the grid. They then look at sentence 1 and find the correct word in the grid. They should not tell the others in the group what they think the missing word is.

4 When they all think they have found the right word, they take it in turns to roll the die and move their counter towards the right word. They can move their counter vertically or horizontally, but they cannot move it diagonally. They cannot cross any of the shaded spaces.

5 The first student who lands on what he/she believes is the correct answer writes his/her initials and the number of the sentence in that space.

6 Steps 4 and 5 are repeated for the other sentences.

7 Let them play for about 15–20 minutes, then stop the game and check their answers. They should give themselves 1 point for each correct answer, but remove 1 point for each incorrect answer. The student with the most correct answers is the winner.

1 materialism **2** bothered **3** reel **4** pressure **5** ban **6** role **7** market **8** nothing **9** confide **10** beneficial **11** preoccupied **12** relent **13** accused **14** no **15** little **16** Not **17** account **18** Hardly **19** sooner **20** isn't **21** now **22** way **23** circumstances **24** So **25** Such

8B Complicated crimes

• Use this activity after Speaking (CB p. 129).

Aim	To extend the theme of crime and punishment
Time	25–40 minutes
Activity type	Whole class + groupwork. Students discuss suitable punishments for some complicated/ambiguous crimes.
Preparation	Make one copy of Activity 8B on pages 167 and 168 and cut into cards.

Procedure

1 Tell students that they are going to hear about some crimes and have to decide on suitable punishments for the people involved. These are all true crimes which have happened in Britain in the last 50–60 years.

2 Give out the cards to nine students at random.

3 Ask the students with cards to tell the rest of the class about their crime case. Students can ask for clarification and make notes,

4 When all of the cases have been read out, ask your students to work in groups of four or five, and discuss what punishments they would give to the people in the different cases. Set a time limit of about 15 minutes for this.

5 After 15 minutes, extend this into a whole-class discussion. Tell the class that they should all try to agree on a suitable punishment for the different people.

Follow-on

Ask students if they know of any crimes in their country which have been slightly ambiguous or not so clear-cut, and what happened to the people involved.

Here are the actual punishments given out for each of the crimes in the activity.

1 The man was made to pay £200 compensation to his neighbour and do 60 hours' community service.

2 She was let off with a warning. The bullies were expelled from the school and made to attend a drugs rehabilitation course.

3 The hijackers were sent to prison indefinitely. Later, they were transferred to an immigrant detention centre. The political system in their country has now changed, and plans are being made to repatriate them. (Students might also like to know that several other passengers also claimed to be hijackers, because they decided they would rather go to prison than return to their country.)

4 The farmer was sentenced to life in prison for murder. This was later reduced to the lesser crime of manslaughter, and he was released after less than four years.
(Students might also like to know that the thief who was wounded then tried to sue the farmer for his injuries.)

5 The 16-year-old was sent to prison and released in his 20s. The 18-year-old was executed for the murder of the policeman, even though he didn't actually do the killing.

6 Nothing happened to the judge. He was allowed to continue working as normal.

7 She was found guilty of the crime of 'perverting the course of justice', and sent to prison. She was released after just two years. She had to change her identity afterwards because public opinion against her was so strong.

8 He was found guilty of causing death by dangerous driving, and sent to prison for seven years.

9 The man was sent to prison, where he spent 24 years.

8C Mixed comparatives and superlatives

• Use this activity after Language development 2 (CB p. 132).

Aim	To review language of comparison and superlatives
Time	15 minutes
Activity type	Pairwork. Students rearrange words to make sentences.
Preparation	Make one copy of Activity 8C on page 169 for each pair.

Procedure

1 Divide the class into pairs, and give each pair of students a copy of the activity.

2 Explain that the first of each pair of sentences can be said in a different way with the same or very similar meaning. The second gives this meaning, but the words in bold are in the wrong order.

3 Working in pairs, students should rearrange the words in bold.

4 Allow them about 10–15 minutes for this, then check their answers.

Follow-on

Working in the same pairs, students use the structures and expressions from the activity and from Language development 2 to talk or write about one or more of the following: football teams / sports people / cars / fashion designers / clothing companies / sports / films / actors / pop/rock stars or groups / towns + cities in their country / mobile phones / food + national dishes / writers / books / artists

1 A holiday in Australia is not nearly as expensive as you think.
2 The new Ferghini Peoplevan is easily the safest car on the road.
3 You're far more likely to be hit by lightning than win the lottery.
4 Ronald is nowhere near as intelligent as Juliet.
5 In my opinion, Paper 3 of the CAE is by far the most difficult paper.
6 This is just about the worst film I have ever seen.
7 My History exam was a great deal more difficult than my Maths exam.
8 Travelling by train is one of the most pleasant ways of travelling.
9 It's not such a good idea to use a telephone during a thunderstorm.
10 The more I study, the less I know.
11 There's a British saying which means 'the faster you work, the more mistakes you make'. (Students may need a dictionary to check the meaning of *haste.*)
12 You're behaving like a little child!

9A Body-language bingo

• Use this activity after Listening 1 (CB p. 139).

Aim	To extend the theme of communication
Time	15 minutes
Activity type	Classwork. Students listen to descriptions and interpretations of different body gestures in a multiple-matching bingo-style activity.
Preparation	Make one copy of the body-language descriptions (page 170), and cut into cards. Make one copy of the body-language pictures (page 171) per student. These should not be cut up.

Procedure

1 Give each student a copy of the body-language pictures and ask them to work in pairs to briefly discuss what they think each gesture means.

2 Ask each student to choose five pictures and write his/her initials under each picture.

3 Explain that you are going to read them descriptions of the different pictures, and what the gestures mean. Each time they hear a description of one of their pictures, they should put a tick next to it.

4 As soon as they think they have heard descriptions of all their pictures, they should call 'Bingo!'.

5 Pick one of the gesture-descriptions cards at random and read the description aloud, twice, at normal speed. Continue doing this for the other cards until one of the students calls 'Bingo'.

6 If his/her pictures match the descriptions you have read, he/she is the winner.

Instead of reading the cards yourself, you could ask one of the students to do it. Alternatively, they could do this in groups of five or six, with one of them reading the descriptions out (you will need to make one set of cards for each group if you use this method).

Follow-on

Students could work in groups to talk about other conscious and unconscious gestures/body language, or compare those in the activity to those in their countries. (This works particularly well in multi-national/multi-cultural classes, where they may be interested to learn of other gestures in their classmates' countries. Keep it clean!)

The description cards on page 170 are in the same order as the pictures (reading left to right, top to bottom).

9B Complete the sentences

• Use this activity after Language development (CB p. 140).

Aims To review reporting verbs
To review idioms with animals
Time 10–15 minutes
Activity type Pairwork. Students compete to complete gapped sentences with reporting verbs and names of animals.
Preparation Make one copy of Activity 9B on page 172 for each pair.

Procedure

1 Divide the class into pairs. Give each pair a copy of the activity. Tell them to decide who, in each pair, will be Student 1 and Student 2.

2 Explain that they should look at the first sentence in each pair (A), which is direct speech. They should then look at the second sentence (B), which is a reported version of sentence A. There are two gaps in each sentence B. Student 1 has to complete the reporting verb (gap 1), Student 2 has to supply the name of an animal (which completes an idiomatic expression from Vocabulary) (gap 2).

3 Students should take it in turns to try to complete their part of each sentence with an appropriate word. This word should have the same number of letters as there are spaces. They should pay particular attention to the word form in each case.

4 Allow them about 10–15 minutes for this, then check their answers.

Follow-on

In the same pairs, students write five of their own direct-speech sentences, then write a reported version of each sentence, leaving out the reporting verb. They should try to include more of the target language from Vocabulary in their sentences. The direct-speech and gapped reported-speech sentences are then passed to another pair to complete.

1 Joan complained that Harry could talk the hind legs off a donkey.
2 Jane apologised for rabbiting on for so long.
3 Tom confessed to letting the cat out of the bag.
4 Clare advised us to get the facts directly from the horse's mouth.
5 Mark noticed that Jan wouldn't say boo to a goose.
6 The teacher suggested repeating a word parrot-fashion until it stayed in our heads.
7 Emma reminded us that she was the first person to smell a rat.

9C Wrong-word crossword

• Use this activity after Use of English 2 (CB p. 147).

Aim To review vocabulary from Module 9
Time 15–20 minutes
Activity type Pairwork. Students identify wrong words in a sentence and replace them with the correct words, which they put into a crossword.
Preparation Make one copy of Activity 9C on page 173 for each pair.

Procedure

1 Divide the class into pairs, and give each pair a copy of the activity.

2 Explain that sentences 1–25 all contain a wrong word or word form. Students must decide what the incorrect word is, what it should be, then write the correct word in the crossword grid. All the correct words have appeared in Module 9. A lot of the wrong words have similarities in sound or spelling to the correct words.

3 Allow students about 15 minutes for this, then review their answers.

Follow-on

Working with a dictionary, students can look up the meanings of the wrong words, and use some of them in their own sentences. They can then remove the word and hand their gapped sentences to another student to complete (e.g. 'I can't eat this chicken – it's _____' Answer = *raw* (number 21)).

1 hindered→hinted **2** crawl→growl
3 whisper→whistle **4** forgetful→forgettable
5 violently→fiercely **6** exceptional→excessive
7 marketable→marketing **8** belied→believed
9 unappreciated→unprecedented
10 ramble→rumble **11** crumpet→trumpet
12 imitate→intimidate **13** delectable→desirable
14 repudiated→reputed **15** rampant→pungent
16 suggestive→suggested
17 differentiate→discriminate **18** convoy→convey
19 decode→encode **20** augmented→argued
21 raw→roar **22** crack→croak
23 scurry from→scare away **24** graduated→evolved
25 quack→squeak

10A Matching participles

• Use this activity after Language development 1 (CB p. 156).

Aim	To review participle and *to*-infinitive clauses
Time	15 minutes
Activity type	Pair- and groupwork. Students rearrange words to make sentence halves, then match the sentences together.
Preparation	Make one copy of Activity 10A on page 174 per four students. Cut into two parts (Pair 1 + Pair 2).

Procedure

1 Divide the class into groups of four. Divide each group into pairs. Give one pair in each group a copy of the Pair 1 or Pair 2 paper. The pairs should not show their paper to each other.

2 Explain that between them, the two pairs have 12 sentences. Pair 1 has the first half and Pair 2 has the second half of each sentence. The words in each half have been jumbled.

3 Working in their pairs, they should put the words into their correct order, then without looking at each other's paper, work with the other pair to join the sentence halves together (e.g. 1 H). The first group to do this successfully is the winner, provided that they have correctly rearranged their words.

Follow-on

Students look at the first halves of the sentences in their activity and use their own ideas to complete them (e.g. *Arriving in class ten minutes late, I realised I had missed the most important part of the lesson*).

1 H Arriving in class ten minutes late / ten minutes late in class, I discovered everyone had already left.
2 I Barely ten minutes after meeting up with my boyfriend, we had a terrible row and haven't spoken since.
3 C Having spent so much time studying, I was disappointed to find I had failed the exam.
4 L Generally speaking, eating out can be an expensive experience.
5 J Not wanting to oversleep and be late again, I set my alarm clock to go off early.
6 K Despite being rather/extremely overweight and rather/extremely clumsy, he's a great dancer.
7 E To see the shocked look on his face, you would have thought he had just seen a ghost.
8 G Opening the box which was lying in the middle of the room, I was shocked to find a gun.
9 D Dressed entirely in black from head to toe / from head to toe entirely in black, he looked like a villain from a James Bond movie.
10 A Bitterly disappointed with my exam results, I decided to leave college and never return.
11 F To make sure we didn't upset the teacher again, we made sure we were on time and paid attention during the lesson.
12 B Not being particularly fit or healthy / healthy or fit, she found it difficult to keep up with the rest of us.

10B Musical tastes

• Use this activity before Speaking (CB p. 160).

Aim	To provide lead-in stimulus for Speaking To review/pre-teach key vocabulary related to music
Time	15–20 minutes +
Activity type	Pair- or small-group work. Students listen to people talking about different types of music, and identify what kind of music is being described from key words.
Preparation	Make one copy of Activity 10B on page 175 for each student. Make one extra copy, and cut into cards.

Procedure

1 Divide the class into eight groups or pairs. Give one student in each group a card. He/She should not show this to the others in the group.

2 Ask the student with the card to read it through quietly to him/herself. While he/she is doing this, ask the other student(s) in the group to think of different types of music and make a list (e.g. rock, classical, etc.).

3 Ask the students with the cards to take it in turns to read their cards out aloud (twice, at normal speed) to the whole class, who make a list of the key words and expressions that they hear.

4 When all the cards have been read out, students work in their groups or pairs to decide what kind of music was being described in each case. Allow them about five minutes for this, then review their answers, asking them which key words/expressions helped them to decide.

5 Give each pair/group a copy of the complete activity, and let them read through all the descriptions, highlighting key words/expressions that they think would be useful when talking about music (they can use dictionary to look up any words they are not familiar with). They could make a note of these, as they will find them useful in the next part of the Coursebook.

Follow-on

The Speaking section in the Coursebook provides comprehensive follow-on for this activity, and will give students the opportunity to recycle key vocabulary from the activity.

1 Rock, heavy metal or punk metal **2** Jazz (or possibly Blues) **3** Country and western **4** Latin music **5** World music **6** Rap **7** Opera **8** Musicals

10C Hot potatoes

- Use this activity at the end of Module 10. At this stage of the course, you could also repeat the CAE quiz that you will find at the beginning of these activities.

Aim	To review some of the target language from all modules of the course
Time	20 minutes +
Activity type	Groupwork. Students answer questions related to key vocabulary, structures, etc. from Modules 1–10 of the course.
Preparation	Make one copy of Activity 10C on pages 176 and 177 per four students. Cut into cards. Also make one copy of the answers below per four students.

Procedure

1 Divide the class into groups of four. Ask each group to choose a referee.

2 Give each group a set of cards and tell them to place these face down on a desk. Give each referee a copy of the answers. He/She should not show these to the other three in the group.

3 The students without the answers each take four cards from the top of the pile. The aim of the activity is for the students to then get rid of their cards.

4 One student begins the activity by looking at one of his/her cards and answering the questions on it. The referee should tell them whether the answers are right or wrong (the cards are numbered for quick reference). If he/she gets all three of the questions correct, that card is given to the referee. If any of the answers are wrong, that card is placed at the bottom of the pile on the table, and another card is taken from the top of the pile. If this happens, the referee should not give the correct answers.

5 Step 4 is repeated by each of the students in turn.

6 The winning student is the first student in each group to get rid of all his/her cards.

Variations

You can make the game shorter by reducing the groups to three students and/or by reducing the number of cards to three. You can make the game longer by increasing the group size and/or increasing the number of cards to five or six. Alternatively, set a time limit of 15 minutes – the student in each group who has got rid of the most cards in this time is the winner.

Answers (Do not show these to the others in the group)

1 a) achieve b) take c) make
2 a) ideal way b) sheltered upbringing c) positive outlook
3 a) came; have been b) haven't had; have been working c) has spent; will have spent
4 a) expects b) about c) bound/sure
5 a) gregarious b) persistent c) sensitive
6 a) both b) neither c) None
7 a) tapping b) rubbed c) pressed
8 a) since b) result c) whom
9 a) *delete* a b) has→have c) luggages→luggage/baggage/(suit)cases
10 a) attendance b) inconvenience c) neutrality
11 a) in b) on c) to
12 a) allowed b) obligation c) advisable
13 a) humour b) full of c) roses
14 a) unbelievable b) replacement c) irreversible
15 a) nibble b) suck c) pick at
16 a) I'll lend you my car provided that / providing you take good care of it. b) You should get up early, otherwise you'll be late. c) He'll pass the exam whether he does any revision or not. / whether or not he does any revision.
17 a) evidence/sign b) sceptical c) confident/optimistic/hopeful
18 a) meeting b) phoning c) to have
19 a) complimentary→complementary b) affects→effects c) principle→principal
20 a) rather b) supposing/suppose c) though/if
21 a) Not only b) Little did I know c) Under no circumstances
22 a) well b) off c) high
23 a) Having b) Although c) To
24 a) to b) that c) from
25 a) shout to→shout at b) anxious for→anxious about c) heard about→heard of
26 a) entirely predictable b) absolutely hilarious c) vastly overrated

How much do you know about the CAE exam?

1 Match the paper number with the name of the paper and the time allocated for it.

Paper 1 ....
Paper 2 ....
Paper 3 ....
Paper 4 ....
Paper 5 ....

Paper 1	a) Listening	i) 1 hour
Paper 2	b) Speaking	ii) Either approximately 15 or 23 minutes
Paper 3	c) Reading	iii) Approximately 40 minutes
Paper 4	d) Writing	iv) 1 hour 15 minutes
Paper 5	e) Use of English	v) 1 hour 30 minutes

2 Which paper do you do first? Which paper do you do second? Which papers do you do on the same day?

3 How many parts are there in the Reading paper? How many questions do you have to answer in total?

4 True or false? In the Reading paper:

a) you can use a bilingual (= English + your language) dictionary. ☐ T ☐ F
b) you can use a monolingual (= English only) dictionary. ☐ T ☐ F

5 True or false? In the Reading, Use of English and Listening papers, you will have marks removed for getting a wrong answer, or for not answering a question.

☐ T ☐ F

6 How many parts are there in the Writing paper, and how many questions do you have to answer in total?

7 Which of the following would you *not* be expected to write in the Writing paper?

a letter a poem a report an article
a story a competition entry a proposal
a review an information sheet a postcard

8 How many words should you write in each part of the Writing paper?

9 Complete the words in this sentence.

In the Use of English paper, you are tested on your control and knowledge of v _ _ _ _ _ _ _ _ y, g _ _ _ _ _ r, s _ _ _ _ _ _ g, p _ _ _ _ _ _ _ _ _ n, w _ _ _ - _ _ _ _ _ _ _ g, r _ _ _ _ _ _ r and c _ _ _ _ _ _ n.

10 How many parts are there in the Use of English paper? How many questions or items are there in total?

11 True or false? In the Listening paper:

a) there are five parts. ☐ T ☐ F
b) you hear each part twice. ☐ T ☐ F
c) there are approximately 40 questions. ☐ T ☐ F
d) Part 4 consists of three sections which have to be done at the same time. ☐ T ☐ F

12 Complete this paragraph.

In the Speaking test, there will be (a) .... or (b) .... other people in the room with you. There are (c) .... parts to the test. You are tested on your (d) g _ _ _ _ _ _, your (e) v _ _ _ _ _ _ _ _ _, your (f) discourse m _ _ _ _ _ _ _ _ _, your (g) interactive c _ _ _ _ _ _ _ _ _ _ _ _ and your (h) p _ _ _ _ _ _ _ _ _ _ _.

13 What overall percentage do you need to get in order to pass the CAE exam?

Student 1

1 This film has been the worst film I ever saw.
2 Robert lives in Edinburgh for more than ten years.
3 I was far better off financially since I have started working here in April.
4 By the time we had arrived at the station, the train left.
5 I play football for this team for two years, and before that I have played tennis for my school.
6 If he continues to spend money at this rate, he has spent everything before his holiday will end.
7 I usually work in London, but at the moment I spend some time working part-time for a company in Dumfries.
8 Normally I am enjoying going to parties, but now I'd like to enjoy my own company for a while.
9 Last Saturday, I was seeing a really great film at the cinema in the town centre.
10 The film had already been starting by the time we got to the cinema.
11 We played golf when it started raining.
12 I'm going on holiday tonight, so this time tomorrow I lie on a beach in the Bahamas!
13 They're bound be late this evening; they always are.
14 He was on the point leaving the house when suddenly the phone rang.
15 Here you are at last. We were to leave without you.
16 Thanks for being so understanding. I thought you would really angry with me.

Student 2

A We had a great time last summer: we were visiting our family in Australia.
B Good luck in your exam tomorrow. I think of you then.
C Everyone said she would happy when we told her the news, but she was really upset.
D She usually is visiting her parents at the weekend, although this weekend she is seeing her friends for a change.
E When the teacher had finally arrived in the classroom, she realised that half her students left.
F They were so angry that they were on the verge shouting each other.
G While I watched television, I suddenly heard a strange scratching noise at the window.
H Thanks for calling me. In fact, I was just to phone you myself.
I I work here since the beginning of the year, and prior to that I have worked in an office in Canterbury.
J Everyone says he's sure pass his exam tomorrow. He's never failed one yet.
K Normally he works hard, but for the time being he takes things easy because of his poor health.
L They have been the rudest people I ever met.
M When I got to Bob's house, I discovered he had already been leaving for the airport.
N We knew each other since we have met at a party last year.
O I'm taking too much time off work, so if I'm not careful I have used up all my holiday leave by the time summer will arrive.
P Janine works at Pembury and Co. since she got married.

Answers

Student 1	1	2	3	4	5	6	7	8	9	10	11	12	13	14	15	16
Student 2	L															

Down

1 Your degree will be **given to you** by a famous politician. It will be .presented.. in a ceremony at the town hall.

2 The college's selection procedure is **really tough**. It is notoriously .... , and only about 50% of candidates are accepted.

3 The college is really **top class**. It is one of the most .... institutions in the country.

5 Don't try to do too much, or you'll feel **shattered** all the time. You won't learn much if you're .... !

6 Everyone wants to know what the **teaching** is like. Well, the .... here is second to none.

7 The degree will be a **real plus** when you start looking for work. It offers .... that other qualifications don't.

10 The college **will let you know about** the trips and excursions that it offers. Everything is .... well in advance.

11 I think this is the **right** course for you. I believe you will find it ....

Across

4 The syllabus is **well planned**. Everything is perfectly ....

8 Your tutors get a bit **ratty** if you miss classes. They also get .... if you are persistently late.

9 You should try to **go to** all your classes. It's important to .... them whenever you can.

12 Most students **take part** in the social programme. We recommend that you get .... in at least one regular activity.

13 Our students generally get **brilliant** exam results. Up to now, they have mostly been ....

14 The course isn't suitable for **people who don't know anything about the subject**. It wasn't designed for complete ....

15 When **the course is over**, you will receive your degree. You won't actually receive this until six weeks after .... of the course.

Part 1

assertive decisive energetic friendly gregarious patient persistent resilient sensible tactful tolerant

Part 2

Some of the people we deal with are just so bad tempered and unreasonable that you feel like shouting at them! You have to be .... , and accept that it's part of your job.

My job requires me to be reasonable, practical and to show good judgement. Basically, I need to be .... and serious all the time. As you can imagine, this can get a little bit boring at times.

We have to be .... to all our customers, even if we feel like strangling them from time to time. If they think that we like them and are ready to help them, they're more likely to come back to us again in the future.

It's essential in my line of work that we don't upset or embarrass the clients we represent. We have to be .... at all times. It isn't always easy, and sometimes you say something and instantly wish you hadn't!

If you're a quiet person, who prefers to sit at the back and just watch and listen, then my job definitely isn't for you! You need to be .... , to speak up, say what you want and make sure that everyone listens. It definitely isn't a job for wallflowers and wimps!

I think that to do a job like mine, you have to be fairly .... It's not unusual for things to go horribly wrong, so you need the strength of character to accept these disasters, and still feel happy in your work when you return to the office the next day.

I work long hours and I seem to spend most of those running around the factory making sure that everything is going smoothly. It's not a job for couch potatoes – you really need to be .... !

The management in my company is very reluctant to make changes, and rarely listens to its staff if they suggest new ideas. The trick is to be .... and not take 'no' for an answer. If you have a good idea, you have to push it at them again and again until they finally give in.

I need to be quite .... in my job. It's important to allow people to say what they want without criticising them. That way, they are more likely to come up with new ideas and put more back into the company.

I spend a lot of time entertaining customers and making sure that they're having a good time, so it's important that you enjoy being with, and talking to, people. I've always been fairly .... , so the job suits me perfectly.

I need to think quickly in my line of work. I have to make decisions quickly and with confidence. If you're not the .... sort, you probably wouldn't survive five minutes in my job.

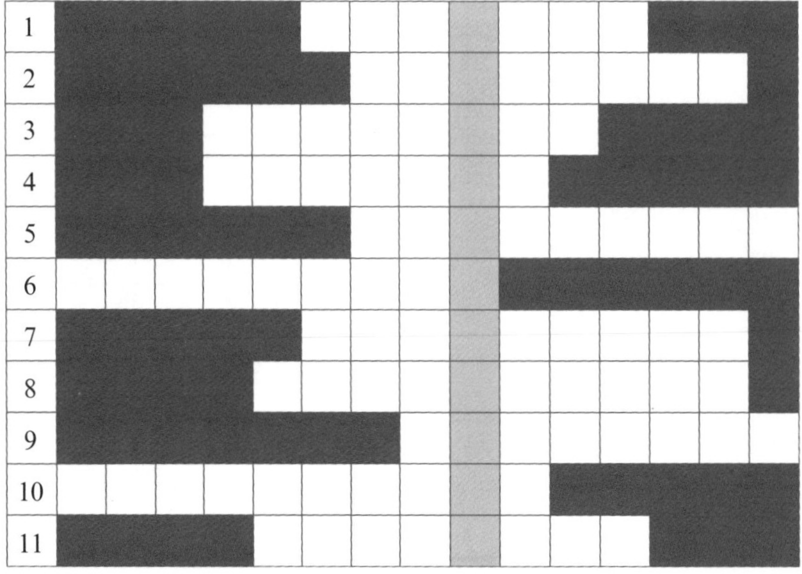

Part 1: Larry the lion

Larry **t**he lion lived in the jungle in the remote Baramungi mountains. Being rat**h**er lazy, h**e** enjoyed nothing better than sleeping unde**r** his favourit**e** tree and taking life easy. Being King of the jungle, he felt that thi**s** was the kind of lifestyle he deserved.

Saturdays, however, were differe**n**t. He would go for a walk, check out the other animals in the jungle, and remind everyone just h**o**w important he was.

This particular Saturday was **n**o differ**e**nt, and so after gorging hims**e**lf on a hearty breakfast, Larry ha**d** a quick bath in a nearby stream, made himself look suitably regal and started out on his rounds.

The firs**t** animal he came across was Molly the mouse, nibbling on a piece of cheese. Puffing himself up to his full height, Larry walked t**o**wards the little creature.

The tiny rodent saw Larry approaching and put down her breakfast. Larry stood over her, looked down disdainfully and asked in a loud, proud voice, 'Who's the kin**g** of the jungl**e**, shor**t**y?'

'You **a**re of course, sir,' replied Molly. 'Everybody k**n**ows that, sir. And may I say, sir, how, er, kin**g**ly, you'**r**e looking toda**y**, sir?'

This was **j**ust what Larry liked to hear. Satisfied, he contin**u**ed on his walk. A hundred metres or so further on, he **s**aw Gerald the giraffe chewing on a few tasty leaves which he had found on a **t**ree. 'Hey, lanky legs!' said Larry. 'Who's the king of the jungle?'

The tall creature looked down from his snack and, recognising Larry, replied hum**b**ly, 'You are, your maj**e**sty.' 'That is the **c**orrect **a**nswer. Well done,' said Larry smugly. He then continued on his way. It wa**s**n't long befor**e** he met Belinda the bear cautiously licking some honey from a beehive.

'Oi, stick**y** nose, who's king of the jungle?' Larry asked her.
'Well, you are,' the sweet-to**o**thed mammal replied. 'And always will be,' she added, anxious to keep the big cat happy. Happy with this answer, Larry decided it was time to go home and get on with the to**u**gh task of relaxing and taking it easy.

He ha**d**n't gone more than fifty metres when he suddenly spotted Gary the gorilla sitting **o**n a tree branch and mu**n**ching contentedly on a bunch of bananas. 'Oh well,' thought Larry, 'I'll do just one more before I go home.' He saun**t**ered up to Gary and said, in a loud, proud voice:

'Hey, you stupid mon**k**ey. Get dow**n** here right n**o**w and tell me **w**ho the king of the jungle is.'

Slowly and casually, the huge ape climbed down from his perch. Then, to Larry's shock, he picked him up, **t**hrew **h**im on the ground, jumped up and down on him and finally boxed his ears before climbing back into his tree and continuing with his br**e**akfast.

Bl**a**ck a**n**d blue with brui**s**es, a **w**ide-eyed Larry look**e**d up at Ga**r**y and said, 'All right, ...'.

Part 2: The punchline

There's ....'.... ....'

Game 1

1 One of Britain's most beautiful cities is Cambridge, ... you will find the country's second oldest university.	2 The ancient city of Petra, ... you will find in the Jordanian desert, dates back over 3,000 years.	3 You shouldn't believe .... you read in the papers.
4 The book was first published in 2001, ... over five million copies have been sold worldwide.	5 Ramton, a village .... situated between Manchester and Liverpool, has a population of only 17.	6 Can you remember the name of the boy .... brother plays professional football?
7 She didn't do any work, ... she failed the test.	8 The box office closed at midnight, ... over 2,000 tickets had been sold.	9 The computers in the library, ... are well over five years old, will be replaced next year.

Game 2

1 Not much is known about the people .... built the city of Machu Picchu in the mountains of Peru.	2 Marilyn Monroe, ... strange death in 1962 remains a mystery to this day, was a famous 1950s actress.	3 The two English schools, ... opened only last year, are already in financial difficulty.
4 I couldn't remember the name of the school .... I spent so many months learning English.	5 I started work here as a junior clerk in 1996, ... I've risen to the post of senior manager.	6 The hottest day in Britain was in 2003, ... a temperature of over 40 degrees centigrade was recorded.
7 *Romeo and Juliet*, a play .... two young lovers meet a tragic end, was written by William Shakespeare.	8 Students .... living in college accommodation can eat in the college canteen for free.	9 All the students, ... had worked extremely hard, failed the exam.

Game 3

1 We waited until half past eleven, ... we gave up waiting for her and went home.	2 He gave us several ridiculous excuses for his absence, ... we believed.	3 The sports centre is going to be opened by Charlie Rigden, ... he is.
4 It might rain, ... you should take an umbrella.	5 I've only got two students in my class, ... turned up today.	6 Who was the first person .... to walk on the Moon?
7 Some people didn't reply to my invitation, ... annoyed me quite a lot.	8 We missed our flight, ... we lost two days of our holiday.	9 I'm trying to remember the name of the town .... I spent my last holiday.

Game 1 (Referee)

1 where
2 which
3 everything/anything that
4 since when
5 —
6 whose
7 as a result of which
8 by which time
9 many/all/most/some of which

Game 2 (Referee)

1 who
2 whose
3 one/both of which
4 in/at which
5 since when
6 when
7 in which
8 —
9 some/many of whom

Game 3 (Referee)

1 at which point
2 none of which
3 whoever
4 in which case
5 both/neither of whom
6 —
7 which
8 as a result of which
9 where

Student A

(Do not show this to Student B.)

1 When he was 32, he inherited a **well-paid** business from his father.
2 Could you give me the 16-number code on your credit card?
3 I really don't visualise how he manages to be so successful.
4 We weren't allowed into the nightclub because there was an intimate function.
5 The directors discovered that the company had been fooled out of over £20 million.
6 He looks rather boring and old-fashioned, but appearances can be deceptively.
7 I refuse to play cards with him because he's such a trickster and never plays by the rules!
8 I had to tap the doorbell several times before she answered.
9 She sat in front of the television, gently patting her cat.
10 If it starts to itch, try not to rub it with your fingernails or you'll make it worse.
11 The police has arrested two men who they believe robbed the jeweller's.
12 She's usually on time, but each few days she turns up late or misses a class.
13 There's few chance of passing your exam unless you work hard.
14 Jane and Bill love going on holiday, but both of them enjoys flying.
15 Within the month of starting the exam, Enrico had already decided to leave.

Student B

(Do not show this to Student A.)

A Although neither of them passed the exam, they weren't accepted onto the course of their choice.
B Don't give her your money. She's a confidence cheat and you'll never see her or your cash again.
C Tom, Annette and Toby aren't really superstitious, but every of them carries a lucky charm.
D The room was small, with a very private atmosphere.
E I arrived on Monday, and over a next week I made a lot of good friends.
F He thought we didn't know his secret, but he was only tricking himself.
G We were given very little opportunities to say what we thought.
H It looks deceptive easy, but it's really rather difficult.
I He didn't want to wake everyone in the house up, so he pressed gently on the window to get my attention.
J 'Good boy,' she said, and stroked him a few times on the back.
K He yawned, scratched his eyes and tried to concentrate on the road ahead.
L The number of people taking the CAE have increased gradually over the last few years.
M I want you to relax, close your eyes and see a beach at sunset.
N Thirteen has always been a lucky digit for me.
O My job isn't particularly **lucrative**, and I'm always short of money at the end of the month.

Missing words

attention tempered arrogance rudeness (x2) side innocence skills far value
devotion perspective political appetite joke convictions scrutiny inconvenience
sorry name clothes array power press

First sentence

Celebrity interview 1

The popular .... describe him as a suave, sophisticated demi-god with charm, personality and intelligence.

Celebrity interview 2

He's made a big-screen living playing a vast .... of violent roles, whether as a gangster, a corrupt policeman or a sadistic army sergeant.

Celebrity interview 3

The public have an insatiable .... for news of their favourite celebrities, and I must admit that I was quite excited to get the chance to interview her.

Celebrity interview 4

Thanks to her role as Peggy on television's *Alimony Street*, hers is a household ....

Second sentence

So to meet him in the flesh really came as a pleasant surprise.

When I met him, however, I began to wonder whether any of these journalists had actually met him, or were just playing a .... on their reading public.

And much like the TV character she plays, she exudes an air of .... and helplessness that has found her a place in the general public's heart and which has taken her so .... in such a short space of time.

Here, after all, was the most glamorous female star to have appeared on our cinema screens for years.

Third sentence

However, initial appearances can be deceptive, and I soon learnt what people mean when they say that we shouldn't take others at face ....

But what we see on film and what we get in reality are often two different things, and this was no exception.

Here is a polite, intelligent young man who, unlike others, hasn't let fame go to his head.

What struck me the most was his complete lack of any of these qualities, with social .... coming right at the bottom of the list.

Fourth sentence

Not for him the intolerable .... and .... that celebrity can bring.

After only ten minutes in her presence, I began to suspect that under the childlike veneer she employs so well on the screen there lurks a mature and determined woman who doesn't suffer fools gladly, enjoys having .... over others and never, ever takes 'no' for an answer.

Although he is in his late twenties, I felt like I was interviewing an adolescent schoolboy.

Not for me the tall, slim, sharp-dressed starlet with a twinkle in her eyes, a toothpaste smile and charming manners.

Fifth sentence

When I was shown into her dressing room, I got a rather short, slightly overweight, bad-......... teenage girl in scruffy casual .... who scowled at me all the way through the interview, refused to answer half my questions and couldn't pay .... for more than a few minutes.

He couldn't sit still, couldn't concentrate for more than two minutes at a time, answered my questions in monosyllabic grunts, and not once made eye contact with me.

So it didn't come as too much of a shock to me when I found myself on her wrong .... by asking about her current boyfriend.

And despite his reputation for avoiding public .... (which presumably means journalists like myself probing into his private life), he seemed quite happy to sit and answer all my questions.

Sixth sentence

True, he has strong religious .... and .... beliefs, but he doesn't ram them down your throat, and he spoke objectively about the things he feels passionate about.

It certainly gave me a different .... on fame.

There was also an air of barely suppressed anger about him, and it seemed that, at the slightest .... or provocation, he would blow his top like a volcano.

Without warning, she dropped the sweet, innocent act and told me, in no uncertain terms, to 'change the subject right now or get out'.

Final sentence

It all made a refreshing change from interviewing the usual self-centred celebrities who fill our screens, newspapers and magazines, and our interview was the most enjoyable two hours I've spent in ages.

I guess .... just goes hand in hand with those like her who achieve celebrity status!

Here, then, is a private wolf in public sheep's clothing.

I guess a more sympathetic person would feel .... for him, but I couldn't see beyond the spoilt child that he really is, and although I have looked hard for his 'appeal factor', I am still thoroughly perplexed at the blind .... he receives from his fans.

You are in your home town, having dinner at a **restaurant** you have never been to before. The **food** is all right, but the **service** is rather slow, and the **waiters** are not very attentive or polite. At the end of the **meal**, you accidentally **leave** the restaurant without **paying**. You are walking home when you realise your mistake. Do you go back and **pay**?

You are doing your **shopping** in a big, busy **supermarket**. It's a hot day, and you are feeling very thirsty. You find yourself by a refrigerator full of **bottles** of ice-cold **water**. You take one of the bottles and drink the water while continuing to do your shopping. When you get to the **checkout** to pay for your shopping, do you also pay for the bottle of water you have **drunk**?

Your **CAE exam** is next week. You have done a lot of preparation, but you don't feel very confident about your chances of getting a good grade. You then hear that somebody has illegally managed to get hold of a copy of the exam that you are going to take, and posted it on the **Internet**. For a small **fee**, you can access the **website** and look at the exam before you take it. Would you?

You are enjoying a meal in a restaurant with your **best friend** (whom you have known for many years) and his/her **girlfriend/boyfriend**. At the end of the meal, your friend goes to pay the bill at the cash desk. While he/she is doing this, his/her **partner** makes a **pass** at you (he/she tells you that you are very **attractive** and that he/she would like to go out with you). You think that he/she is very attractive and lots of fun to be with. You don't have a girlfriend/boyfriend. Would you accept the offer?

Your best friend (whom you have known for many years) is in **trouble** with the **police**. They have accused him/her of accidentally **hitting** someone in the street while riding his/her **motorbike**, and then riding away. The person he/she hit has not been seriously **injured**, but has needed treatment for some minor **injuries**. Your friend has told the police that he/she wasn't on the road when the accident happened, and has asked you for an **alibi**. Would you provide your friend with an alibi?

Your neighbour has gone away on holiday and has asked you to look after her **pet rabbit**, Binkle, for a couple of weeks. While she is away, Binkle **dies**, and you have a horrible feeling that it is your fault (maybe you didn't give him enough food or water). You know that your neighbour is extremely fond of Binkle. You also know that you can buy an **identical substitute** from a local **pet** shop. Do you buy the substitute rabbit and **pretend** to your neighbour, when she returns from her holiday, that it is Binkle?

You need some **money** for a night out with your friends, so you go to your local **bank** to use the **cash machine**. You put your **card** into the machine, and ask it to give you **£100** and a **receipt**. The money comes out. To your surprise, you are given **£200** instead of £100. You check your receipt: it says you have only **withdrawn** £100. Do you inform the bank that their cash machine has given you too much money?

You are relaxing in your local **park** when you see some **children** repeatedly **hitting** a smaller child. The child is crying and clearly frightened, but his parents are not there. The parents of the other children, however, are. They are watching their children **picking on** the smaller child and not doing anything to **stop** them. Do you **intervene**?

Driving home late one night, you pull into your road and accidentally **hit** your neighbour's **car** which is **parked** by the side of the road. The **damage** to your car is only superficial, but your neighbour's car has more serious damage: a broken **headlight**, a cracked **bumper**, a smashed wing **mirror** and a big dent in the **driver's** door. You don't think anyone has seen the **accident**. You are also not very fond of your neighbour. Do you tell her what you have done?

You desperately need a new **computer**, since your old one has stopped working and you don't think it's worth **repairing**. Unfortunately, you can't **afford** a new one. Then one day at work, a **colleague** offers to sell you a brand-new, top-of-the-range computer for a fraction of the shop **price**. However, you strongly **suspect** that this computer has been **stolen**. Do you buy it from your colleague?

Team 1

1 He must have been really angry.	2 It's advisable leave early if you don't want to be late.	3 OK, so we don't need to make so much food.
4 Sorry, you can't smoke in here.	5 You'd better get down to some hard work if you want to pass.	6 Well, you don't need to hand it in until Monday.

Team 2

1 You should have withdrawn some more.	2 You didn't have to, but thank you anyway.	3 All right, maybe we should get going.
4 Oh great, so I didn't need to stay up half the night revising.	5 You better book early to avoid disappointment.	6 I need to ask my teacher for permission to leave early first.

Team 3

1 Yes, but it's advisable to arrive a couple of days earlier.	2 He must have been out of his mind!	3 You are not under an obligation to attend the classes.
4 He can't have done. He's been away all week.	5 It could be anything up to two million.	6 She said she might, but wasn't sure.

Team 4

1 You must have known he would; he always does.	2 There could be as many as 50.	3 She might have, but there's no sign of her.
4 No, and he should have arrived half an hour ago.	5 He could telephone yesterday to let me know, but he didn't.	6 Right, but you are required to arrive at least 15 minutes before.

Team 5

1 We felt oblige to give him plenty of warning.	2 No. I could have come, but I felt a bit tired.	3 She might have been busy.
4 No, you're under no obligation to attend.	5 I've been able to save about ten thousand pounds.	6 Oh well, you should have told me you would be absent yesterday.

Team 6

1 Strange. She should have called hours ago.	2 In that case, he must have gone out.	3 Yes, unfortunately, but we were able to get a seat on a later departure.
4 I felt obliged to because she's an old family friend.	5 You could have told me before I went to the supermarket.	6 I'm afraid smoking is forbidding in the building.

	Student 1	Student 2
1 *You shouldn't count your chickens before they're hatched.*		
2 **The more the merrier.**		
3 *No news is good news.*		
4 **It's never too late to mend.**		
5 A miss is as good as a mile.		
6 *It never rains but it pours.*		
7 Actions speak louder than words.		
8 ***Rome wasn't built in a day.***		
9 When the cat's away, the mice will play.		
10 *When in Rome, do as the Romans do.*		
11 *You shouldn't make mountains out of molehills.*		
12 **A bad workman always blames his tools.**		
13 *More haste, less speed.*		
14 **One good turn deserves another.**		
15 Two's company, three's a crowd.		
16 *His bark is worse than his bite.*		
17 Many hands make light work.		
18 ***Too many cooks spoil the broth.***		
19 You should strike while the iron is hot.		
20 *Pride comes before a fall.*		

A He said that the accounts figures were wrong because the computer had miscalculated the sales figures. And if you believe that, you'll believe anything. I reckon he's just looking for excuses for his own ineptitude.

B It's so typical. Things can go really smoothly, then suddenly everything goes spectacularly wrong at once. And then there are a lot of heated discussions and temper tantrums which create bad feelings all round.

C This is quite a big job, and there aren't enough of us here to do it in time. However, if we ask some of the workers from the shop floor to help out, we'll get it done in no time at all.

D I don't know why he got so upset. I made a trivial mistake, that's all, and he blew it out of all proportion. I don't know why he can't get things in perspective. Maybe he's just too thin-skinned.

E The company has just started, so we can't expect to make a great deal of money yet. We'll need to wait a bit longer, be patient and just make sure we all work towards the same objectives.

F We've got some great ideas and we've got the ways and means of putting them into action, so we might as well get started straight away, no delays.

G The office was fine when it was just me and Andrea in there, working to a common purpose, bouncing ideas off each other. But now they've moved Timpkins in, there's just not enough room.

H Everyone in the office is congratulating one another because we nearly got the contract. What's the point? 'Nearly' isn't good enough, is it?

I We don't need all these people working on the project, treading on one another's feet. The fewer people we have, the quicker the job will get done.

J Our boss is a bit of a slave driver and makes sure he gets his money's worth out of us. So as you can imagine, as soon as he leaves the office, everyone relaxes, and we don't get much work done.

K He's got a good eye for a deal, and I'm sure he could be beneficial to the company, but until he puts his ideas into practice, he might as well just keep his mouth shut.

L People are a bit frightened when they first meet him because he can be abrupt and bad-tempered, but they soon find out that he's really a very fair-minded person with an excellent sense of humour.

M If you're having problems with writing that report, let me help you. It's only fair. After all, you gave me a hand with writing my speech for the annual general meeting.

N Our boss is so self-important and full of himself, always telling everyone how wonderful he is. Well, I reckon that very soon he'll be cut down to size.

O They've promised to give us the contract, but I don't think we should get too excited just yet. Personally, I won't be happy until I see it in black and white.

P He joined us from another company, and keeps telling us that their way of working was better. I really think he needs to adapt his methods to suit this company.

Q I like it when we don't get any post, as it's usually a bill or invoice or a customer complaining or cancelling an order.

R The office was practically dead when it was just the two of us there, but we were joined by three others last week, and there are two more joining us on Tuesday, so things should be much more lively.

S I know that she's been the worst, most unreliable member of staff here, but in view of her breadth of experience I think that we should give her a second chance. She might improve.

T The boss has asked me to give him the report and sales figures by lunchtime. Well, if I try to do it in that amount of time, it's bound to be full of mistakes.

START ➤	**1** She's a very influence woman who knows all the right people.	**2** He isn't very energy, and would rather sit in front of the TV all day.	**3** We held a festival to celebrate the diverse of cultures in the city.
7 Participate in the afternoon activities programme isn't compulsory.	**6** There are several history buildings in the town.	**5** Your persist amazes me, Mr Bond.	**4**. It's believe! He went out in the snow in shorts and a T-shirt.
8 He's a very depend person who will always do what you ask him.	**9** Oh dear. I think you must have understand the instructions I gave you.	**10** Jealous is a terrible and irrational thing that eats away at you.	**11** I told my boss I thought I was pay, and demanded a bigger salary.
15 Do you think that accurate is more important than fluency in spoken English?	**14** It was easy to guess the deep of feelings at the meeting.	**13** I think that cruel to animals is absolutely despicable.	**12** You'll find several good biographies of Beckham in the fiction department.'
16 We plan to large the school building in order to accommodate the new courses.	**17** The weather was terrible, so fortune we had to cancel the trip.	**18** The police tried to prevent a violent confront between the two groups.	**19** The road is closed because the council are strong the bridge over the river.
23 It was a real please meeting you last week.	**22** We walked the length and broad of the Woodstock Road looking for a chemist.	**21** The police had to release him because they didn't have enough prove.	**20** Once the damage has been done, I'm afraid it's reverse.'
24 She's very tolerate of lazy people, and gives them a really hard time.	**25** No one has seen him for days. His appear is a complete mystery.	**26** The best way to rich your knowledge of the world is by travelling.	**27** She has several responsible, and doesn't take any of them seriously.
31 He's a bit of a conform, and prefers to do things his own way.	**30** The city is terribly populate, with over 12 million people.	**29** When I checked my bank balance, I discovered I was draw.	**28** The area is rather develop, with poor transport, education and health facilities.
32 I went to a education school, where both boys and girls were taught together.	**33** His stupid behaviour on the trip danger the lives of everyone.	**34** The castle date the other buildings in the town by at least a hundred years.	**35** He was unable to persuade his work in the factory to join him on the strike.
FINISH	**38** British food is not all horrible. You shouldn't general like that.	**37** I had to speak to him about his aggression behaviour in class.	**36** I think this chicken is a bit cook. It's fine on the outside, but still a bit red on the inside.

Student 1

1 My mother was a warm and .... person who made sure we were well looked after.
2 Although they were in their 20s, they were both .... on their parents for everything.
3 Lacking any real parental care, the twins relied on .... other for help and support.
4 I was always absolutely .... by my elder brother's stories of his travels around the world.
5 The modern nuclear family is rapidly becoming more common than the more traditional .... family in many parts of the world.
6 As children, my sister and I were .... , and did absolutely everything together.
7 Ours was a tightly .... family; we always supported one another whenever anything went wrong.
8 I admire my father a great deal, but I .... up to my mother even more.
9 My grandfather had a big nose, my father has got a big nose and I've got a big nose. I guess it's something that .... in the family.
10 As the eldest child, I was held .... for anything that went wrong.
11 Unlike a lot of my friends, I didn't have any brothers or sisters, but I never missed having ....
12 There was always a .... bond between me and my brothers.

Student 2

1 Despite being very different from each other, my brother and I had a .... relationship.
2 She was completely .... to her children, and would do anything for them.
3 I tried hard at school and always made a conscientious effort not to .... my parents.
4 I used to feel that my sister was always rather .... of my achievements.
5 My parents never really had high .... of me, so were amazed when I announced I was going into politics.
6 When I told my parents I was dropping out of university, they were understandably .... with me.
7 My sisters and I were close in many ways, but we never really .... it off with one another.
8 In my opinion, some parents can be over-......... towards their children, with the result that they never learn about the real world.
9 My brother was always more successful than me, but I never felt .... towards him.
10 My parents, my brothers and my sister all work in education, but I want to be an actor. I suppose I was always the black .... of the family.
11 From the very .... she was a problem child who ran wild and always got into trouble.
12 My brother .... after my father, but I think I'm more like my mother.

Trouble in Paradise

There's more than meets the eye on the tropical island of Babarrie, as Brendan Buxton discovers.

Situated in the South Pacific, the island of Babarrie is one of the most beautiful islands in the world. The coastline offers miles of white-sanded, palm-fringed beaches, lapped by crystal-clear waters abundant with fish and other marine life. The interior has thousands of hectares of beautiful, unspoilt rainforest with some of the rarest, most exotic flora and fauna on the planet. The climate is a pleasant 28°C for most of the year, with wall-to-wall sunshine from October through to May.

The island, which is about 42 kilometres long by 29 wide at its widest point, is home to 42,000 people. The capital, Tompolia, is a quiet, easy-going place where life has barely changed in the last 60 years. The atmosphere is laid-back and relaxed. The town only comes alive in the evening, when everybody heads to their favourite restaurant for a sumptuous, candlelit seafood meal. Late at night, the only sounds are the breeze rustling the coconut trees, the lapping of the waves on a moon-kissed beach and the occasional twittering or hooting of a nocturnal animal hidden in the shadows of the forest.

Visually, Babarrie is certainly beautiful. But beauty is only skin deep. Scratch the surface of this tropical paradise and you get a different picture. People seem relaxed, but this is only because there's nothing to do. Unemployment is on the rise (over 40%), and many are turning to crime to make a living. Night-time muggings and other violent crimes in Tompolia are becoming more common. Drugs are becoming a problem. Inflation is running at 60%. Services and amenities are limited. Transport infrastructure is poor. The roads haven't been maintained for years, and are very dangerous. Road-related accidents, injuries and deaths are high. Access to the island is difficult (the only way to get there is on a 15-hour ferry ride from the island of New Caledonia, which itself is difficult to get to).

Healthcare is practically non-existent. Leisure facilities are non-existent. Education is not high on the government's list of priorities: the schools (and there are only five on the whole island) lack just about everything you care to mention, from chairs and tables to pencils and writing paper. Only about 20% of children receive any formal education, and most of these leave school before they are ten. Consequently, illiteracy is high.

There are very few natural resources to exploit – even fresh water is limited. There are frequent power cuts. The government cannot invest in anything because it has nothing to invest. The people live on a diet of fish and coconuts, which might sound nice in the short run, but gets very tiring after a few days. Average life expectancy is a shocking 46 years. Basically, Babarrie has big problems . . .

Easipower Oil

You are the director(s) of Easipower Oil and you want to build a refinery on the island of Babarrie.

You can offer several benefits to the island by building this refinery. First of all, you will create jobs for 2,000 people. You will also be able to provide cheap, much-needed oil for the local community. Furthermore, in addition to paying all national taxes, you will donate 2% of your net profits to local health, educational and social organisations, and 3% of net profits to local environmental groups who are working to protect the island's fragile ecosystem. Finally, you will minimise any negative visual impact on the environment by building much of the refinery underground.

Maurice-Filippson Tobacco Company

You are the director(s) of the Maurice-Filippson Tobacco Company, and you want to build a cigarette factory on the island of Babarrie.

You can offer several benefits to the island by building this factory. First of all, you can provide jobs for 1,500 people. These are guaranteed jobs-for-life. You will also provide free accommodation and meals for all of your employees. You have estimated that your factory will bring the island $150 million a year in tax, and you have agreed that on top of this, you will invest 1% of your net profits in local healthcare, and an unspecified percentage of profits in improving local transport and other communications infrastructure.

Club Yoyo Holiday Resorts

You are the director(s) of Club Yoyo Holiday Resorts, and you want to build a high-profile luxury holiday resort on the island of Babarrie.

You can offer several benefits to the island by building this resort. First of all, you will be providing jobs for 1,800 people. Your employees will all get free meals. You will provide a nursery and school for their children. Overseas tourists (up to 50,000 a year) will bring foreign capital to the island, as they will be spending money in local shops and restaurants. You have also offered to build an airport to make access to the island easier, and will invest 2% of your net profits on improving roads on the island. You are also considering building a sports and entertainment complex on the island for local people to use.

F.C. Pharmaceuticals

You are the director(s) of F.C. Pharmaceuticals, which produces drugs and medicines, and you want to build a factory on the island of Babarrie.

You can offer several benefits to the island by building this factory. First of all, you can provide jobs for 1,600 local people. You will also build a hospital for the island, and pay for its upkeep (staff, equipment and medicine) for the first five years of your operations on the island. During this time, you will also provide free medicine to anyone on the island who needs it. You will also invest $50 million in a comprehensive health awareness and education programme which you believe will benefit the whole island. Furthermore, you will invest up to 3% of your net profits on improving transport and communications infrastructures on the island. You are an environmentally aware company, and your factory will have minimum impact on the local environment (your factory will not require the destruction of any area of ecological importance or beauty).

Notes for Babarrie Investment Committee

1 The company Easipower Oil want to build a refinery on Babarrie, and have offered you some incentives for letting them do this. However, you have uncovered some facts that they do not really want you to know.

First of all, you know that they pay their employees very low salaries and make them work long days (up to 12 hours) in conditions that could be considered dangerous (their health and safety record in other countries is very bad). Secondly, the refinery will need a lot of water, which is very scarce on the island. Thirdly, there is the risk of water, air and (in the area immediately around the refinery) noise pollution: Easipower has already received international criticism for polluting other areas in which they operate. Furthermore, the construction of the refinery will require the destruction of 400 hectares of forest and the destruction of two villages (displacing 600 people). The port will also need to be enlarged at your expense.

2 The Maurice-Filippson Tobacco Company wants to build a cigarette factory on Babarrie, and has offered you some incentives for letting them do this. However, you have uncovered some facts that they do not really want you to know.

First of all, you have discovered that the company prefers to employ children, because they can pay them very little and make them work long hours (up to 14 hours a day). Secondly, the accommodation that they provide their employees in other countries is of very poor quality and badly maintained, with shared facilities (up to 100 people using one bathroom, toilet and kitchen). Furthermore, the presence of the factory may be an encouragement for local people to smoke, and your country has a strict anti-smoking policy (you are particularly concerned by their advertising slogans 'Smoking is cool' and 'Real men smoke MFs!'). You are also worried about the visual impact that the factory will have on the local environment.

3 The company Club Yoyo Holiday Resorts wants to build a high-profile luxury holiday resort on Babarrie, and has offered you some incentives for letting them do this. However, you have uncovered some facts that they do not really want you to know.

First of all, although they can provide jobs for 1,800 people, this is only seasonal: the rainy season from June to September means that the resort will be closed during that time. Secondly, the building of the resort will require the destruction of over 1,000 hectares of forest, the redevelopment of two beaches (one of which is a seasonal nesting ground for rare turtles) and the destruction of a two-kilometre stretch of coral reef. The resort and its grounds (including gardens, four swimming pools, tennis courts, fountains and a golf course) will require a huge amount of your limited fresh water supplies. Three villages along the coast will need to be destroyed, displacing almost 1,000 people. You have also heard that Club Yoyo resorts in other countries discourage their guests from using local shops and restaurants, and instead encourage them to remain in the resort grounds for their meals, etc. Furthermore, you are concerned about the social and cultural impact that so many foreigners will have on the islanders and their traditional way of life.

4 The company F.C. Pharmaceuticals, which makes drugs and medicine, wants to build a factory on Babarrie, and have offered you some incentives for letting them do this. However, you have uncovered some facts that they do not really want you to know.

First of all, the factory will include a laboratory which uses animals, including cats, dogs, rabbits and monkeys, for medical experiments. Secondly, after an initial five-year period in which they will provide free medicine to the islanders, they will then sell it at its 'retail' price, which you know is well beyond the amount most islanders can afford. Furthermore, the company was recently the subject of a major international scandal when it was discovered that managers in several of their factories were selling drug-making equipment and raw materials to drug dealers around the world. There have also been reports that, despite their claims to be environmentally aware, they have been caught dumping chemicals in the sea or burying them underground, with disastrous environmental results.

1 There was so much traffic on the road. It was a p............y .... journey home.

2 Everybody knew that he was telling lies. It was t............y .... that he wasn't telling the truth.

3 You must make sure you're not late for your exam. It's e............y .... that you arrive on time.

4 Nobody can agree what to do about her. Opinion is d............y ....

5 She made a fortune from her first book. Now she's s............y ....

6 She went purple with rage when I told her. She was a............y ....

7 Everybody is against the plans to build the new airport. They're c............y .... to the proposal.

8 Without their support, this project would be almost impossible. We're h............y .... on them to make it a success.

9 A lot of doctors say that eating red meat now and again is good for you. They say that it can be p............y .... to eat it once in a while.

10 Everybody wants to live here. It's a h............y .... neighbourhood.

11 We all told her how concerned we were for her. We were t............y .... that she wasn't looking after herself properly.

12 I really expected to pass the exam. Naturally I was b............y .... when I failed.

13 Everyone says he knows more about farming and agriculture than anybody else. He's w............y .... to be outstanding in his field.

14 Jennifer and Carol look a lot like each other. In fact, they're v............y ....

15 A few years ago, digital cameras were beyond the price range of most people. Nowadays, they are r............y ....

16 Ian and Laurence both look at life the same way. They are f............y .... in their outlook.

17 I woke up in the middle of the night because I could hear someone moving around downstairs. As you can imagine, I was r............r ....

L	S	O	O	P	P	O	S	E	D	O	K	A	B
T	L	T	H	E	D	E	S	I	R	A	B	L	E
S	O	B	V	I	O	U	S	I	Z	L	E	R	N
W	W	S	I	M	I	L	A	R	O	A	F	E	E
O	T	H	E	P	G	A	Z	O	C	R	N	C	F
R	G	A	S	O	O	N	T	H	H	M	A	O	I
R	T	F	U	R	I	O	U	S	E	E	I	G	C
I	D	E	N	T	I	C	A	L	A	D	B	N	I
E	E	T	S	A	H	E	G	O	P	E	S	I	A
D	E	P	E	N	D	E	N	T	L	I	K	S	L
E	R	A	R	T	A	M	P	A	N	T	R	E	H
D	I	V	I	D	E	D	I	N	O	W	H	D	N
S	C	H	E	S	I	N	T	H	E	R	O	M	S
X	H	D	I	S	A	P	P	O	I	N	T	E	D

The adjectives you want can be found by reading horizontally (left to right) and vertically (top to bottom).

Student ___

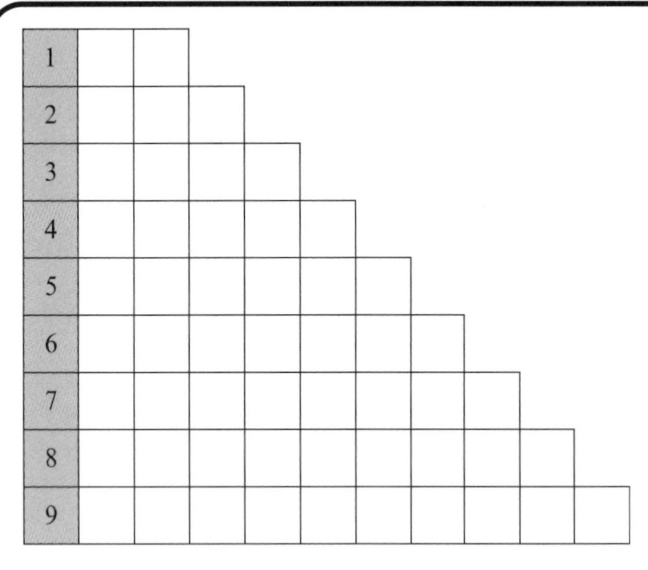

Student 1

1 Our teacher says she'll let us leave early .... condition that we work really hard.
2 .... for all the extra help he got from his teacher, he would have failed the exam.
3 I'll lend you my camera as .... as you promise to take great care of it.
4 He was arrested for disturbing the .... with his loud music and riotous behaviour.
5 We'll go for a picnic .... it rains.
6 Local schools are being encouraged to .... a healthy lifestyle by banning burgers and other fast food from their menus.
7 When it comes to writing essays, remember that it is quality, not .... that counts.
8 We have decided to .... the committee's recommendations as soon as we receive funding.
9 He's a photographer, so when his camera equipment was stolen and he couldn't afford to replace it, he lost his ....

Student 2

1 The government expected tourists to visit the country .... droves, so they were heartily disappointed when there was just a slow trickle.
2 I found the exam a .... difficult, but to be honest, it could have been much worse.
3 Let's get away .... it all and take a long holiday.
4 I said take a little sip, don't .... the whole bottle. (*Do not use* drink *here.*)
5 If you feel travel sick, a good trick is to slowly .... a dry biscuit.
6 Despite his unhealthy diet and heavy smoking, I'm sure he'll .... the rest of us.
7 .... that I make enough money, I'll join you on holiday.
8 She leads a very busy .... , so I rarely get to see her.
9 His habit of spending money on all his friends really .... the kind of generous person he is.

Student 3

1 I think you've got a long way .... go before you can take the exam.
2 I can't decide whether or .... to apply for a place at university or spend some time travelling when I finish school.
3 Despite the huge number of tourists visiting the island, the locals' way of .... hasn't changed in almost 500 years.
4 You need to strike the .... balance between work and relaxation.
5 Switzerland has one of the highest standards of .... in the world.
6 Can you .... living rough on the streets without a penny to your name?
7 When she missed yet another of my lessons, I was .... annoyed. (*three possible answers*)
8 There were very few people there. The place was .... deserted.
9 She was .... furious when I told her I was seeing someone else.

1 Complete this sentence with an appropriate word:

Nutritionists recommend that we eat a/an .... diet.

a) even b) steady c) stable d) balanced

2 Complete this sentence with an appropriate expression:

If you decide to eat less food because you want to lose weight, you .... a diet.

a) turn to b) take on c) go on d) set up

3 Complete this sentence with an appropriate expression:

If you eat less of a certain kind of food because it is not very good for you, we say that you ....

a) cut into it b) cut it out c) cut down on it d) cut it up

4 Complete this sentence with an appropriate expression:

If you want to become more healthy, it is recommended that you .... a sport.

a) take up b) take to c) take on d) take in

5 What do we call somebody who doesn't eat:

a) meat?
b) any product derived from or produced by animals?

6 Which of these methods of cooking is the most healthy? Which is the least healthy?

frying roasting steaming barbecuing grilling boiling

7 Physical exercise can turn fat into muscle. True or false?

8 Meat, cheese and eggs contain a lot of fibre. True or false?

9 Complete this sentence with an appropriate word:

People who are very overweight are obese; they suffer from ....

a) obeseness b) obesity c) obesinatis d) obesatia

10 Rearrange the letters in bold to make an appropriate word:

The opposite of an active lifestyle is a ***tedsernya*** *lifestyle.*

11 Correct the spelling mistakes in this sentence:

Make sure you get a good balance of vitamines and minnerals in your diet, eat foods which are rich in protien, and avoid foods which are high in callories.

12 Which of these forms of exercise is considered to be the best for all-round physical well-being?

walking jogging swimming skiing

13 What do you think a *couch potato* does?

14 *............ food* is another expression for *fast food.*

15 What's the difference between a health club and a health centre?

16 Cholesterol is bad for you. True or false?

17 Which of these illnesses and other disorders can be caused or made worse by eating a poor diet?

cancer diabetes heart disease
strokes thrombosis stress
skin problems / poor complexion
halitosis (bad breath) flatulence
high blood pressure sleeping disorders
premature ageing

18 Rearrange the letters in bold to make an appropriate word:

Lack of regular exercise, a poor diet and unhealthy habits like smoking can reduce your life ***cneptayxce****.*

19 Complete this expression with an appropriate word:

You are what you ....

a) consume b) eat c) have d) devour

20 Complete this proverb with an appropriate word:

An apple a day keeps the .... away.

a) doctor b) fat c) ageing d) illness

First sentence of first pair ➤	He was surprised and more than a little disappointed when I turned down his offer of a job.	I have my own reasons why I refused .... , but I don't intend to explain why.	When I was young, my father, who was rather old-fashioned, made sure that I had a very sheltered upbringing.
.... intentions were good, but I don't think it's right to be protected from difficult or unpleasant situations.	It has always been my dream to walk across the USA from the east to the west coast.	If I ever went .... , however, I would probably spend my time hanging out on a beach somewhere.	Are you one of those people who are always thinking about having plastic surgery to look younger?
If .... , perhaps you should consider something a little less dramatic – like changing your diet, for example.	There are several recommended ways of staying young and beautiful for as young as possible.	One of .... is to take regular exercise or take up a sport or other physical hobby.	In my opinion, overseas travel is a wonderful way of learning about other cultures.
I love it, ... is why I never seem to save up enough money to buy myself any luxuries.	Photography is a great hobby, because it gets you out and about, and keeps you fit in a relaxed sort of way.	.... is golf, although not everyone can do this unless they have easy access to a golf course.	I prefer the comforts of home to the rigours of travel, probably because I've never been keen on discomfort.
.... has my wife, which probably explains why we spend so much time sitting in front of the TV and moaning about the weather.	I bought my first computer back in 1987, when they were large, slow and ridiculously expensive.	Since .... , there have been some remarkable technological developments and improvements.	The ability to get on a plane and be on the other side of the world in hours is something we take as a matter of course.
To us, doing .... is completely normal, but it would have been almost inconceivable to our grandparents.	After almost ten years teaching in a school in Oxford, I decided to take the plunge and become a freelance writer.	.... a move was risky, of course, but I gave the matter a lot of thought and careful consideration.	Lying on a beach soaking up the sun is, in my opinion, one of the most boring things you can do.
What's .... , it's been proven that it can be terribly dangerous for you.	A lot of people seem to complain about everything and anything whenever they get the chance.	I'm not sure if it's .... upbringing that's to blame, or just the fact that they are plain miserable and bad-tempered.	First of all, it is highly unlikely that a miracle cure will be found for obesity.
.... , I'm sceptical of claims made by certain food manufacturers that their products can reduce cholesterol.	Sometimes it seems that the whole world is inexplicably obsessed with looking beautiful.	.... the reasons, a lot of companies are making a lot of money out of people.	◀ Second sentence of last pair

1 If you ask me, I'm confident about a cure for cancer will be found in the next few years, provided that there is enough investment in research.

2 I don't hold out much hope that people in my country will all be healthier within the next few years unless they make radical changes to their lifestyle.

3 I don't see any evidence that people will be persuaded to change their eating habits over the night, especially while fast food and ready-prepared TV meals are so easily and cheaply available.

4 In my opinion, it is highly likely that the majority of people in developed countries will be obese at some stage in the not-too-distant future, thanks to poor eating habits and a sedentary lifestyle.

5 For years, fast-food companies have, metaphorically speaking, been pushing their products down our throats, but now we are seeing a reversal in this trend.

6 Governments are punching a huge amount of money into campaigns to promote a healthier lifestyle, but with very little success.

7 In order to attract new customers and help to revitalise the national economy, a lot of banks are offering debt-free loans of up to £10,000.

8 The country which consumes the most fish per head of the population is generally considered to be Japan, where life expectancy has been shown to be the highest in the world.

9 It wasn't until I had been cooking dinner for two hours I remembered the guests I was expecting were vegetarians and wouldn't touch the chicken casserole I had decided to make for them.

10 It's been four long years since I last took a break from work, and then I only managed to spend a cold, wet week in Scotland with my parents.

11 All I ever have wanted to do is be famous and make lots of money, but I really don't think this is going to happen in the near future, or indeed at any time in the future.

12 What I really want to do is to spend a year travelling around the world, so I'm working hard and saving up as much money as possible, and with a bit of luck I'll do it next year.

13 I enjoy my job, I like the people in my office and of course it's good to get a regular salary cheque, but recently I've been considering to spend some time working for a charity in a developing country.

14 You've been a very enthusiastic employee and you've done your best, but unfortunately I regret to tell you that the company hasn't been particularly satisfied with the quality of your work.

15 I tried to apply for several jobs with different companies, and was even invited to a few interviews, but unfortunately I wasn't accepted anywhere, so now I'm thinking about setting up my own business.

Opening credit: 100 points

If you are 100% sure you have chosen the correct answer, you may risk 15 points.

If you are 75% sure that you have chosen the correct answer, you may risk 10 points.

If you are 50% or less sure that you have chosen the correct answer, you may risk 5 points.

	Points risked	Points won	Points lost
1 Right / Wrong			
2 Right / Wrong			
3 Right / Wrong			
4 Right / Wrong			
5 Right / Wrong			
6 Right / Wrong			
7 Right / Wrong			
8 Right / Wrong			
9 Right / Wrong			
10 Right / Wrong			
11 Right / Wrong			
12 Right / Wrong			
13 Right / Wrong			
14 Right / Wrong			
15 Right / Wrong			
	Your total score:		

seirujni evitaicerppa boj gnihctaw dekcehc revoerom yal esitcarp esiar suounitnoc egamad luftcepser ~~teeffa~~ seitilibissop noitpmussa esivda ylfeirb gnirud sselhtrow noitca noissecorp tnarelotni dettimrep snoitcejbo suoitneicsnoc etaredisnocni gnirevocsid diova ediseb

Example: *We often have to make important decisions which* ***effect*** *our lives.*
effect ➤ affect

1 I haven't managed to get any work done; I've had an endless process of visitors all day.
2 I've had six continual hours of meetings today, without so much as a five-minute break.
3 There's no way anybody will buy this; it's priceless junk.
4 We wanted to make money, so began to explore the chances of opening a club in the city.
5 While the speeches were being made, everyone listened in respectable silence.
6 The university is working to rise the number of students from state schools.
7 This activity gives students the opportunity to practice their speaking skills.
8 Before you start to cut it up, lie the material flat on the table.
9 A lot of people are intolerable of other people's political and religious beliefs.
10 The rent on my new house is reasonable and, however, the location is perfect.
11 The earthquake caused extensive structural harm to several buildings in the town.
12 Our school is regularly controlled by a fire-safety officer, who makes sure that it is safe.
13 As a punishment, she was not permission to attend any school activities.
14 Most parents don't know what their kids are looking at on TV.
15 He slept calmly while the early part of the night.
16 It's a temporary work, but I'm hoping it will be made permanent.
17 When I wanted to build an extension on my house, local residents raised strong criticisms to the building application.
18 It was very inconsiderable of you to keep us waiting for so long.
19 Carlos is a conscious and hard-working student who always gets good results.
20 She was appreciable of Greg's concern for her health.
21 We stopped off shortly in London on our way to Geneva.
22 The driver of the truck sustained only minor damages to his legs and arms in the accident.
23 The Curies are best known for inventing radium.
24 Road safety is taught to young children so that they can prevent road accidents.
25 When the accident happened, I was standing right besides her.
26 A lot of people make the presumption that poverty only exists in the Third World.
27 Environmental groups want tougher activity on pollution from cars.
28 We strongly advice you to take out medical insurance when visiting China.

Student 1

- Today is Laurence and Lisa's **wedding** day. They have just got **married**.
- Rick has been **working** for the same company for 40 years. Today is his **retirement** day. He has just **retired**.
- Ian and Stephanie have lived all their lives in a big **city**. They have just moved to a quiet **village** in the **countryside**.
- Janet is an **actress** and recently made her first **film**. It has become an international success, and she has just become very **famous**.
- Alison has lived with her **parents** for 19 years. She has just **left home** for the first time.
- Andy has been **married** to Julie for 12 years. They have two **children**. Unfortunately, they have just got **divorced**.
- Timothy recently finished **school** after finishing his **exams**. He has just started at **university**.

Student 2

- Claire has **worked** for the same **company** for ten years. She has just left to start her own **business**.
- Elizabeth and Tom have lived all their lives in **Britain**. They have just **moved** to **Italy** to start a new life.
- For years, Harriet has lived in a **rented flat**. She has just bought her first **house**.
- Mike had a well-paid **job** in the city. He has recently been made **redundant** and is now **unemployed**.
- Edward has been in full-time **education** for 16 years and recently left **university**. He has just started his **first job**.
- Sarah's extremely **rich** grandfather died recently. She has just **inherited** a lot of **money** from him.
- Harry **robbed** a bank recently. He was **caught**, and has just been sent to **prison** for ten years.

Team 1

I've taken these exams three times, and each time I get grade Ds or below.

This room smells absolutely disgusting.

Everyone keeps telling me that I've got a really wonderful job.

Tony is having a really horrible time in London.

Maria didn't expect a brilliant grade in the test, but she didn't expect to do so badly.

Your work has been very poor, and your results correspondingly bad.

My neighbours are really beginning to annoy me with their loud music.

I'm really fed up with the way you treat me.

Why are you always telling your boyfriend to get lost?

Team 2

Her idea is excellent in theory.

He asked me to marry him.

This must be one of the worst hotels in the city.

I didn't question your actions at the time.

I'm not sure if my plan will work.

Frederico has the potential for being an excellent student.

You should always take a mobile phone with you when going on a long car journey.

I doubt I'll get the job, but I'm going to apply anyway.

Your computer is old, slow and unreliable.

I wish you wouldn't ...
It's high time you got ...
If only they knew ...
I'd prefer it if they didn't ...
However, I'd rather run ...
It's not as if I was risking ...
If only she had ...
I'd sooner stick ...
It's high time you started ...
You looked as if ...
Suppose he did ...
If only I had ...
But what if it doesn't ...
I wish I could ...
If only he could ...
He really wishes he had ...
Supposing you broke ...
You talk to me ...

... the risk of getting it wrong ...
... down in the middle of nowhere ...
... as if I were ...
... rid of it and bought ...
... stayed at ...
... making more of an effort ...
... smoke ...
... you knew what ...
... just that one day ...
... gone to an English school ...
... my head in ...
... play their Coldplay albums ...
... studied harder and paid ...
... how difficult it can be ...
... get better ...
... work ...
... be more ...
... life and limb ...

... by trying.
... in Oxford instead.
... and coming on time.
... a child.
... in practice?
... the Carlington instead.
... than doing nothing at all.
... and needed help?
... a new one.
... in here all the time.
... at times.
... grades in my exams.
... more attention in her class.
... until three in the morning.
... and walked out on you?
... a food processor!
... you were doing.
... motivated and industrious.

1 Nowadays, everybody wants the best car, the most modern home entertainment system, the best designer clothes. This .... has made us forget what life is really about.
2 You're such a couch potato. You can't even be .... to go for a walk these days.
3 He's got a terrible memory, but he can .... off the names of every player who scored in the last World Cup.
4 You shouldn't give in to everyone's demands all the time. If you always bow to popular .... , people are bound to take advantage of you.
5 When the government made it illegal to smoke in public places, there were so many complaints that they were eventually forced to lift the ....
6 He was very important in getting the company started, and he played a crucial .... in making people aware of their products.
7 A handful of phone networks have succeeded in capturing the .... for mobile users, with smaller companies going out of business.
8 She's so greedy, she'll think .... of eating a whole packet of biscuits in one sitting.
9 I thought I could .... in you, but now I found out you've gone and told everyone what I said!
10 I bit of exercise now and again can be very .... for your health.
11 I wasn't ignoring you. I was a bit .... with a few problems I'm having at work, that's all.
12 For years I resisted getting a mobile phone, but then I decided to .... and buy one.
13 A lot of major international companies stand .... of exploiting workers in their factories.
14 He's a really good friend, but at .... time has he ever been my boyfriend.
15 I thought that working as a tour guide would be great fun, but .... did I know how hard it would be.
16 .... only is our new teacher very strict, but he is also extremely rude.
17 On no .... are you to miss a lesson without informing me first.
18 .... had the lesson began than half the students started falling asleep.
19 No .... had he arrived than he started getting on everybody's nerves.
20 Beautiful she may be, but intelligent she most certainly .... !
21 Originally we didn't think it was important, and only .... are we beginning to realise how crucial it really was.
22 I'm supposed to start work at six in the morning, but no .... am I going to be able to get up in time!
23 Under no .... are you to come late again.
24 .... good were his exam results that he was offered places at five universities.
25 .... was her enthusiasm that everyone else on the course was motivated as well.

Start ↓									
	confide		little		isn't		so	speak	beneficial
ban		anything		account		bothered	events		
								no	confidence
	sooner	reason	place		thinking	immediately	customers		
benefit				role					nothing
	can't	materialism			possibility	preoccupied		way	
now	force		repent	such			hardly		later
		circumstances			pressure	charged		very	prohibition
reel			accused				because	relent	
	market	then		able	never	materialistic			not

You can move vertically or horizontally, but you cannot move diagonally. You must not cross the shaded spaces.

Case 1

A man living in a remote village steals a car from his neighbour. He is arrested outside a hospital with his wife, who is about to have a baby. He explains that he needed the car to get his wife to the hospital. When asked why he didn't call for an ambulance, he explained that he didn't think the ambulance would arrive on time. He had asked his neighbour if he would take them to the hospital, but his neighbour had refused. What should happen to the man?

Case 2

A 16-year-old girl is arrested for stealing things from a shop. She tells the police that she is being bullied at school. The bullies told her to steal for them because they needed money for drugs. They threatened to hurt her if she refused to help them, or if she told the police about them. What should happen to the girl? What should happen to the bullies?

Case 3

A jet airliner on an internal flight in another country is hijacked, and the pilots are made to fly to your country. Once the plane lands, all the passengers are released unharmed, and the hijackers surrender without a fight. Their one and only weapon, a gun, turns out to be a replica. They explain that they hijacked the plane to escape from their country, which has a very repressive political system. What should happen to the hijackers?

Case 4

Two thieves break into a farmhouse. The farmer, who is upstairs, hears them and comes downstairs to confront them. He has a shotgun (legally, as he is a farmer). He shoots. One of the thieves is killed, and one of them is seriously wounded. The farmer tells the police that he was defending himself and his property. He claims that he was 'frightened for his life'. What should happen to the farmer?

Case 5

Two young men are robbing a shop, when they are interrupted by the police. One of the policemen approaches them. The youngest thief, who is 16, has a gun. The oldest, who is 18 but has a mental age of eight, tells his friend to 'Let him have it'. His friend fires the gun and kills the policeman. What should happen to the 16-year-old? What should happen to the 18-year-old?

Case 6

A man is arrested for murder. At his trial, the judge ignores some of the evidence that could prove the man is innocent. The man is found guilty and is executed. A short while later, it is discovered that he was innocent after all, and the evidence ignored by the judge could have saved him. What should happen to the judge?

Case 7

A man is arrested for murder. His girlfriend is convinced that he is innocent, but she also knows that he has been in trouble with the police before. In order to protect him, she gives the police an alibi for him which she knows is not true. The police discover that the man is guilty and that his girlfriend has told them a lie. What should happen to the man's girlfriend?

Case 8

A man loses control of his car, which goes off the road and lands on a railway line. He escapes unhurt, but a train hits his vehicle and several people on the train (including the driver) are killed. The police discover that the man had fallen asleep while driving the car. He was extremely tired because he had spent the whole of the previous night chatting to his girlfriend in an Internet 'chat room'. What should happen to the man?

Case 9

A man is seriously depressed after his parents die and he becomes homeless. He sets fire to a curtain in a church, then calls the police and tells them what he has done. He explains that his actions were a 'cry for help'. Damage to the church is minimal, and nobody is hurt. However, arson is considered to be a very serious crime. What should happen to the man?

1 A holiday in Australia is a lot cheaper than you think.
A holiday in Australia **think not as as is you nearly expensive**.

2 The new Ferghini Peoplevan is much safer than other cars on the road.
The new Ferghini Peoplevan **road the is car on easily the safest**.

3 The chances of winning the lottery are much smaller than being hit by lightning.
You're **lottery far likely to hit win by lightning more the be than**.

4 Compared to Juliet, Ronald is really quite stupid.
Ronald **as nowhere intelligent is Juliet near as**.

5 In my opinion, Paper 3 of the CAE is much more difficult than the other papers.
In my opinion, Paper 3 of the CAE **most paper far the is by difficult**.

6 I don't think I've seen any other film that's as bad as this.
This **seen about the is film just I worst ever have**.

7 My Maths exam was much easier than my History exam.
My History exam **exam more Maths a great deal was than my difficult**.

8 I can't think of many more pleasant ways to travel than by train.
Travelling by train **travelling one of ways the is pleasant of most**.

9 You shouldn't really use a telephone during a thunderstorm.
It's not **thunderstorm good such a telephone use to a during idea a**.

10 The amount I know decreases with the time I spend studying.
The **study know more I, the I less**.

11 There's a British saying which goes 'More haste, less speed'.
There's a British saying which means **'mistakes make the you faster you, the more work'**.

12 You're 18 years old, not eight!
You're **child little a behaving like**!

This gesture gives the impression that the person is praying for something to happen, or not to happen. We also use this when begging somebody to do or not to do something, in which case it can be semi-humorous. A variation of the clenched fists seen here is that of the palms gently pressed together, and the thumb and fingertips touching those on the other hand. When raised and shaken above the head, it usually means victory, or 'I've won!'

In this gesture, we are creating a shield or barrier that we are prepared to use when we don't want to see something. Often seen at very tense moments in important football matches, it means 'I can hardly bear to look'. The hands don't make contact with the face, but are held in position just in case they're needed.

This gesture is particularly common with young children when they have done or said something wrong. There are two possible meanings to the gesture, one being 'I've embarrassed myself, so I don't want to look anyone in the eye' and the other being 'I'm hiding because I don't want anyone to see me'.

This gesture provides us with a virtual knife or dagger that we use when angry or are shouting at someone, and emphasises what we are saying. When thrust towards the person we are speaking to, it takes on a threatening meaning, as if to say 'I'm going to kill you'. Parents will use the less threatening version with their children, which has the forefinger pointing vertically up, accompanied by a rapid shaking movement.

Not a conscious gesture, but one that we often use when in a situation where we feel vaguely uncomfortable in someone's company. A protective barrier is being formed, the vulnerable chest being almost totally covered in preparation for a virtual physical attack. It is also used when we simply want to feel more comfortable.

This is an unconscious gesture, although why we do it is unclear. It is often understood as meaning that someone is telling us a lie. The fingers can either rub, or tug, the lobe, and it is often accompanied by excessive eye contact, or by avoiding eye contact altogether.

A gesture we usually use when we hope that the person we are referring to is not listening. The hand movement, which imitates the one we politely make when we are yawning, is usually accompanied by a slightly tilted droop of the head and is basically saying 'This person is so boring!'

Often employed when we are talking and want to stop someone interrupting us, this means 'Hang on, let me finish'. We are using the palm to act as a barrier, metaphorically bouncing the speaker's words back at him. Also used to mean 'Stop' when someone comes towards you, and is often used by the police. This gesture can be inappropriate in some countries, where it can be taken to mean 'I curse you'.

The open palms of this gesture are basically saying 'Look, I have nothing to hide', but it can also mean 'It/He/She could be anywhere'. It is a lazy, non-verbal response, often accompanied by a shrug of the shoulders, to mean 'I don't know' or 'Search me'.

We sometimes use this to emphasise a point or a number of points that we are making, the chopping gesture being used to highlight the important issues and make the listener aware of how serious we are. Often, but not always, used when we are angry about something and we want to make sure it doesn't happen again.

A
B
C
D
E
F
G
H
I
J

1 A 'You're always talking, Harry,' said Joan, 'It's so annoying. I reckon you would go on talking for ever if nobody told you to shut up!'

B Joan **1** _ _ _ _ _ _ _ _ _ _ _ that Harry could talk the hind legs off a **2** _ _ _ _ _ _ _.

2 A Jane said, 'I'm so sorry I spoke for so long.'

B Jane **1** _ _ _ _ _ _ _ _ _ _ _ _ for **2** _ _ _ _ _ _ _ _ _ _ _ on for so long.

3 A 'All right,' said Tom, 'It was me who gave your secret away.'

B Tom **1** _ _ _ _ _ _ _ _ _ _ to letting the **2** _ _ _ out of the bag.

4 A 'If I were you,' said Clare, 'I would make sure you get the facts from him directly rather than relying on rumours.'

B Clare **1** _ _ _ _ _ _ _ _ us to get the facts directly from the **2** _ _ _ _ _ _ ' _ mouth.

5 A 'I've just realised how terribly timid Jan is,' said Mark.

B Mark **1** _ _ _ _ _ _ _ _ that Jan wouldn't say boo to a **2** _ _ _ _ _ _.

6 A 'A good way to remember a word is to say it again and again until it stays in your head,' the teacher said.

B The teacher **1** _ _ _ _ _ _ _ _ _ _ repeating a word **2** _ _ _ _ _ _ _ -fashion until it stayed in our heads.

7 A 'Don't forget that it was me who first became suspicious of her,' said Emma.

B Emma **1** _ _ _ _ _ _ _ _ _ us that she was the first person to smell a **2** _ _ _.

1 It's been hindered that he's going to resign.
2 The dog gave a low crawl and started coming towards me.
3 She gave a sharp whisper, and everyone stopped talking.
4 It was an instantly forgetful film.
5 He's violently competitive and hates to lose.
6 The residents all complained about the exceptional noise coming from the factory.
7 It was such a successful marketable strategy that demand rapidly exceeded supply.
8 The company is belied to be making plans to terminate its contract with us.
9 In the last five years, crime has increased on an unappreciated scale.
10 From our house, you can hear the steady ramble of traffic on the motorway.
11 He loves to crumpet the fact that he is better than everyone else.
12 She likes to imitate her employees with her aggressive behaviour and unreasonable demands.
13 Companies try to make their products more delectable by excessive use of hype.
14 She's repudiated to be the richest woman in the United Kingdom.
15 The house was full of the rampant odour of burnt garlic.
16 It has been suggestive that the school starts doing classes on Saturdays.
17 I think it's wrong to differentiate against people because of their age.
18 Advertisements often convoy the message that 'thin is beautiful'.
19 If you want to make the message secret, you should decode it so that no one else understands what it says.
20 It has been augmented that computers cause more problems than they solve.
21 We could hear the raw of the crowd from almost three kilometres away.
22 Her throat was so sore she could do little more than crack.
23 Some animals have their own defense mechanisms to scurry from other animals. (*two words*)
24 Mobile phones have graduated from the bulky, expensive status-symbols that they used to be.
25 I woke up in the night to the quack of rubber-soled shoes walking across the floor.

Pair 1

1 late minutes Arriving in ten class,
2 with ten boyfriend after meeting Barely up minutes my,
3 spent studying so Having time much,
4 speaking Generally,
5 late to again and wanting oversleep Not be,
6 and being overweight extremely clumsy Despite rather,
7 shocked see face look the on To his,
8 middle the the the was which Opening in of box room lying,
9 head in toe black Dressed from to entirely,
10 results with my disappointed Bitterly exam,
11 didn't To make teacher sure we again the upset,
12 fit particularly Not healthy or being,

Pair 2

A never decided to return leave and I college.
B the keep rest difficult found it to she up with us of.
C failed had I was exam to find I disappointed the.
D like James movie he a from villain a Bond looked.
E thought just you ghost would have he seen had a.
F the made we were lesson on we time and attention sure paid during.
G was find I gun to a shocked.
H already had discovered I left everyone.
I row haven't had a we and since spoken terrible.
J go early I my alarm to set off clock.
K dancer great a he's.
L be experience out an can eating expensive.

1 I have always liked my music to be exciting, to have a real kick, something that would shock granny and scare the cat. People say it's a bit childish, but I like to hear something with screaming vocals, deafening electric guitars, thumping drums, primitive lyrics, that kind of thing. In fact, the louder and more aggressive, the better. Basically, unless it's the kind of music that will blow up my amplifier or make me prematurely deaf, I'm just not interested.

2 It's relaxing, laid back, what I call midnight music. In my opinion, the saxophone must be the coolest instrument in the world. However, I think you need to be in the right place for it. A large concert hall or the living room of a semi-detached in Orpington doesn't feel right for something so smooth and mellow. The best place to enjoy it and get a feel for the music would probably be a dimly lit, smoky cellar in somewhere like Chicago or New Orleans. People who say it's boring have probably never listened to it properly.

3 My friends think I'm a bit odd to like music like this. They joke about it, ask me where my cowboy hat is or how my horse is, say 'Yee-ha!' and make strange twanging noises with their tongue. I suppose for me it evokes memories of when I was young and growing up in the American Midwest. Over there, that kind of stuff was on the radio all the time, and I guess it's just stayed with me.

4 I love it, it's so passionate, so physical, so lively. I can't listen to it without wanting to put on a flamenco dress, down a margarita, grab a pair of maracas and tango or cha-cha-cha my way around the room. And I'm not the only one, as you can see from all the salsa classes that are springing up everywhere. It's so evocative of hot and sultry nights on the beach or some side street in Mexico City, Buenos Aires or Rio.

5 I think that these days young people place too much emphasis on pop music from the USA or Britain. There's so much more out there which is so much better than the bubblegum pop you hear on the radio. And if you listen to tribal music from Africa or traditional Irish folk songs or Aboriginal chants, you realise the influence they have had on modern music. There's really nothing original about contemporary pop music; it's all been borrowed or stolen from other cultures.

6 People criticise it, say it's trashy and promotes violence, gun culture and drugs, or that it's sexist. However, you can't deny that the lyrics are sharp and clever, and the music is lively, with a real beat. It's what I call 'outsider' music, ideal for young people who feel alienated. It makes them realise that there are lots of other people out there who feel the same anger with the world and the way we are all supposed to conform.

7 It's been described as highbrow elitist music for snobs and millionaires. With ticket prices for a live performance as high as £250, this would certainly appear to be the case. However, you can't deny the passion and the emotions that composers like Mozart or Wagner could arouse. What's more, you just can't beat the visual and aural impact of seeing and hearing these people sing accompanied by a full orchestra; it's pure magic.

8 Nowadays, it seems like every play, every novel, every film, is being turned into one of these. If it's not singing, dancing cats, then it's an anguished Quasimodo or Phantom or Heathcliff singing about love and loss and all that sickly sweet sentimental nonsense. It really gets on my nerves. Believe it or not, there are people out there who have gone to the same show 20 or 30 times, bought the album, the T-shirt and the souvenir key ring, and know the story and the lyrics backwards. All I can say is they must have more money than sense, or indeed taste.

1 Complete the sentences with one of these words:

feel make achieve do win take earn

a) You can .... a goal, a target or an ambition.
b) You can .... a risk, a chance or the plunge.
c) You can .... decisions, sacrifices or mistakes.

2 Complete each sentence with two of these words:

outlook upbringing positive way ideal sheltered

a) Writing to an English-speaking penfriend is a(n) .... of practising your written English.
b) As a child, I had a very ....
c) She's very cheerful, and has a really .... on life.

3 Complete these sentences with the correct form of the verbs in bold.

a) I **come** here in 2002, and I **be** here ever since.
b) I **not / have** a break yet because I **work** hard all morning.
c) He **spend** most of his money already, so by the time he gets back, he **spend** everything.

4 Complete these sentences with an appropriate word.

a) She's a good student, so everybody .... her to do well in the exam. (*Don't use* wants.)
b) Thanks for calling. In fact, I was just .... to call you. (*Don't use* going.)
c) She's so unpunctual, so she's .... to be late again. (*Don't use* going.)

5 Complete these sentences with an appropriate word beginning and ending with the letters in bold.

a) He loves being with people. I don't think I've ever met anyone who's so **g**............**s**.
b) She never gives up; she's so **p**............**t**.
c) Don't upset her; you know how **s**............**e** she can be.

6 Complete the sentences with one of these words:

no neither each none every both all

a) My two brothers are similar in that they .... love socialising.
b) They're both very intelligent, but .... of them enjoys studying.
c) I hate History, Maths and Biology. .... of them interest me much.

7 Complete the sentences with one of these words. You will need to change the word forms:

feel scratch pat press tap hold rub stroke

a) 'You're mad!' he said, gently .... the side of his head for emphasis.
b) She yawned, ... her eyes and poured herself another coffee.
c) I .... the doorbell several times, but nobody came to the door.

8 Complete these sentences with an appropriate word.

a) He started gambling ten years ago, ... when he's lost thousands of pounds.
b) He came late, as a ....of which he missed the most important part of the lesson.
c) A lot of people, many of .... are intelligent and rational, believe in ghosts.

9 Correct the mistake in each of these sentences.

a) I feel absolutely terrible; I think I'm getting a flu.
b) The police has investigated the crime, but haven't found any clues.
c) On my last holiday, all my luggages went missing.

10 Change the words in bold into a noun.

a) Your **attend** has been very poor this term.
b) We apologise for any **inconvenient**.
c) She tried to maintain a position of **neutral** during the argument.

11 Choose the correct word in italics to complete these sentences.

a) Everybody seems to have decided what to do, but I haven't had a say *in / on / to* the matter.
b) They put a lot of pressure *on / at / to* me to pass the exam.
c) Would you like to contribute *with / to / for* the discussion?

12 Rearrange the letters in bold to make an appropriate word.

a) I'm sorry, but you're not **wadloel** to smoke in here.
b) You're under no **anligoboti** to attend afternoon classes.
c) It's **esavibdla** to arrive ten minutes before the exam begins.

13 Choose the correct word in italics to complete these idioms.

a) He's got an excellent sense of *amusement / fun / humour* and always makes people laugh.
b) She's so *filled with / full of / complete* in herself, always telling people how wonderful she is.
c) You need to accept that life isn't always a bed of *roses / daffodils / tulips*.

14 Put the words in bold into their correct form by adding a prefix and a suffix.

a) His behaviour was absolutely **believe**. It was so embarrassing.
b) When he resigned, they had to find a **place** for him.
c) Once it goes wrong, the process is completely **reverse**.

15 Choose the correct words for the definitions:

munch suck swig wolf down nibble (at) pick at guzzle chew drain

a) to eat something slowly with very small bites
b) to drink something through a straw
c) to eat something very slowly, without really wanting to eat it at all

16 Rephrase these conditional sentences using the words in brackets.

a) I'll lend you my car if you take good care of it. (that)
b) You should get up early, because if you don't, you'll be late. (otherwise)
c) He'll pass the exam, even if he doesn't do any revision. (whether)

17 Complete each sentence with a word that can be used to express an opinion about the future.

a) I can't see any .... that things will get better.
b) I'm .... of claims that things will get better.
c) I'm .... that things will get better.

18 Choose the correct verb form in each of these sentences.

a) I'm sorry, but I don't remember *to meet / meeting* you at the conference last year.
b) If Carrie isn't at home, try *to phone / phoning* her office to see if she's there.
c) I was really busy, but I stopped *to have / having* a cup of coffee.

19 Change one word in each of these sentences.

a) The school runs general English classes in the morning, and complimentary studies classes in the afternoon.
b) The film wasn't particularly good, but the special affects were amazing.
c) The principle at the college will address the students at ten o'clock.

20 Complete each sentence with a word that can be used when talking about hypothetical situations.

a) She's suggested playing tennis, but I'd .... go for a picnic.
b) I can't say it to her face, but .... I were to put it in a letter?
c) He's only a salesman, but he acts as .... he owned the company.

21 Complete each sentence with these words:

no know little circumstances only I under not did

a) .... was it hot, but it was also very humid.
b) I thought it would be a piece of cake. .... how difficult it would be.
c) .... are you to enter the office unaccompanied.

22 Complete these sentences with an appropriate word.

a) Nobody has heard of him here, but he's very ....-known in my country.
b) I wasn't expecting any problems, so I was caught completely ....-guard when he told me there was a delay.
c) We knew it was a ....-risk investment, but we were still surprised to lose so much money.

23 Complete these sentences with an appropriate word.

a) .... been fooled once, I wasn't going to let it happen again.
b) .... disappointed with my lack of success, I felt that things could have been much worse.
c) .... hear him boast about how rich he was, you would think that money was the only important thing in life.

24 Complete these sentences with an appropriate word.

a) He objected .... the way everyone treated him.
b) I disagree .... the clothes you wear say what kind of person you are.
c) She tried to discourage me .... joining the drama club.

25 A preposition in each of these sentences is incorrect. Identify and correct it.

a) Please don't shout to me; I'm trying to do my best.
b) Laurence has been missing for days, so I'm rather anxious for him.
c) Everybody keeps talking about someone called Beckham, but I've never heard about him.

26 Rearrange the letters in bold to make words. The first letter of each word has been underlined.

a) I guessed how the film would end because the whole thing was **nyetirle capritleeb**.
b) We couldn't stop laughing; the whole thing was **tubasyolel oiulrihsa**.
c) People say it's a brilliant film, but personally I found it **syytla readtorev**.

Exam practice 1: Reading

Paper 1 Part 4: multiple matching

You are going to read an article in which six people talk about the part that travel plays in their lives. For questions **1–15**, choose from the people (**A–F**). The people may be chosen more than once.

Which person

currently has very little time for holidays?	**1**	
believes that the destination often compensates for a tedious journey?	**2**	
dislikes being kept waiting on arrival at a destination?	**3**	
prefers to let someone else organise the travel arrangements?	**4**	
finds that travelling provides inspiration for work?	**5**	**6** .....
says that feelings of homesickness can spoil the enjoyment of travelling?	**7**	
is reluctant to give up creature comforts when travelling?	**8**	
is inclined to overpack when going on a journey?	**9**	
visits so many places that it's hard to recall much about them individually?	**10**	
regards travel as an educational experience?	**11**	
does not spend holidays in the way some people might expect?	**12**	
welcomes the culture-change that long journeys make possible?	**13**	
appreciates the opportunity for reflection that travel can offer?	**14**	**15** .....

Exam practice 1: Reading

My life in travel

A Anita Roddick, founder of the Body Shop

As kids, we spent the summers helping out in my mum's café, sneaking off when we could to the nearby beach or amusement park. Since then, the insights I've gained from extensive travelling have been ploughed back into my work with the Body Shop. I think of travel as a university without walls, and for me there has always got to be a purpose to it. Lounging around on a beach doesn't grab me at all, but I love soaking up the environment; walking, hiking and rafting with family and friends in remote places. Long train journeys are sublime because they give me time to think, but air travel is abysmal .When I do have to do it, I make sure that I just take hand luggage, so I'm not hanging around the baggage reclaim at the end of the journey.

B Laura Bailey, model

I've clocked up lots of miles through my work since I started modelling.When I was small, we mainly rented cottages on the coast in the holidays, although I do remember going camping in Europe. These days, I regard long-haul travelling as a necessary means to an end; I tolerate it because the place itself generally more than makes up for the tedium of getting there. When I travel for work, I have no choice but to let my agent organise the trip, but I'll very often stay on afterwards or take a detour to somewhere I want to explore. Over the years, what I've learned is just to go with the flow – sometimes the wrong turn can turn out to be the right turn.

C Bear Grylls: explorer

People assume I must be a nightmare to go on holiday with because I'll always want to be going up a mountain, but actually that's the last thing I'm up for. I get my adrenaline rush from work, so on holiday I want to chill out properly, preferably on a beach. For me, travelling is all about arriving at a place. On my lecture tours to the USA, I always try to squeeze a lot into as little time as possible so that I can fly there and back in one day. If you're away from your family, then the fun part of travelling is eliminated.

D Matthew Williamson: fashion designer

When I go on holiday, I tend to go abroad because I want to escape from my daily routine. Even though the reason for going is to recharge my batteries, I take my sketchbooks with me so that I can draw new ideas for the next season's collection. However, I don't enjoy the procedure of getting from A to B at all – in fact, I would probably travel twice as much as I do if I could just magic myself there.

When I was younger, I used to rough it a bit on cheap package holidays, but there's no way I would do that now. For me, the biggest luxury if I go away now is time, and because I like everything to go smoothly and pleasantly, it's safer to let my personal assistant sort out the whole thing for me.

E Nelly Furtado singer

I grew up in Canada, and remember many holidays just swimming in the lakes and going cycling. But one of my best memories is of backpacking around Europe with friends when I was around 18, and waking up and not knowing where we were. Even now, I'm quite happy with this kind of wandering, rootless lifestyle, although I still haven't learned to travel light. I've done so much touring in the last three years that it all blends into one for me these days, and grabbing a day or two off to look around a city counts as a vacation. The process of getting there means you can just sit with your memories – flying back to my old home town I get swamped by all these feelings.

F Jamie Cullum, singer

With short-haul flights, it hardly feels like you have left home, so I'm quite happy to do long-haul travel if there's plenty of time to soak up the atmosphere at the other end. Although I'm not averse to the occasional package holiday with friends – it certainly takes away the hassle – I've had some of my best times travelling alone, perhaps because it's a more intense experience. I used to take loads of books and stuff with me, but I've learned that sometimes you can get so immersed in reading that you could be anywhere instead of appreciating what is around you.

Exam practice 1: Use of English

Paper 3 Part 4: gapped sentences

For questions **1–5**, think of **one** word only which can be used appropriately in all three sentences. Here is an example **(0)**.

Example:

0 Sarah is a good teacher and her students find her really .... to talk to.

Carl wasn't a fan of hard work and was always looking for an .... life.

It's .... to see why so many couples choose that lovely island for their honeymoon.

Example: **0** | E | A | S | Y |

1 Estelle suffered a .... in income when she moved to a less responsible post.

There has been a .... in the number of people applying for seasonal work in the retail sector this year.

After the long hot spell, there was hardly a .... of water left in the reservoir.

2 On the preparatory course, all subjects are covered, but none in great ....

The actor managed to convey a great .... of feeling in the final scene of the film.

The water in the swimming pool is of a constant .... , so there is no shallow end.

3 Fiona's tutor .... her to believe that she was well prepared for her first work placement.

After the film on recruitment procedures, Sarah Brown .... a short discussion on the issues that it had raised.

It was Simon's poor timekeeping that .... to his dismissal from the company.

4 The newspaper is .... a series of articles on employment opportunities in the hotel and catering sector.

The works manager was just .... through the firm's safety policy for the benefit of the new recruits when the fire alarm sounded.

If you want to apply for a teaching job, time is .... short, as applications have to be in by the end of the week.

5 At the end of the heated negotiations, the staff .... for a 3 per cent pay increase.

Graham found a spare seat in the corner of the library and soon .... down to work.

There had been a fall of snow overnight, but it had only .... on the higher ground.

Exam practice 1: Use of English

Paper 3 Part 5: key word transformations

For questions **1–8**, complete the second sentence so that it has a similar meaning to the first sentence, using the word given. **Do not change the word given**. You must use between **three** and **six** words including the word given. Here is an example **(0)**.

Example:

0 Do you think I could borrow your pen, please?

WONDERING

I .... I could borrow your pen, please?

Example: **0** | W | A | S | | W | O | N | D | E | R | I | N | G | | I | F | | | |

1 It was the most expensive restaurant I've ever been to.

MORE

I've .... expensive restaurant.

2 It's two years since I first went to dancing classes.

BEEN

I've .... for two years.

3 I don't think Yasmin is likely to call round tonight.

THAT

I think .... will call round tonight.

4 People say it is the world's most polluted city.

SAID

It .... polluted city in the world.

5 By booking in advance, students can get a 10 per cent discount.

ENTITLED

Students who .... to a 10 per cent discount.

6 You have to hand in your assignment by the end of the month.

DEADLINE

The .... is the end of the month.

7 There's not much chance of our winning the match.

LITTLE

We .... winning the match.

8 Who left the room in such a mess?

RESPONSIBLE

Who .... the room in such a mess?

Exam practice 1: Writing

Paper 2 Part 2

Write an answer to one of the questions **1–3** in this part. Write your answer in 220–260 words in an appropriate style.

1 An internationally renowned art gallery is looking for people to work there as temporary assistants over the summer. The job involves acting as an English-speaking guide for groups of tourists from many different countries. You are interested in applying for the job. Write a formal letter to the director of the gallery, asking for further details and giving information about:

- any relevant previous experience
- your suitability for the job
- your level of English.

Write your **letter**.

2 A group of young people (18–21) from various countries is coming to your area as part of a cultural exchange scheme. They will have an organised programme of daytime activities, but will be free in the evenings. The local tourist office has asked you to produce an information sheet in English which will give members of the group information about:

- social activities available in the area in the evening
- how to take part in them
- an indication of the cost.

Write your **information sheet**.

3 A website for English language students in your country has decided to include a page of book reviews written by students themselves. You have been asked to provide a review of a book in English which you have read for pleasure recently. It should be a book which you would recommend to other readers.

In your review include:

- brief information about the plot and characters
- an indication of the level of English needed to read it
- the type of person who you think would enjoy reading it.

Write your **review**.

Exam practice 1: Listening

Paper 4 Part 2: sentence completion

You will hear a university student talking about some research she has been involved with on the subject of superstitions in the United Kingdom. For questions **1–8**, complete the sentences.

SUPERSTITIONS IN THE UK

The researchers intend to use a [**1**] as a way of collecting future data.

The exact number of people taking part in the initial research was [**2**]

The researchers found that various types of [**3**] were used by 28% of people.

The commonest saying used by respondents to bring good luck was [**4**]

The objects most widely regarded as unlucky by respondents were [**5**]

The researchers were surprised to find that people involved in [**6**] were quite superstitious.

The most superstitious age group turned out to be [**7**]

When talking about the future, the speaker says the fact that superstitions are always [*and* **8**] is important.

Exam practice 2: Reading

Paper 1 Part 3: multiple-choice questions

You are going to read a newspaper article. For questions **1–7**, choose the answer **(A, B, C or D)** which you think fits best according to the text.

A TALE OF TWO EVERESTS

I gazed up at the peak as I unpacked my picnic. The sun shone and there was not a cloud in the sky. I told my companions that I would climb to the summit of Everest immediately after lunch. They smiled 5 indulgently and began cracking open hard-boiled eggs. I could hardly believe that, after all the preparation, I was here at last. Nor could I believe my luck – to have found a Mount Everest I could scramble up without ice pick or oxygen, while 10 looking at the house of my ancestor, after whom the other, *real* Mount Everest is named. Not many people know about this gentler peak, five miles from the hill station of Mussoorie in the foothills of the Himalayas – and roughly 500 miles from Mount 15 Everest proper. I must admit that until we reached this idyllic base camp, neither did I. But there it was, an attractive grassy peak, wafted by the scent of wildflowers and grazed on by a herd of cattle.

Everest was my maiden name, and I was told as 20 a child that Sir George Everest was my distant ancestor. He had been in charge of map-making in India at the time Mount Everest was first measured. I was also told that he was a brilliant mathematician and surveyor, whose calculations were so accurate 25 that they named Mount Everest after him as a reward. For many years, whenever I gave my name people would say, 'Did you get your name from the mountain?' and I would be unable to conceal my sense of pride as I replied, 'No, it was the other way 30 round actually. The mountain got its name from us.'

As a young adult, however, I was taken aback to discover that when Mount Everest was named in 1856, not only had Sir George been back in England for more than a decade, the mountain 35 already had several ancient local names including the beautiful Tibetan 'Chomolungma', meaning 'mother goddess'. It was actually an Indian pundit who had calculated the height and established that this was the world's highest peak. Renaming it to 40 honour a British mapmaker caused controversy even then. The issue was hotly debated in the newspapers, and it was only the outbreak of the political crisis known as the Indian Mutiny that quelled the row. By the time the mutiny was over, the debate had fizzled out and Sir George's name was 45 firmly affixed to the mountain. When I discovered this, I felt ashamed of the way in which the British had gone round the world imposing names on places that had perfectly good ones already. I half considered changing my name to Charlotte 50 Chomolungma as some kind of a gesture.

It is only recently, after several trips to India, that I have again become interested in my ancestor's activities and come to appreciate the sheer scale of what he achieved. He had successfully mapped the 55 vast and varied terrain that is the Indian subcontinent, probably braving flood and fever, traversing hill and jungle, and surviving tiger, snake and scorpions along the way. Most of what is known about Everest, the man, however, is somewhat sparse and 60 mundane. After 25 years of map-making, he left India and settled down to married life in England. But when I read that Park Estate, his home in India still existed, albeit in ruins, I felt compelled to visit.

What I found was a long, low, ruined mansion 65 offering spectacular views over the plains in one direction and a breathtaking panorama of snow-capped Himalayas in the other. That's when my guide had pointed out what he called 'Everest Peak' – just nearby. I stared at the hill overlooking Sir 70 George's garden. 'Does it have another name?' I asked cautiously. 'Maybe,' he replied, 'but here everyone calls it Everest Peak.' After lunch, I climbed to the top of the hill. I stood at the top thinking about Sir George in his old age, treading the pavements of 75 London beneath chilly grey skies, but dreaming of this wonderful spot.

When I turned and began to descend, I could scarcely believe my eyes. All across the plateau below me, soldiers in light khaki were spreading out 80 drawing boards and setting up theodolites. It was if I had conjured up ghosts. I went over to greet them and learnt that the survey branch of the Indian army had chosen that afternoon to arrive and update their maps of the area. They were friendly, but could cast 85 no further light on my ancestor. I couldn't help but notice, however, that Everest Peak was marked 'Hathipaon' on their maps.

Exam practice 2: Reading

1 From the first paragraph, we learn that the writer

A had always wanted to climb a peak such as this one.
B had been preparing to climb this peak for a long time.
C had brought various pieces of climbing equipment with her.
D had previously had no idea that there was a peak here to climb.

2 As a child, how did the writer feel about her name?

A embarrassed that it attracted so much attention
B proud to be named after such a prestigious place
C keen to explain her connection with a famous mountain
D disappointed when people didn't recognise its significance

3 The verb 'fizzled out' (line 45) suggests that

A people had lost interest in the issue of the name.
B the debate about the name had been settled amicably.
C open debate about the name was no longer permitted.
D people realised they had been wrong to oppose the name.

4 Why did the writer's attitude towards her ancestor change as she got older?

A She disapproved of his changing the name of a mountain.
B She realised that he'd lied about measuring the mountain.
C She found out about his involvement in a political event.
D She discovered that stories she'd been told were untrue.

5 What impresses the writer now about Sir George Everest?

A the size of the task he managed to complete
B the fund of stories circulating about him
C the long-term significance of his work
D the romantic nature of his life story

6 On top of Everest Peak, the writer felt

A renewed pride in her family's achievements.
B surprised that Sir George never returned there.
C relieved that her family is still remembered there.
D glad that her ancestor had lived to an old age.

7 What did the writer learn from the Indian mapmakers she meets?

A New names are being given to places in the area.
B The hill she had climbed has a number of names.
C Maps of the area are in the process of revision.
D Sir George's influence is still strongly felt locally.

Exam practice 2: Use of English

Paper 3 Part 2: open cloze

For questions **1–15**, read the text below and think of the word which best fits each gap. Use only **one** word in each gap. There is an example at the beginning **(0)**.

Example: **0** | A | S |

Doctors without borders

Médecins Sans Frontières (MSF), also known **(0)**.............. Doctors without Borders, is a non-governmental, non-profit-making charity which was **(1)**.............. up in 1971 by a small group of French doctors. They believed that all people, whatever **(2)**.............. race, religion or political beliefs, have a right **(3)**.............. medical care, and that the needs of the sick go beyond national borders. So, in **(4)**.............. event of an earthquake, a famine or a civil war, the charity, **(5)**.............. consists mainly **(6)**.............. doctors and health workers, aims to get a team into the country promptly and deliver emergency aid. Getting there quickly enough to **(7)**.............. a difference is the main priority, so medical kits are prepared **(8)**.............. advance and kept **(9)**.............. standby.

The charity, **(10)**.............. network extends over 18 countries, operates in some of the most remote and unstable parts of the world, and is determined to remain completely independent **(11)**.............. all political, economic or religious powers.

(12).............. sometimes attracts controversy is when a team returns from a mission and tries to **(13)**.............. the situation and its causes to the attention of the world's media or the United Nations. **(14)**.............. not always popular with national governments, the organisation's achievements **(15)**.............. recognised in 1999, when it won the Nobel Peace Prize.

Exam practice 2: Use of English

Paper 3 Part 4: gapped sentences

For questions **1–5**, think of **one** word only which can be used appropriately in all three sentences. Here is an example **(0)**.

Example:

0 Sarah is a good teacher and her students find her really .... to talk to.

Carl wasn't a fan of hard work and was always looking for an .... life.

It's .... to see why so many couples choose that lovely island for their honeymoon.

Example: 0 | E | A | S | Y |

1 After listening to both sides of the argument in the debate, the audience was left to .... their own conclusions from the evidence presented.

Felicity could see that her father was trying to .... her into a discussion about politics, so she changed the subject.

I was able to .... on some excellent research papers when I was doing that assignment on charity fundraising.

2 If you need to .... access to the building after hours, you have to find the caretaker.

You can .... great insight into the values of other cultures by studying their literature.

I can't see what Helen hoped to .... by telling everyone that secret.

3 At first, Frances was .... to understand that fame would bring problems as well as benefits.

Traders in the market said that business had been .... since the new shopping mall opened.

Trevor found the pace of life in the village rather .... compared to what he was used to.

4 Harry has just .... a role in a new soap opera on TV.

The best time to buy fish at the harbour is just after they've .... the day's catch.

Simona's flight .... in Turin rather than Milan, where the airport was closed because of fog.

5 Paul and Joanne have reached the .... in their relationship where they trust each other completely.

I can't see the .... in arguing with Gloria because she never listens to what you say to her.

The .... that the director was trying to make in the film was rather lost in the publicity given to some controversial scenes.

Exam practice 2: Use of English

Paper 3 Part 3: word formation

For questions **1–10**, read the text below. Use the word given in capitals at the end of some of the lines to form a word that fits in the gap in the same line. There is an example at the beginning **(0)**.

Example: **0** | D | O | C | U | M | E | N | T | A | R | Y |

Reality TV

For many years, the boundary between what is fiction and what is	
reality on television has been rather blurred. For example,	
(0) .... programmes in which we watch people as they	DOCUMENT
go about their **(1)** .... lives, either at work or at home, have	DAY
long been popular. Then there are the 'candid camera' style	
programmes in which ordinary people suddenly find themselves	
in strange and **(2)** .... situations and, supposedly	EXPECT
(3) .... that they are being filmed, react in an amusing way.	AWARE
That these programmes can be an **(4)** .... for the victims	EMBARRASS
goes without saying, yet people seem happy to take part in them.	
What was new about the concept of reality TV, as epitomised by	
Big Brother, was the idea that a group of **(5)** .... would	VOLUNTARY
imitate the challenges of real life whilst never moving outside the	
'house' or TV studio. How **(6)** .... people can be in this sort	SPONTANEITY
of situation is **(7)** .... , and why people rush to take part is	DEBATE
something of a mystery, given that most end up being	
(8) .... ridiculed in the media or rejected by the audience.	PUBLIC
When asked, these people talk about the challenge or the money,	
but they are not very **(9)**	CONVINCE
In truth, what makes them do it seems to be the **(10)** .... of	ATTRACT
achieving what's known as 'celebrity status' – their five minutes	
of fame.	

Exam practice 2: Writing

Paper 2 Part 1

You are a member of a voluntary organisation that arranges exchange visits between families in different countries aimed at encouraging international understanding.
The organisation recently held an Information Day for local families interested in joining the scheme, which you helped to arrange. You have been asked to write a report about the day for the next meeting of your local branch of the organisation.

Read the note below from the local manager of the organisation, the programme of activities for the Information Day and two extracts from feedback sheets filled in by people who came to the event. Then, **using the information appropriately**, write a report as outlined below.

The Information Day was fairly successful, but one or two things didn't go as well as we expected. We need to think about which activities went well and which didn't, with a view to putting on a better day next time. I'm attaching some comments from families who came to the event, and I'm sure you've got some ideas of your own. Could you write a report, summarising the feedback we've received and your own impressions, and including some ideas for how we could organise a better event next time.

Thanks

INFORMATION DAY FOR FAMILIES

10.30 Opening speech by local manager — Too early - people still arriving

11.00 Video diaries made by families who've taken part — Poor quality - check equipment?

11.45 Question-and-answer session with families who've taken part in the scheme — Why these two families? - one of them had lots of problems on their trip!

13.30 Lunch

14.30 The practical details (led by Admin. Manager) — Good - better earlier?

15.30 Closing speeches — Too long - people drifted away

Lovely day. We're really fired with enthusiasm for the scheme as a result of coming.

The question-and-answer session worried us a bit. We're not so sure about joining the scheme now. The video diaries were fun - but one would've been enough.

Now write your **report** for the next meeting, commenting on the positive and negative aspects of the Information Day and making suggestions for improvements (180–220 words).
You should use your own words as far as possible.

Exam practice 2: Listening

Paper 4 Part 3: multiple-choice questions

You will hear a radio interview with a woman called Megan Turner, who runs a company which makes ethically produced clothes. For questions **1–6**, choose the answer (**A**, **B**, **C** or **D**) which fits best according to what you hear.

1 Why did Megan decide to leave her job in London?

A She was persuaded to do so by an old friend.
B She realised that her real interest was in fashion.
C She lacked the time to develop her leisure pursuits.
D She had grown tired of her rather routine existence.

2 What gave Megan greatest satisfaction in the early days of *Ethically Me*?

A the chance to work closely with Minisha
B the fact that she could make her own decisions
C the opportunity to design her own range of clothes
D the knowledge that she was doing something worthwhile

3 What worried Megan most about setting up the business?

A risking her financial security
B having to ask people for help
C convincing people of her principles
D needing to find out about unfamiliar things

4 For Megan, the key aspect of 'ethical' production is

A behaving fairly towards the people who make the clothes.
B supporting charities in the areas where the clothes are made.
C being honest with the people who choose to buy the clothes.
D paying a fair price for the materials used in making the clothes.

5 What does Megan say about her suppliers?

A They are inspected by specialists in health and safety issues.
B They are ranked according to how 'ethical' they are.
C They are changed each year in order to maintain standards.
D They belong to a group which checks on them annually.

6 What disadvantage of running her own business does Megan admit to?

A She wishes she had more of a social life.
B She doesn't like the idea of employing staff.
C She finds she misses the company of colleagues.
D She can't get used to living without a regular salary.

Exam practice 3: Reading

Paper 1 Part 2: gapped text

You are going to read an article about zoos. Six paragraphs have been removed from the article. Choose from the paragraphs **A–G** the one which best fits each gap **(1–6)**. There is one extra paragraph which you do not need to use.

The last menageries

The departure of the elephants from London Zoo will mark a turning-point in the history of animals in captivity. Michael McCarthy reflects on centuries of tradition.

5 So farewell then, Dilberta, Mya and Layang-Layang. Cultural change is a gradual process, and only rarely can its meaning be crystallised by a single event. There can be no doubt, however, that such is the case with London Zoo losing its elephants.

1	

10 What is it about this statement that draws us up short, grabs at our feelings in some obscure way? To try and pin it down, imagine that the news was instead that the zoo was to lose, say, the lemurs or its hydraxes (the elephants' rat-sized closest relatives). Would we notice? 15 If we were not animal enthusiasts to a high degree, would we care? Hardly.

2	

The age-old and traditional version of the zoo is a menagerie, defined as a collection of wild beasts in captivity for exhibition. Know something? Its day is done, 20 and the departure of the elephants from the heart of London symbolises and marks its passing, powerfully, incontestably and finally. It's the exhibition bit that is finished as a justification for zoos all over the world.

3	

The conservation argument in favour of zoos is even 25 stronger, however, and in a world where extinction is becoming increasingly common, we need havens for our endangered species. Leading zoos all recognise now that cramped quarters and exploitation are no longer acceptable, and that conservation must be their rationale if zoos are to have a future. 30

4	

The capture of these animals and keeping them, rather than eating them – the original menageries – came with the kings and queens. The pharaohs of ancient Egypt were prominently addicted. Antelopes and similar animals are depicted wearing collars on Egyptian tomb 35 pictures, and most monarchs followed with their own collections.

5	

Other societies have always valued such creatures for their fascination, and throughout the Middle Ages, kings and emperors continued to create their menageries when 40 the animals could be found. In Britain, the great age of the zoo was throughout the first three-quarters of the 20th century, and zoos everywhere were enjoyed, often beloved places to take families. Zoos as menageries seemed great, but nobody ever asked how the animals 45 felt.

6	

London-loving columnist Simon Jenkins, of the city's daily newspaper the *Evening Standard*, is one such voice. 'Call yourself a zoo? Where's your elephants then?' was very much his argument this week. But spare 50 a moment to think of the creatures themselves. It is hard not to imagine that once the narrow and bare walls of the modernist elephant house, designed by the famous architect Sir Hugh Casson in the 1960s, have been left behind, the rolling green acres of Whipsnade Zoo will 55 come as a welcome change. Over indeed, the age of the menagerie. We might regret it; the animals surely won't.

Exam practice 3: Reading

A The reason for this has been the growth of concern for animal welfare in the last 25 years. The rise of the animal-rights movement has led people to look at these places from the animals' point of view, and dislike what they see. The signs of depression and deviant behaviour observed in many creatures in close confinement has been confirmed as real. Indeed, an animal welfare case can be made against having zoos at all.

B For 170 uninterrupted years, these noble beasts and their predecessors have curled their trunks and swished their tails under the endlessly fascinated gaze of millions of visitors. The Zoo has now announced that the remaining three on the cramped Regents Park site are moving to rural Whipsnade Zoo, which has nearly 20 times as much space. They will not be replaced.

C This has necessitated a big shift in attitude. Until recently, menageries and zoos were one and the same thing, and menageries have been with us for thousands of years. Their fascination seems to lie in two profound human emotions: firstly, the desire of great and powerful men to display dominance over great and powerful animals; and secondly, the residual awe we all feel in the presence of big wild beasts.

D It is immensely to London Zoo's credit that it is sacrificing three of its prime public attractions for the sake of their own welfare. There is no doubt that many visitors will be disappointed, and strong criticism has already been aired.

E Zoos first became popular in Europe in the 1400s, when explorers returned with strange creatures from the New World. Over the years, larger collections appeared, which were also centres of research. In 1907, the practice of displaying animals began, which vastly improved their conditions, as barred cages were replaced with larger, more natural enclosures.

F Yet something about the elephants, and their final departure, touches us all, and not just because of any childhood nostalgia. It's rather that this astonishingly shaped mixture of strength and gentleness is somehow at the heart of what many of us think a zoo is – a place of wonders to be gazed at. If you take the elephants out of it, is it really a zoo?

G The Ancient Romans kept wild animals for something more sinister than display: slaughter in the arena. Even the gentle giraffe was a victim of their bloodthirsty killing.

Exam practice 3: Use of English

Paper 3 Part 1: multiple-choice cloze

For questions **1–12**, read the text below and decide which answer (**A**, **B**, **C** or **D**) best fits each gap. There is an example at the beginning **(0)**.

Luxury on the high seas

The ship named *The World* is quite **(0)**....B..... one of the largest and most luxurious sailing the seas today and is to be seen **(1)**.............. off at top international events such as the Carnival in Rio and the Monaco Grand Prix as it makes its way round the world. Strikingly beautiful, the ship **(2)**.............. the best aspects of cruising with every home **(3)**.............. and the quiet intimate atmosphere of a **(4)**.............. exclusive country club.

This **(5)**.............. remarkable ship has 110 luxury apartments and 88 studio residences, some of which **(6)**.............. as their owners' permanent home, whilst others are hired out by the night. The ship is **(7)**.............. out with everything that the well-off retired person could need, including a library, a theatre, a full-sized tennis court and a golf driving range where residents can **(8)**.............. instruction from top professional golfers. One **(9)**.............. of the latter is golf balls made out of fish food, designed in response to the **(10)**.............. of environmentally conscious residents. For the gourmet on board, the ship has plenty to **(11)**.............., everything from top-class restaurants to casual eateries, although residents also have the option of **(12)**.............. themselves a snack in their own apartments.

0	**A** purely	**B** simply	**C** basically	**D** essentially
1	**A** dropping	**B** calling	**C** drifting	**D** stopping
2	**A** combines	**B** links	**C** ties	**D** attaches
3	**A** relaxation	**B** ease	**C** comfort	**D** relief
4	**A** deeply	**B** strongly	**C** vastly	**D** highly
5	**A** truly	**B** comprehensively	**C** fully	**D** abundantly
6	**A** perform	**B** serve	**C** provide	**D** represent
7	**A** supplied	**B** stocked	**C** equipped	**D** fitted
8	**A** receive	**B** accept	**C** retain	**D** collect
9	**A** issue	**B** character	**C** point	**D** feature
10	**A** troubles	**B** nerves	**C** concerns	**D** stresses
11	**A** attract	**B** offer	**C** satisfy	**D** tempt
12	**A** fixing	**B** fetching	**C** managing	**D** settling

Exam practice 3: Use of English

Paper 3 Part 5: key word transformations

For questions **1–8**, complete the second sentence so that it has a similar meaning to the first sentence, using the word given. **Do not change the word given.** You must use between **three** and **six** words including the word given. Here is an example **(0)**.

Example:

0 Do you think I could borrow your pen, please?

WONDERING

I .... I could borrow your pen, please?

Example: | 0 | W | A | S | | W | O | N | D | E | R | I | N | G | | I | F | | | |

1 I don't eat fast food if there is an alternative available.

NO

I only eat fast food .... available.

2 The aim of the meeting was to promote trade between the two countries.

HAD

The meeting .... trade between the two countries.

3 If we continue to burn fossil fuels, climate change will get worse.

STOP

Unless we .... climate change will get worse.

4 We will soon run short of water if it doesn't rain.

BE

There will .... if it doesn't rain.

5 'Why don't we share a taxi,' suggested Maria.

SHOULD

Maria .... a taxi.

6 I'd rather read a book than see a film.

PREFERABLE

For me, reading .... seeing a film.

7 'Don't forget to post my letter, Tom,' said Mary.

REMINDED

Mary .... letter.

8 You can choose to stay either in a hotel or in a guesthouse.

HAVE

You .... either in a hotel or in a guesthouse.

Exam practice 3: Writing

Paper 2 Part 1

You are studying at an international college in London. Like many of the students, you make use of the college's gym facilities which are managed by a private leisure company called Gymwise. After reading an article about the gym in the college magazine, your class conducted interviews and did a survey amongst students at the college. You have decided to write a letter to the editor of the newspaper, responding to the article, briefly summarising the information from the survey and presenting your conclusions.

Read the newspaper article and look at the chart below, together with the comments from some of the college students you interviewed. Then, **using the information appropriately**, write a letter as outlined below.

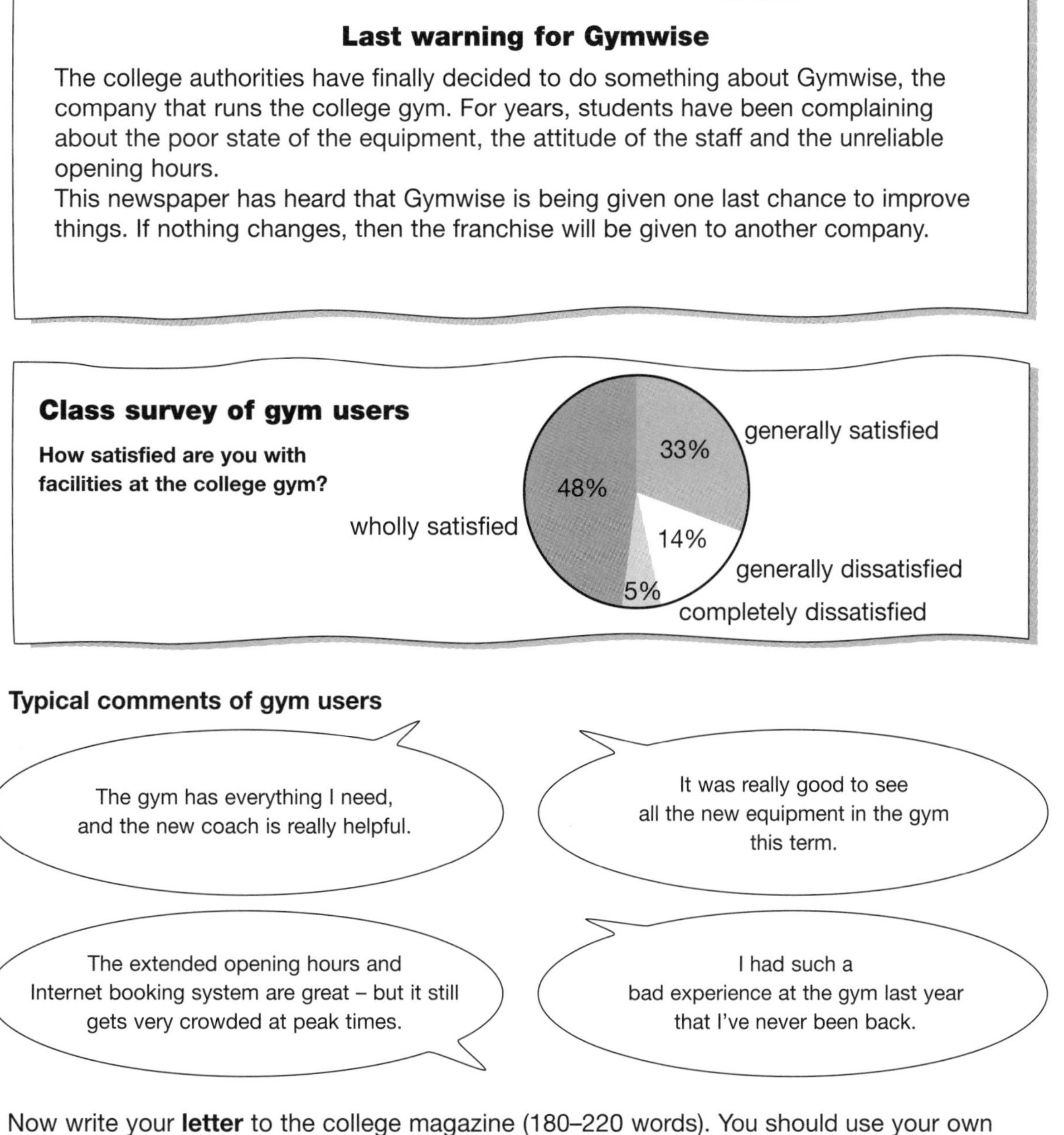

Last warning for Gymwise

The college authorities have finally decided to do something about Gymwise, the company that runs the college gym. For years, students have been complaining about the poor state of the equipment, the attitude of the staff and the unreliable opening hours.
This newspaper has heard that Gymwise is being given one last chance to improve things. If nothing changes, then the franchise will be given to another company.

Now write your **letter** to the college magazine (180–220 words). You should use your own words as far as possible. You do not need to include postal addresses.

Exam practice 3: Listening

Paper 4 Part 1: multiple choice

You will hear three different extracts. For questions **1–6**, choose the answer (**A**, **B** or **C**) which fits best according to what you hear. There are two questions for each extract.

Extract One

You hear part of a radio debate about wildlife film-making.

1 The two speakers agree that for viewers

A the photography may have become more interesting than the wildlife.
B the technology used in wildlife film-making has become too predictable.
C the technical advances in equipment are of no great consequence.

2 What is the woman suggesting about computer-generated images in wildlife films?

A They are only likely to appeal to children.
B People will generally realise that's what they are.
C Environmentalists could regard them as more desirable than photographs.

Extract Two

You hear part of an interview about a website called ARKive.

3 According to the interviewee, the idea for the website

A predated the technology to produce it.
B didn't originally come from scientists.
C wasn't initially popular with naturalists.

4 What view about the project does she express?

A It should be given more investment.
B Anybody should be able to use it for free.
C Endangered species should be given priority on it.

Extract Three

You hear part of a talk given to members of a wildlife group.

5 What is the male speaker describing?

A trying to prevent a new development being built
B investigating the effects of a new pipeline
C working for a group of conservationists

6 What aspect of her work does the woman like least?

A writing reports
B the unsocial hours
C physical contact with animals

Exam practice 4: Reading

Paper 1 Part 2: gapped text

You are going to read an article about mobile phones. Six paragraphs have been removed from the article. Choose from the paragraphs **A–G** the one which best fits each gap **(1–6)**. There is one extra paragraph which you do not need to use.

How the mobile conquered the world

Sadie Plant follows the invention that has changed the way societies work

In the early 1990s, the mobile phone didn't feature greatly in debates about the so-called 'communications revolution', but since then, it has leapt from obscurity towards ubiquity. Once considered a toy for the elite, today it has crossed social and geographical boundaries to find its way into the hands of the young, the old, the rich and the poor; even into communities largely untouched by other modern technologies.

1 ______

Students in Beijing, for example, explained the importance of maintaining contact with families in which they are the only children. Somali traders on dhows showed me how their mobiles allowed them to keep up with the movements of goods between Mogadishu and the Middle East.

2 ______

In spite of the high incidence of phone theft, this group values the security of knowing that assistance – often a lift home – is only a call away. They were also the first to see the potential of text messaging. More than two billion text messages were sent around the world in 2000 alone, and this traffic has since intensified, with an estimated eight billion messages in 2004, with much of this growth coming from users under the age of 25. In Japan, the teenage generation has become known as *oya yubi sedai*, the 'thumb tribe', on account of the dexterity with which they text, unaccountable to an older generation. But teenagers and text messaging are only part of the mobile story – so how do we account for the rest of it?

3 ______

The effects of this extend beyond these travellers; even people who go nowhere face new instabilities as traditional structures of employment, family, community and cultural life are disturbed by unprecedented movements of information, money and commodities.

4 ______

Even in the West, where phone ownership was once only available to those with a fixed address, a regular income and a large amount of cash, almost anyone can now buy a mobile phone off the shelf. The mobile also adds and answers to the more subtle senses of mobility which mark so many contemporary lives: the restless, non-committal feeling that all plans are contingent and might change at any time; an awareness that life is unpredictable and insecure.

5 ______

Does that sound like an exaggeration? Think about it – carried on the person, often all the time, a mobile phone is something to which people grow genuinely attached. It alters the experience of solitude, providing a stream of ways to fill dead time and constant reminders – not always welcome – that one is never quite alone.

6 ______

But it is in the developing world that the mobile phone's impact has been the most immediate. Bangladesh is one of several countries in which mobiles are used as public village telephones, sometimes powered by solar energy, and often offering access to the latest digital services. For whole communities, they offer access to a new world and new horizons.

Exam practice 4: Reading

A Indeed, in Thailand, many students told me that they could move south to Bangkok only when their parents were assured that they could keep in touch by mobile phone. Meanwhile, in the West, joggers in the park can be observed conducting their personal banking on a mobile handset.

B In the light of this phenomenal growth in usage, I was commissioned last year to study the sociological impact of the mobile phone through several regions of the world. I was amazed to see how fast, how far and with what diversity the mobile phone has spread. Because it extends a most basic human quality – the ability to communicate – there are few aspects of life that it fails to touch.

C Historians and sociologists see clues to this in a parallel development. The telephone arrived at the exact period when it was needed for the organisation of great cities and the unification of nations. The mobile, in its turn, arrived to suit a time of mobility. Unprecedented numbers of people are now on the move, whether as commuters, nomadic workers, backpackers, freelancers, exiles or migrants.

D Mobiles have changed the parameters of public space, too, blurring the edges of that private world. Visible and audible to all, their usage has rewritten many social rules about where, when and what one should communicate.

E Mobile phones encourage and respond to this. In China, which is witnessing vast migrations of people to the cities from the countryside, the mobile has become a crucial part of migrant life: a way to keep in touch with families back home and also a means of establishing oneself in a new social environment.

F Because it connects individuals rather than locations, the mobile phone alters people's expectations about such things; about what is possible and desirable, and changes the parameters of their social lives. It affects their perceptions of themselves, their boundaries and capacities, and becomes part of who they are.

G Meanwhile in Birmingham, England, adolescent girls convinced me that because mobile phones 'make it cool to talk', even their most taciturn male friends are becoming more communicative. Indeed, teenagers have become the conduits through which mobile phones have found their way into the wider society. For the young throughout the world, the sense of freedom of movement and the privacy afforded by the mobile are highly valued.

Exam practice 4: Use of English

Paper 3 Part 3: word formation

For questions **1–10**, read the text below. Use the word given in capitals at the end of some of the lines to form a word that fits in the gap in the same line. There is an example at the beginning **(0)**.

Example: | 0 | D | A | I | L | Y |

Can you be addicted to Coffee?

Until recently, I was a coffee addict, and coffee was part of	
my **(0)** .... life. Looking back, I can see I was utterly	DAY
(1) .... on a regular dose to get me through the day.	RELY
After the first cup at breakfast, I'd feel **(2)** .... and	VIGOR
ready to face the world; but unless I had another mid-	
morning, my energy level would slump and I'd become	
(3) .... Then one day my boss mentioned that he'd	IRRITATE
given up coffee. I nodded **(4)** .... , but the very	POLITE
thought of missing just one of my morning coffees made me	
feel **(5)** .... It suddenly occurred to me that I might	EASY
have become **(6)** .... on the stuff.	DEPEND
The definition of an addiction is 'taking a drug **(7)**	EXCESS
and being **(8)** .... to cease doing so without adverse	ABLE
effects'. If I was looking for evidence of the latter, it wasn't	
hard to find. Like many people, I sipped coffee all through	
the working week, only to find that at weekends when my	
(9) .... dropped, I ended up with a massive headache.	CONSUME
I tried vainly to give up on several occasions, but within two	
days was back on it again. **(10)** .... , I discovered that	EVENT
it takes a full couple of weeks to get the caffeine out of your	
system. But it's worth persevering, because providing you	
break the habit completely, the benefits are enormous.	

Exam practice 4: Use of English

Paper 3 Part 4: gapped sentences

For questions **1–5**, think of one word only which can be used appropriately in all three sentences. Here is an example **(0)**.

Example:

0 Sarah is a good teacher and her students find her really .... to talk to.

Carl wasn't a fan of hard work and was always looking for an .... life.

It's .... to see why so many couples choose that lovely island for their honeymoon.

Example: **0** | E | A | S | Y |

1 Ralph is always ready to .... his opinion about things, even when he is quite ignorant of the issues.

I may have to .... up guitar lessons as I don't really have enough time to practise.

We asked the teacher if he would .... us an indication of what topics we should revise for the exam.

2 Jennie's manager thought that she .... very well with one particularly awkward customer.

My uncle has .... in second-hand cars all his life.

The knee injury .... a severe blow to Tom's hopes of captaining the football team in the cup final.

3 The police are .... a lot of progress in the case of the missing work of art.

The top boy band of a decade ago is thinking of .... a comeback.

The prime minister is .... a speech in parliament today on the subject of crime prevention.

4 Let's .... a date for the next meeting of the committee.

I'm afraid my car's out of action until I find someone to .... a problem with the lights.

Tony tried unsuccessfully to .... the two pieces of metal together with superglue.

5 I've got a .... ache in my leg, which is uncomfortable but not really painful.

It was a .... day with lots of cloud and no sun whatsoever.

Sandra painted her walls in very .... colours, such as dirty pink and slate grey.

Exam practice 4: Writing

Paper 2 Part 1

You are studying at an international college. You are the student representative on the committee which organises adventure sports activities for the students. This year, students went to the Mulvane Outdoor Pursuits Centre for the first time, and the college principal has sent you a memo asking for a report on this trip.
Read the memo from the principal and a leaflet giving information about the centre, together with comments from some of the students who took part. Then, **using the information appropriately**, write your report according to the principal's instructions.

MEMO
To: Student Rep, Adventure Sports Committee
From: Niall Sanchez, Principal
Re: Mulvane Outdoor Pursuits Centre

We need to make sure that we're getting the right level of service and value for money from the Mulvane Centre. I need to know:

- the quality of the facilities and tuition;
- any problems;
- any suggestions for improvements on future trips.

Could you write me a report covering these three points, please?

Mulvane Outdoor Pursuits
Spring Break programme (college group)

The centre will provide:
- shared rooms (private bathroom);
- choice of activities including watersports and climbing;
- fully trained tutors;
- all equipment.

Some days, the only watersport available was white-water rafting. The water-skiing was great, but I only got two days of it.

The climbing tutor certainly knew his stuff, and the equipment was all top quality.

I didn't realise that 'shared room' meant four other people. A bathroom between five of us was hardly 'private'!

Now write your **report** for the principal (180–220 words). You should use your own words as far as possible.

Exam practice 4: Listening

Paper 4 Part 3: multiple choice

You hear part of an interview with Tom Westfield, who is an adventure sailor, and Alison Nunn, who works as his personal assistant or PA. For questions **1–6**, choose the answer (**A**, **B**, **C** or **D**) which fits best according to what you hear.

1 Tom feels that the main advantage of having a PA is that

A he can make better use of his own time.
B he has someone to discuss the work with.
C he can leave her to sort out any problems.
D he has been forced to develop new skills.

2 Alison says that when she first met Tom

A she was already familiar with his book.
B she was impressed by his charity work.
C she suggested that he might employ her.
D she'd recently given up working as a PA.

3 Why did Tom decide to sail around the world?

A He was determined to prove that he was fit enough.
B He thought it would boost his self-confidence.
C He hoped it would lead to further adventures.
D He'd lost interest in other extreme sports.

4 Why did Tom start giving talks on the subject of motivation?

A He feels he should pass his enthusiasm on to others.
B It was a way of getting finance for forthcoming trips.
C It was a chance to think through his own feelings about it.
D He realised that he had the power to inspire young people.

5 How does Alison see her role in Tom's next trip?

A keeping in touch with those supporting him
B standing by in case he needs her
C giving advice on the best route
D being the centre of operations

6 How is Tom feeling at this stage in the preparations for his next trip?

A frightened by the prospect of what might go wrong
B aware that he will need good luck in order to succeed
C worried about whether he's got the minor details right
D calm in the knowledge that everything's under control

Exam practice 5: Reading

Paper 1 Part 4: multiple matching

You are going to read a magazine article about five authors. For questions **1–15**, choose from the authors **(A–E)**. The authors may be chosen more than once.

A Louis de Bernières
B Hanif Kureishi
C Iain Banks
D A.L. Kennedy
E Helen Simpson

According to the article, which writer

has still not published a novel?	**1**	
has produced a remarkable quantity of material since the nomination?	**2**	
achieved great success shortly after receiving the nomination?	**3**	
was already well known in 1993 for writing material other than novels?	**4**	
finds it hard to survive financially on the income from writing?	**5**	
feels that life has not changed much since the nomination?	**6**	
has discovered that being famous has its drawbacks?	**7**	
finds that writing is limited by domestic constraints?	**8**	
didn't expect to be nominated for this type of award?	**9**	
enjoys the material rewards that writing can bring?	**10**	**11** .....
had previously been unaware of the existence of the award?	**12**	
was encouraged by the award to continue along the same path professionally?	**13**	
feels less emotionally committed to novel writing than in 1993?	**14**	**15** .....

Exam practice 5: Reading

Turning the page

Sally Wilson interviews five authors who were short-listed for a 'best young fiction writer' award in 1993, asking them what it meant to them at the time, and whether they feel they have fulfilled their promise.

When **Louis de Bernières** was nominated for the award in 1993 he observed, 'It was like having rockets attached to your back. It gave me a tremendous psychological boost.' At that time, de Bernières, an ex-teacher who didn't start writing until he was 35, was almost completely unknown. He had won an award for one of his three early novels, and each was praised for its 'comic brio', but sales were minimal. All this changed the following year with the publication of *Captain Corelli's Mandolin*, which has sold more than two million copies and been turned into a Hollywood blockbuster. De Bernières, who 'has a wonderful life, thanks to *Corelli*,' moved from a flat in London to a huge house in the country. But his celebrity also means that he now feels he is under a lot of pressure to prove that *Corelli* was not a lucky accident; the only story he has published since is a very short book, *Red Dog*. Now three-quarters of the way through his next novel, he finds that the 'fierce obsession' that kept him hunched over his computer into the early hours has gone.

In 1993, with the struggle to be a writer and start a career behind him, **Hanif Kureishi** had a far more significant achievement to celebrate than the nomination: the birth of his twin sons. Kureishi has been writing professionally since 1980, when he was a playwright at a leading London theatre, but it was his screenplays for films such as *My Beautiful Laundrette* that sealed his reputation in the 1980s. Although he hasn't 'made a substantial living' since the films, he proved he could also write fiction, and he has continued to write novels. He says these days he doesn't care so much. 'The pride and pleasure you get from your first novel are so great, now it's just not the same. But if my son wrote a novel, it would be much more important to me.'

'I thought, they must have got this wrong', remembers **Iain Banks**, of his nomination. 'I'm here with all these literary writers. I've always had this ambiguous status – am I literary or am I popular?' Either way, even in 1993 Banks was rich. Since his million-selling debut novel *The Wasp Factory* was published in 1984, Banks has gone from scribbling rejected science-fiction stories whilst doing dead-end office jobs to a basic annual salary of a quarter of a million pounds. Both populist and experimental, Banks has written an impressive total of eleven novels and eight science-fiction books, each taking only around three months to complete, although he has now slowed down to one book every two years. At 48, he spends his hefty royalties on expensive music-making technologies and fast cars. He's also managing director of an explosives company 'I guess, several years on, it's just more of the same – more money, more cars,' he says.

A.L. Kennedy, meanwhile, had 'never heard of the award, so I didn't know if it mattered or not.' At 27, she was the youngest of the nominees, with a soon-to-be-published first novel, written while working as a teacher. It bought her publicity, which was 'useful' and turned what was then a hobby into a job. She has written award-winning short stories, non-fiction and two novels, seems to provoke extreme reactions from critics and is still subsidising her books, having to write for newspapers to make ends meet. 'What you learn as a literary novelist is that you'll never earn a living out of your novels.' She finds writing prose fiction a 'slog' and extraordinarily isolating. Now, at 37, she is dogged by back problems, a legacy of days spent typing, and a few years ago with 'no life to speak of' was tempted to give it all up.

Helen Simpson, whose writing career took off in her early twenties when she won a magazine competition in the 1980s, remembers: 'I thought, this nomination will help keep things in print and I'll be allowed to write short stories as long as I want. The common wisdom is they don't sell.' Although a novel is yet to appear, her reputation is none the worse for that. In her collections of short stories, she has given a sharp poetic, funny spin to issues of pregnancy and motherhood, traditionally considered too dull to write about. Her stories are not directly autobiographical, but looking after her children has meant that each of her three books took five years to write. 'I get furious with myself for not managing more. I've got all these ideas. But I find it hard to write at the expense of other people. Trying to be a good mother and staying married 'takes up an awful lot of time'. Now the children are older, she thinks she should be able to work faster.

Exam practice 5: Use of English

Paper 3 Part 1: multiple-choice cloze

For questions **1–12**, read the text below and decide which answer (**A**, **B**, **C** or **D**) best fits each gap. There is an example at the beginning **(0)**.

Earworms

According to recent research, certain songs **(0)**....A..... in our minds because they create what's called a 'brain itch' which can only be 'scratched' by singing them. This kind of song, sometimes known as an 'earworm', has an upbeat melody and **(1)**.............. lyrics. Women tend to be most susceptible to earworms, and musicians more **(2)**............. to them than non-musicians. Interestingly, even the greatest musicians have suffered from earworms. Take Mozart for example; when his children **(3)**.............. to finish playing a tune on the piano, he would feel compelled to complete it for them.

It goes without **(4)**.............. that this research will be of **(5)**.............. interest to the pop-music industry which is always looking to boost **(6)**............. of CDs. One of the key **(7)**.............. of an earworm is its simplicity, since a song with lots of detailed content is not so easily assimilated by the brain. Earworms need to be **(8)**............. very quickly, so that people can reproduce them in **(9)**.............. while walking down the street, simply because they can't **(10)**............. them out of their heads.

Some experts argue that if you listen to such an infectious **(11)**.............. of music several times, it will go away – but others are not so **(12)**...............

0	**A** stick	**B** block	**C** hold	**D** trap
1	**A** returning	**B** constant	**C** compulsory	**D** repetitive
2	**A** favourable	**B** inclined	**C** receptive	**D** liable
3	**A** lacked	**B** missed	**C** skipped	**D** failed
4	**A** speaking	**B** saying	**C** telling	**D** talking
5	**A** exact	**B** particular	**C** proper	**D** typical
6	**A** sales	**B** markets	**C** deals	**D** outlets
7	**A** items	**B** issues	**C** features	**D** matters
8	**A** taken in	**B** caught on	**C** called up	**D** settled down
9	**A** full	**B** entirety	**C** whole	**D** complete
10	**A** send	**B** bring	**C** get	**D** have
11	**A** bit	**B** portion	**C** piece	**D** slice
12	**A** convinced	**B** influenced	**C** converted	**D** resolved

Exam practice 5: Use of English

Paper 3 Part 2: open cloze

For questions **1–15**, read the text below and think of the word which best fits each gap. Use only **one** word in each gap. There is an example at the beginning **(0)**.

Example: **0** | A | S |

Cirque Éloise

The circus has traditionally been regarded **(0)**.............. entertainment for children, with acts involving animals accounting **(1)**.............. a large proportion of the acts. These days, things have changed, however, and some circuses, **(2)**.............. as Cirque Éloize from Montreal, clearly aim to appeal more to adults. **(3)**.............. of the animals, Cirque Éloize offers a breathtaking mix of human athleticism, daring and strength. In **(4)**.............. to juggling, trapeze acts and bicycle stunts, all of **(5)**.............. are executed perfectly, the circus also has a clown. But this is a clown **(6)**.............. a difference.

(7).............. clowns in most other circuses, who simply tell jokes and play tricks, this one is able to walk on the high wire and juggle as well. What's **(8)**.............., he doesn't rely solely **(9)**.............. a big red nose and funny clothes to get laughs, but gives **(10)**.............. highly polished comic performance.

In common with other circuses, this one has a ringmaster firmly in control, but, unusually, he has a guitar slung **(11)**.............. his shoulder and barks only a few gruff words of dialogue. **(12)**.............. seen the show recently, I can testify that one of the best moments comes at the end when all the performers pile, one on top of the **(13)**.............., onto a single unicycle. **(14)**.............. they have settled into position, the stage is bathed in a gentle, golden light and **(15)**.............. is complete silence. It is a truly magical moment.

Exam practice 5: Use of English

Paper 3 Part 5: key word transformations

For questions **1–8**, complete the second sentence so that it has a similar meaning to the first sentence, using the word given. **Do not change the word given.** You must use between **three** and **six** words including the word given. Here is an example **(0)**.

Example:

0 Do you think I could borrow your pen, please?

WONDERING

I .... I could borrow your pen, please?

Example: | 0 | W A S | W O N D E R I N G | I F |

1 Stella is well-known for the attention she pays to detail.

FAMOUS

Stella .... paying attention to detail.

2 I wasn't able to enter the building because my teacher stopped me.

PREVENTED

My teacher .... the building.

3 This door must be kept closed at all times.

CIRCUMSTANCES

Under .... opened.

4 I used to spend a lot of time reading novels as a child.

FOREVER

As a child, ... reading novels.

5 The machine is harder to use than I thought it would be.

EASY

The machine is .... I thought it would be.

6 Ian managed to both lose his job and his girlfriend in one week.

ONLY

In one week, Ian .... , but he also lost his girlfriend.

7 Angela regrets getting angry with Peter.

WISHES

Angela .... angry with Peter.

8 Gina didn't know that the date of the meeting had been changed.

UNAWARE

Gina .... in the date of the meeting.

Exam practice 5: Writing

Paper 2 Part 2

Choose **one** of the following writing tasks. Your answer should follow exactly the instructions given. Write 220–260 words.

1 Your local English language club is running a competition to find the most interesting article to publish in its next newsletter. The theme of the article is: 'The best live music event I have ever been to'
Write an article describing the type of live music event you went to, what you particularly liked about it and why it is better than others you have attended.

Write your **article**.

2 You have received this letter from an international chain of fast-food restaurants. You visited one of their branches whilst on holiday in Britain.

Thank you very much for completing our customer feedback form. We are sorry that you were less than satisfied with certain aspects of your meal and the response of or staff. We would be grateful if you could provide further details about the following points you rated as 'unsatisfactory':
- speed of service
- attitude of staff
- quality of meal.

Write your **letter of complaint** to the restaurant.

3 You are planning to start an English-language film club at your college or workplace. Write an information sheet that explains:
- the aims of the club
- when and how often it will meet
- the range of films and other associated activities
- the advantages of becoming a club member.

Write your **information sheet**.

4 An international magazine has asked its readers to send in a review of two different English-language websites they would recommend, commenting on:
- ease of navigation
- range of information/products available
- any special features.

The review should also say which website would be most appropriate for people of all ages.

Write your **review**.

Exam practice 5: Listening

Paper 4 Part 4: multiple matching

You will hear five short extracts in which people who work as directors in the performing arts are talking about their work.

While you listen, you must complete both tasks.

TASK ONE

For questions **1–5**, choose from the list **A–H** what each speaker directs.

A feature film
B pop-music video
C documentary film
D radio comedy show
E classical music festival
F touring theatre company
G television drama series
H modern dance company

Speaker 1	1
Speaker 2	2
Speaker 3	3
Speaker 4	4
Speaker 5	5

TASK TWO

For questions **6–10**, Choose from the list **A–H** what advice each person gives to young people entering their profession.

A set your own goals
B keep on good terms with others
C don't compromise on quality
D don't expect financial security
E rely on your own judgment
F only work with the best people
G don't be put off too easily
H keep your opinions to yourself

Speaker 1	6
Speaker 2	7
Speaker 3	8
Speaker 4	9
Speaker 5	10

Audio Scripts

Module 1

Listening 1, page 11, Exercise 2a

Hi. Now, I know a lot of you are planning to go off to university or college next term, so I've been asked to talk about how studying there is different from studying at school. Well, one big difference is that you have to manage your own time; on average, only ten to 12 hours a week is actually timetabled teaching on a university course, the rest is up to you to organise. Another new challenge is that you'll have to set yourself learning objectives, and of course it's your responsibility to make sure that you meet them. So it's very different from school. There are various ways of approaching all this, however, and I'm going to go through some of them with you. I've also come along with a few tips I've picked up along the way which it may be helpful to pass on.

Listening 1, page 11, Exercise 3a

Firstly, make sure you know exactly what the course requirements are. You know, how many pieces of work you have to complete by when, and all that. One way of doing this is to get a diary or a wall planner – something that will help you set the year out visually – so that, as term progresses, you can see at a glance how you're getting on.
Then, it's also important to know what's expected of you in terms of the quality of the work. For instance, how to present your work and what you should include. It's a good idea to get hold of some previous students' work that has got high marks. Read it and think about why it was good. If you can't find any in the library, ask your course tutor. Don't be shy – these people are there to help you, and it's their job to make sure you know what you're supposed to be doing.
Finally, once you know exactly what is expected of you, you should start to set yourself deadlines and learning targets. These have to be realistic – try to do too much too soon, and you'll just be disappointed when you fall behind. So, think about things like: how many hours a week you're going to spend studying, or how you can best use the time available to meet your goals. I always find it useful to build in a safety margin – things will sometimes take longer than you think, and you don't want to end up feeling you're always under pressure to catch up.

Listening 1, page 11, Exercise 4a

N = Nick, A = Anne

N: So, what did you think of Rita's talk. Would you have found it useful?

A: Yeah. In my experience, most university students find they have about ten to 12 hours of timetabled teaching each week, and on top of this they will spend from 20 to 30 hours studying in their own time. So I think she had a point when she was talking about the wall planner – you need to get yourself organised, make the best of the available time – because it's also important not to overdo it. Time spent discussing issues with friends isn't necessarily wasted – you're learning key life skills, such as how to debate, think on your feet and use logical reasoning and the art of persuasion.

N: I agree that spending all day, every day with your nose stuck in a book or at a computer isn't necessarily your best use of time, but for me, the most relevant part was when she talked about *what* is expected of you. To my mind, to be a successful university student, you need to be open-minded – ready to try out new ways of thinking. You do your reading, you evaluate the information, and you come up with your own ideas and opinions about it. That's what really marks it out from what you're used to doing at school, and I think that was really what Rita helped them to grasp.

Writing 1, page 14, Exercise 1b

Yeah, the training was really hard, wasn't it? There was no messing about. Before Jamie'd let us get down to work in a real kitchen, we had to go through a demanding course at college – just to cover the basics. After that, he got us into work placements in some top-class restaurant kitchens to learn the hard way – in real life! To round off our training, we all cooked in Jamie's restaurant, as a team. Jamie said it was going to be one of the best places to eat in London, so we had to come up with the goods. Sure, we were shattered half the time, but what a fantastic experience!

Listening 2, page 15, Exercise 2

Speaker 1
When my husband was badly injured at work, I helped him win a court case against his employers. It struck me I could make a useful contribution to society by helping other people win law suits. I was in advertising at the time, had a good salary, but found it a rather superficial atmosphere to work in. With the family's backing, I decided to go back to college and study law. The course was great, but what I didn't realise was quite how much is expected of you as a lawyer. Since qualifying, I've ended up working all hours just to keep on top of things. Still, my colleagues are terrific, and in many ways it suits me very well.
Speaker 2
After five exciting years in the navy, I'd seen the world – but the actual work didn't stretch me. As I didn't seem to be in line for promotion, at the end of my contract I went for some vocational guidance. They said I'd make a good teacher, so I went to college. I felt a bit out of my depth there and very much under pressure – I almost gave up. But my wife was very supportive, and that saw me through. And it was worth it – the kids in my class are really demanding, but I love every minute. Colleagues complain about salary scales or the demands of the curriculum, but I can't get worked up about all that, I'm afraid.
Speaker 3
The farm had been in the family for over 100 years, but it wasn't large enough to remain commercially viable, and I had to give it up. Suddenly, I was not only without a regular income, but, um, I had time on my hands, too. Luckily, I'd always enjoyed a variety of activities such as t'ai chi and meditation, so I took the opportunity to train as a therapist. It was only once I'd qualified that I decided to take it up full time. I run sessions in five different centres on different days, so the commuting came as a bit of a shock after life on a farm. But, er, apart from that, it's been a very positive move for me.
Speaker 4
After a degree in natural sciences, I ended up in accountancy. It paid well, but wasn't the most exciting of jobs. I'd probably be still there today if I hadn't gone to stay with a friend who was working as a doctor in Ghana. One day I just happened to be present when she had to perform an emergency operation. I suddenly realised what I wanted to do with my life, and as soon as I got home from my travels, I signed up for a retraining scheme. I'm now in my second year as a nurse, and although I can't afford to do half the things I used to do, which is a pain, I'm happier than I've ever been.
Speaker 5
I was in computer sales, but also a member of an amateur drama club. Once you've been on stage, you long to act professionally – but my boss got fed up with me taking time off for auditions, so I wasn't actively looking. Then, out of the blue, I was offered a part in this play. It was too good to be true. The only drawback was having to give up my regular job for a six-month contract. My parents were a bit alarmed, but they soon calmed down once they'd seen the show. Um, we're on tour at the moment, which can be tiring, but I was always on the road in sales, so I can hardly complain about that.

Speaking, page 16, Exercise 1d

Words with two syllables:
friendly, friendly
patient, patient
tactful, tactful
Words with three syllables:
assertive, assertive
creative, creative
decisive, decisive
efficient, efficient
fair-minded, fair-minded
flexible, flexible
persistent, persistent
sensible, sensible
sensitive, sensitive
tolerant, tolerant
Words with four syllables:
energetic, energetic
gregarious, gregarious
resilient, resilient

Speaking, page 17, Exercise 3b

In this part of the test, I'm going to give each of you three pictures. I'd like you to talk about them on your own for about a minute, and also to answer a question briefly about your partner's pictures.
Robert, it's your turn first. I'd like you to compare and contrast these three photos, saying what personal qualities these jobs would require and why. Don't forget, you have about one minute for this..

Speaking, page 17, Exercises 3c and 4a

Right ... Well, these photos show a teacher, a doctor and a hairdresser. Looking first at the teacher and the doctor, I think these jobs are similar because they both require sensitivity and tact. To do either of these jobs, you must need a lot of patience. I suspect neither of these people would survive without a good sense of humour because their work must get quite stressful at times. In my opinion, these jobs are both more important than being a hairdresser, but the hairdresser must also make people feel good about themselves and be patient if they have difficult customers!
The most obvious difference between the jobs is that a doctor needs a lot of knowledge and technical skills, whereas for a teacher of young children, I imagine the most important thing would be creativity and energy. A hairdresser is somewhere in the middle. You would also have to be creative to be a hairdresser but also have technical skills, although I think they are easier skills to learn than a doctor's. And I suppose being a good listener might not be quite as essential for a teacher as for a doctor. To be a hairdresser, you must listen a lot too, or you might give people haircuts that they don't like!

Module 2

Listening 1, page 27, Exercise 2

The topic of my talk today is superstition. People have always been superstitious. In the days before science and education, people looked for simple explanations for things happening which they didn't understand and ... and couldn't influence. But as human knowledge has increased, you'd think that there would be less need for superstition – surprisingly, however, superstitious behaviour seems, if anything, to be on the increase. For example, a recent survey in the USA established that only 25 per cent of people there regard themselves as superstitious, yet 75 per cent of those questioned admitted to possessing a good-luck charm of some kind. So, why is superstition still so prevalent?

Listening 1, page 27, Exercises 3b and 4

Section 1
The main explanation for this seems to be that, although popular superstitions may seem like senseless rituals, most of them have a long history in cultural beliefs or ... or religion. It's difficult for people to throw off the influence of such deep-rooted ideas on their lives. In other words, no matter how sceptical they are, they hold on to the lucky charm or ritual rather than running the risk of inviting bad luck. That's why people in some cultures throw salt over their shoulder if they spill it, or touch wood when mentioning the possibility of tragic events. It's also why some people cross their fingers when hoping for good fortune, and why in some languages there is a set phrase which you must use in reply when someone wishes you good luck.
Section 2
Another reason is what we might call social tradition. An example of this is when British people say the words 'Bless you' when somebody sneezes. Similar ... similar traditions exist in most European languages, but where do they originally come from? Well, centuries ago, people believed that sneezing was a sign of dangerous ill-health, so saying kind words to the sneezer was meant to help that person combat the illness. People no longer really believe that their words will help, but the custom persists as a politeness.
Section 3
What is hardest to understand, however, is the power of superstition even when it's obviously unsuccessful. If your team doesn't win, what good did the lucky mascot do? The answer seems to lie in the fact that humans tend to hope for the best, no matter how irrational that appears. So, if things go well, it reinforces our belief in our mascot or ritual; if they go badly, we generally blame ourselves – perhaps ... perhaps we didn't perform the ritual in the right way.
Section 4
So what kind of people are most superstitious? Although lots of ordinary people have their own personal superstitions, such as using a lucky pen for exams, it is probably people working under stressful conditions who are the most superstitious. Sportspeople and actors are a case in point. Superstitions for them often revolve around fixed routines, most noticeably with food and clothes. One famous actor always eats fish, for instance, before an important performance; a top footballer always puts his socks on in the same order – that kind of thing.
Section 5
Finally, we must ask what benefits people from following these rituals. Most sociologists or anthropologists agree that when people are anxious or under stress, they feel out of control, and anything that helps them feel more in control – doing something active rather than just being in the hands of fate – helps them to cope. So, from that point of view, superstition can be seen as a positive force.

Listening 2, page 31, Exercise 2

Good morning. Although I didn't myself believe in the existence of luck, I started doing research into the subject over ten years ago. I was keen to investigate why it is that some people believe themselves to be lucky in life and others don't, and indeed whether it's possible to have any control over how lucky one is.
The first thing I did was to recruit some volunteers. I was interested in studying people who actually believed themselves to be either lucky or unlucky, and so I set out to find them. I did a few radio interviews and advertised on a student noticeboard and in the local press, but without much success – so, in the end, I put an advertisement in a national magazine. Around 400 people responded, and over the years, these extraordinarily patient men and women voluntarily completed questionnaires and kept diaries, as well as participating in various experiments and tests.
So, what did I learn from the research? Well, most people seem to have very little idea as to why they're lucky or unlucky. However, I've now come round to the view that it's the way people think and behave at any given time which is mainly responsible for what happens to them – rather than, for example, any particular intelligence or psychic ability.
In my first experiment, I put together a group containing both 'lucky' and 'unlucky' people – *their* definitions, remember – and gave each person a newspaper. I told them to ignore the articles, but to look through the pages and tell me how many pictures were inside. On average, this task took the unlucky people about two minutes; but the lucky people only took a few seconds. Why? Because in huge letters on page two of the newspaper was a message which took up half the page and said, 'Stop counting – there are 43 in total'. The interesting thing was that although it was staring everyone straight in the face, the

unlucky people tended not to notice it.
So why was this? Well, personality tests revealed that the unlucky people were more nervous types compared to the lucky ones, which maybe prevented them from spotting unexpected opportunities. Lucky people, being more relaxed by nature, tend to see what is there – rather than just what they're looking for.
I had discovered the first and most important principle of luck: the ability to notice and act on chance opportunities. The second, I soon realised, was that making lucky decisions isn't just about approaching a decision logically. It was clear that the luckier people were also following their instincts when they weren't absolutely sure which path to follow.
I've also noticed that 'unlucky' people tend to be downcast by their misfortunes, and so come to expect more bad luck in the future, whereas 'lucky' people have a positive outlook on life and remain positive when things go wrong – imagining instead how things could have been worse.
So perhaps luck does really exist, and our attitude to life is the key. In my book, I point out how lucky people …

Speaking, page 32, Exercise 1

E = Examiner, S = Student

1

E: How long have you been studying English?
S: I've been studying English for six years.
E: What kind of music do you enjoy listening to?
S: I enjoy listening to all modern music.

2

E: What do you like doing in your leisure time?
S: Well, what I enjoy most is playing football. I play regularly for the local team, and in fact we've got a match coming up this weekend.
E: How important is sport to you?
S: Oh, very. Apart from football, I only play squash myself, but I really enjoy watching most sports on TV – as well as football, I watch rugby, tennis and even snooker!

Speaking, page 33, Exercise 3a

En = Enrico, C = Cécile, Ex = Examiner

Ex: First of all, we'd like to know a little about you. Cécile, where do you come from?
C: I was born in France 19 years ago.
Ex: And you, Enrico?
En: Well, originally from a little village in the north-west of Brazil, though I've been living in Portugal for the last ten years.
Ex: Thank you. And could you tell me how long you've both been studying English? Enrico?
En: Well, I started learning English at school when I was about eight, but I've been coming to this language school for the past four years.
Ex: And you, Cécile?
C: I have studied English since 1998.
Ex: Thank you both very much.

Speaking, page 33, Exercise 3b

En = Enrico, C = Cécile, Ex = Examiner

Ex: Cécile, what are your earliest memories of school?
C: Er … mm, er, I was six when I started. It was a very small school and I cried on the first day because I had no friends.
Ex: And you, Enrico?
En: I remember taking a toy with me, and refusing to let it go. I used to keep it on my desk, and I'd scream if anyone tried to move it.
Ex: And who do you think has had the most influence on your life so far?
En: Er … that's a tricky question … it might … I'll have to think about that … it might be my older brother, Paulo – I used to look up to him because he was four years older than me, and we're still very close.
Ex: And what about you, Cécile?
C: I don't know …
Ex: OK. What do you hope to achieve in the future?
C: To pass this exam!

Exam practice 1

Listening, Paper 4, Part 2

I've come along this evening to give you an update on the research that we've been doing in my university into the levels of superstition you find in the United Kingdom today. We called our research the National Superstition Survey, and it was timed to coincide with the event known as National Science Week.
Although most of our findings relate to information we gathered via a questionnaire and through a series of interviews which we conducted in that week, we have continued to collect data via our website, and we'll be carrying on with that in the future.
So, what form did the research take? Our aim was to ask around 2,000 people across the country about their beliefs and attitudes toward superstitions. In the end, a total of 2,068 people actually took part. We asked these respondents – that's what we call the people we asked – to rate just *how* superstitious they thought they were. And we found that 77% said they were at least a little superstitious, while 42% admitted to being very superstitious.
We also asked these people whether they ever did anything superstitious, um, like saying words and phrases to bring them luck, or doing things to avoid bad luck. Twenty-eight per cent of people said that they carried lucky charms of one sort or another, and 26% felt that the number 13 was either lucky or unlucky. By far the most common superstitious saying across the country was 'Touch wood', an expression used by 74% of respondents to ward off bad luck, whilst around 65% used the expression 'Fingers crossed' when they were hoping for good fortune in the future. Interestingly, both sayings were far more commonly used by women than by men – or at least that's what they told us!
When we asked about objects associated with superstitions, we expected to hear about things like black cats and mirrors that people often associate with good or bad luck, but in fact, although 39% of the respondents did mention mirrors, the objects most often mentioned in the context of bad luck were ladders. We did find that this varied in different parts of the country, though.
Some of the results of the survey came as a bit of a surprise to us. For example, we asked all our respondents whether or not they had a background in science. We thought that people like that would be much less superstitious than people involved in the arts or commerce. But this wasn't, in fact, the case.
We also expected to find superstitions being kept alive by parents and grandparents; that young people would be more rational. In fact, our findings showed that people actually get less superstitious with age, and rather than the over-50s, it was teenagers who were very much keeping the traditions alive.
So, what were our conclusions? The range of superstitious behaviour we observed led us to conclude that superstitions are constantly developing and evolving, and so there's no reason to think they'll be declining or dying out in the near future.
If you'd like to know more about common superstitious phrases …

Module 3

Vocabulary, page 42, Exercise 1c

scrutinise, scrutiny
arrogant, arrogance
rude, rudeness
inconvenient, inconvenience
attend, attention
devote, devotion
imagine, imagination
neutral, neutrality
innocent, innocence

Listening 1, page 43, Exercise 2b

I = Interviewer, C = Charlie

I: And next on Celebrity Watch, we have Charlie Lane, who collects autographs of famous people – he'll snap up signed photographs, letters, anything with a signature on. And it's not just a hobby, it's how he makes a living. Charlie, um, what started you off as an autograph collector?

C: Well, I've been collecting something or other ever since I can remember. First, it was plastic dinosaurs, then a bit later my friends were all into model aeroplanes, so I started collecting them, too. Then when I was 11, for some reason my dad gave me a signed photograph of the film star Cary Grant. I wasn't a particular fan of his, but I realized that just the fact that I had this one signature made people – even grown-ups like my parents' friends – look up to me somehow. So I started building up a collection.

I: So you took it all quite seriously, right from the beginning?

C: Well, it wasn't that hard. I'd find the names of famous actors and writers in the library, and just fire off letters to them, asking for signed photographs. I had it all worked out. I'd use the same letter over and over again, just changing a couple of details to suit the person. Today, looking back, it makes me blush when I think of the sort of thing I used to put in them – grovelling statements like, 'I think all your work is fantastic. Nothing would make me happier than to have your autograph.'

I: So, did you do it all through letters?

C: No, by the time I was 12 or so, I was hanging about outside stage doors waiting to ask actors for their autographs. I once waylaid Ray Charles, the jazz musician – he had to pretend he couldn't write to get away from me.

I: Would you say you were obsessed by your collection?

C: Well, I didn't think of it in that way … I mean, it was just a hobby. All my friends had hobbies of some sort – boys do at that age. But then I discovered it was possible to buy autographs at sales and auctions. I would just turn up, and once I realised there was a financial angle to it all, collecting became that much more entertaining. But I must have come across as an odd little kid in the middle of all those professional collectors, because it's all a very retro culture, very backward-looking, you're often dealing with the autographs of film stars and musicians from the 1930s and 40s.

I: So, by the time you left school, and went to university, you were already not just a collector but a trader? Didn't that set you apart from the other students a bit?

C: Well, I was only buying and selling the odd thing, making a bit of money here and there. But I was studying music, and I thought that was where my future lay.

I: Mm-hm. What happened to make you change your mind?

C: Basically, I did a big concert which turned out to be an absolute disaster. That was more or less the end of music for me, at least as a performer. But, actually, I never really chose autographs, I just kind of fell into it because I knew enough to buy and sell, and I suppose I'm a natural businessman. I see it more as a way of surviving than as a sensible career move. I don't feel like I have a job … it's just my life, I mean, I can justify a trip to Paris by visiting second-hand bookshops … or to New York to go to an auction. And I still get excited – reverential almost – about some of the things I handle.

I: It sounds like a nice life.

C: Well, I like the fact that when I wake up and think I don't want to do this today, I don't have to. But I suppose, actually, yes, it's not a bad life.

Listening 1, page 43, Exercise 3a

I = Interviewer, C = Charlie

I: And next on Celebrity Watch, we have Charlie Lane, who collects autographs of famous people – he'll snap up signed photographs, letters, anything with a signature on. And it's not just a hobby, it's how he makes a living. Charlie, um, what started you off as an autograph collector?

C: Well, I've been collecting something or other ever since I can remember. First, it was plastic dinosaurs, then a bit later my friends were all into model aeroplanes, so I started collecting them, too. Then when I was 11, for some reason my dad gave me a signed photograph of the film star Cary Grant. I wasn't a particular fan of his, but I realized that just the fact that I had this one signature made people – even grown-ups like my parents' friends – look up to me somehow. So I started building up a collection.

Listening 1, page 43, Exercise 4a

I = Interviewer, C = Charlie

Section 2

I: So you took it all quite seriously, right from the beginning?

C: Well, it wasn't that hard. I'd find the names of famous actors and writers in the library, and just fire off letters to them, asking for signed photographs. I had it all worked out. I'd use the same letter over and over again, just changing a couple of details to suit the person. Today, looking back, it makes me blush when I think of the sort of thing I used to put in them – grovelling statements like, 'I think all your work is fantastic. Nothing would make me happier than to have your autograph.'

I: So did you do it all through letters?

C: No, by the time I was 12 or so, I was hanging about outside stage doors waiting to ask actors for their autographs. I once waylaid Ray Charles, the jazz musician – he had to pretend he couldn't write to get away from me.

Section 3

I: Would you say you were obsessed by your collection?

C: Well, I didn't think of it in that way … I mean, it was just a hobby. All my friends had hobbies of some sort – boys do at that age. But then I discovered it was possible to buy autographs at sales and auctions. I would just turn up, and once I realized there was a financial angle to it all, collecting became that much more entertaining. But I must have come across as an odd little kid in the middle of all those professional collectors, because it's all a very retro culture, very backward-looking, you're often dealing with the autographs of film stars and musicians from the 1930s and 40s.

Section 4

I: So, by the time you left school, and went to university, you were already not just a collector but a trader? Didn't that set you apart from the other students a bit?

C: Well, I was only buying and selling the odd thing, making a bit of money here and there. But I was studying music, and I thought that was where my future lay.

I: Mm-hm. What happened to make you change your mind?

C: Basically I did a big concert which turned out to be an absolute disaster. That was more or less the end of music for me, at least as a performer. But actually I never really chose autographs, I just kind of fell into it because I knew enough to buy and sell, and I suppose I'm a natural businessman. I see it more as a way of surviving than as a sensible career move. I don't feel like I have a job … it's just my life, I mean, I can justify a trip to Paris by visiting second-hand bookshops … or to New York to go to an auction. And I still get excited – reverential almost – about some of the things I handle.

I: It sounds like a nice life.

C: Well, I like the fact that when I wake up and think I don't want to do this today, I don't have to. But I suppose, actually, yes, it's not a bad life.

Listening 2, page 47, Exercise 1

Extract 1

I = Interviewer, J = Jane

I: Jane, Animal Aid is one of the longest established animal rights groups in the world. What are its goals?

J: Our campaigners investigate laboratory experiments, factory farming and trading in wildlife, amongst other things, and protest peacefully against any form of animal abuse we come across, as we actively encourage a way of living in which no creature has to suffer needlessly. Our inquiries are usually undercover, sometimes using videos, and any evidence we obtain is often used by newspapers and television to bring these issues to the attention of the public.

I: And in what way do you think that you personally have made a difference?

J: Well, I started off as a fund-raiser, which is an essential job but not really up my street. I'm now working with the education department, which is an area I believe in passionately. As well as providing a range of resources for teachers and students to use, we also offer free talks to schools on a wide range of topics from humane research to horse-racing, and talk about simple ways we can all do something to help, whether it's by becoming a vegetarian or not hunting or whatever. It's a particularly worthwhile thing to do because ...

Extract 2

B = Barbara, J = John

B: If the government does agree to the airport having another runway, it'll just be a disaster. I mean, the place will go from being a small village to a really busy town. It'll have a really negative effect on the people who live nearby.

J: Like you, you mean?

B: Like me, yes. I know it's selfish, but just think of the noise. I mean, even so, I probably wouldn't have felt strongly enough myself – it was just that one of my friends talked me into keeping her company.

J: On the march?

B: Yes. Personally, I didn't – and still don't – think anything we do makes a great deal of difference – the government has already made up its mind. So I felt it was rather a pointless way to spend the day. Anyway, I went.

J: Did it go well?

B: The actual arrangements left a bit to be desired, I must say, and some of the speeches went on a bit but actually it was a good opportunity to make contact with other people who are interested in protecting the environment. Before that I'd had the impression they would all be students, but in fact the age range was quite wide.

Extract 3

I = Interviewer, H = Howard Canning

I: You will remember that yesterday we reported on Howard Canning, the man who climbed a mobile telephone mast to protest about the firm not offering services in the Welsh language. Today he's got his feet back on the ground and is with us in our Birmingham studio. Howard, what was all this actually about really?

H: Well, I don't kid myself that doing this kind of thing will change anything overnight but the object of the exercise was mainly to advertise my cause – which is that Welsh is still not the official language in Wales!

I: It's rather a strange way to make your point, though, isn't it?

H: I suppose it is rather unusual and not particularly sensible, either, especially for someone like me who hasn't got a head for heights. As it happens, the company turned a blind eye anyway so I got no publicity whatsoever, which rather destroys the objective!
But I still think that by fighting for the things you believe in, in your own part of the world, you can contribute to global justice.

Speaking, page 49, Exercise 3b

Now, I'd like you to discuss something between yourselves, but please speak so that we can hear you. These pictures illustrate different methods of showing your feelings about important issues. Talk to each other about the advantages and disadvantages of each method and then decide which one would be the most effective. You have about four minutes for this.

Speaking, page 49, Exercises 3c and 4a

S1 = Student 1, S2 = Student 2, E = Examiner

S1: OK, shall I start? Well, I've been on several demonstrations, and I think they can be very effective.

S2: Yes, providing there are enough people there. When people come from all over the country, that must have a huge impact. Of course, if not many people turn up, it's a bit of a let down.

S1: That's true, and besides that, I suppose it must be quite difficult to organise and co-ordinate a big demo.

S2: The great thing, though, is if you get enough people, it gets reported in the papers, so people sit up and take notice, don't they?

S1: Not only that, it puts pressure on the government.

S2: Mind you, if there's any trouble or fights, the publicity can go against you. And sometimes troublemakers join in just for the fun of it.

S1: Mm. Anyway, let's move on to petitions, shall we? I don't think they're used much, are they?

S2: Well, actually, they are. They wanted to close down the local library last year, and some people drew up a petition.

S1: I'm not sure anyone takes any notice of them, though, do they?

S2: As a matter of fact, they do if there are enough signatures. They can be used to put pressure on your local Member of Parliament, for a start. And as well as that, they're quite easy to draw up and organise.

S1: Mm, maybe. I think most people just cross the street if they see someone with a petition. Anyway, what about leaflets?

S2: The problem is that we get so much junk mail these days that leaflets can get lost. I'm afraid I tend to throw them in the bin if they come through my door. I shouldn't, I know.

S1: I do the same most of the time. Having said that, I think they can be effective, you know, if they're designed well, and eye catching, and if it's a good cause.

S2: Oh, we haven't talked about meetings yet ... But if lots of people write in about the same subject, people take notice. Anyway, what do you think? What's the best method?

S1: Well, personally, I think a mass demonstration or march has the most impact, don't you?

S2: Yes, I do. Mainly because of the publicity – it generates a lot of publicity.

E: Thank you. So, which method have you decided [FADE] would be the most effective?

Module 4

Listening 1, page 59, Exercise 2

H= Helena, T = Tom

H: In *Moral Issues*, we look at difficult choices or moral issues that ordinary people face in their lives. Tom Wilkins is general manager of a print company in the north of England. The company's been losing sales, and Tom has unwillingly decided that redundancies are inevitable.

T: The only way we can survive as a company is to cut costs, and at the same time we need to think about the structure of the company ... I think, in fact, we've become a little too top-heavy and we'd benefit from some organisational restructuring. We've got four sales departments – corporate printing, packaging, digital supplies and labels – each with its own sales manager, and what we've decided is to merge these into two larger departments. But what this will mean is that we're going to have to lose two of the sales managers and shift some of the others around to cope with the new structure. So what I have to decide is which two of these sales managers we have to make redundant, and it's a very hard decision, I'm afraid.

Listening 1, page 59, Exercise 3a

H = Helena, M = Mike, Jo = Joanne, Ja = Jason, C = Carol

H: One of the people who could be affected by Tom's decision is Mike Brant, sales manager for corporate printing.

M: I've been with Tom almost from the beginning ... I actually set up corporate printing sales, and I've built up the client base from nothing ... everyone who's anyone knows me in the business. And one thing I've learned over that time is that selling's all about building up good relationships. We've always had good sales in my department; the team's figures have taken a bit of a knocking in the last few years, but we're still 20 per cent above the rest. And you've got to take the rough with the smooth in this business, to be frank.

H: Joanne Farmer is in charge of packaging sales. She feels a little less confident about her chances.

Jo: Well, at my appraisal interview last year, Tom said he appreciated the fact that I am ready to try out new ideas, and I think I could turn this section of the company around if I had the chance. But the proposed reorganisation is a worry to me personally, yes. I only joined the company two years ago, and I feel in quite a vulnerable position because packaging's always been at the bottom of the league as regards sales – but I have to say, my sales team have done a fantastic job in the last year, they've really pulled together, and the figures have definitely perked up. I'd be sorry to go.

H: Jason Pearl was the last to join the digital supplies team, three years ago, and became manager of the team last year.

Ja: Well, I've not been doing the job very long – I joined straight from university, so if it's a 'last in first out' scenario, I'm definitely for the chop. I expect I could find another job, with my experience and qualifications, but I really like it here, even though it's a big change from student life. And things are just beginning to take off with digital supplies – it was a new area for the company when I took it on, and I've pretty well got it at my fingertips now, and I think my clients all recognise that; they're very knowledgeable themselves, so they want someone who can come out with the information they need.

H: The sales manager for the labels division is Carol Spalding, who has been with the company for over ten years.

C: Well, I've been in labels for over ten years now, and I've got all my systems in place, so the whole department just gets on with the job; it all works like clockwork, and everyone's happy. That's always the way, though – you just get something working smoothly, and then someone decides to turn it all upside down. I really dread the thought of what will happen with these changes. But I know Tom thinks we have to do something, though in fact the sales figures for labels have been holding their ground quite well, considering.

Listening 1, page 59, Exercise 4c

H= Helena, T = Tom

H: Tom didn't find the decision easy.

T: There's no easy way of making a choice like this. On a 'last in first out' basis, Jason was the obvious choice for redundancy, but he's a bright young man with a lot of expertise in digital supplies, which is our fastest-growing area, and he's got lots of enthusiasm, so we decided to hang on to him. We weren't quite so sure about Joanne; when we looked at her sales figures, there was a marked upward trend since she'd taken over, and she's shown a lot of initiative, but on the other hand, we weren't sure if she was ready to cope with a sudden increase in responsibility. Carol's fantastically organised, she doesn't like change, but once she's got used to the idea, she'll cope well with the extra responsibility. And then Mike ... he's been here forever, but his figures have been declining for years, and he doesn't really seem to be aware of the way the industry's heading. So, in the end, we decided to make Mike and Joanne redundant. But as I say, it was a tough decision, and I think it'll raise a few eyebrows.

Listening 2, page 63, Exercise 1

I = Interviewer, K = Kate

I: My guest today is the psychotherapist, Kate Holt, and I'm going to be asking Kate about personality. Kate, what I've never understood is how brothers and sisters can be so different. Surely, if siblings have the same upbringing, you'd expect them to be similar in personality?

K: Well, this is why some people are convinced their children are born with different characters already in place. Either that, or they must be under the sway of external influences such as school or friends. But you have to remember that no two children are ever going to be brought up in exactly the same way. Things change radically between the births of different children – a new child may be a much anticipated event or a big upheaval, and at different stages of their lives, parents may be more or less financially secure, emotionally content or whatever. My own view is that at the time of each birth, parents are in a new situation, which means that inevitably they relate to the new child in a very different way to previous ones – even if that's not their intention. And as a result, each child develops their own way of responding, which is what makes them who they are.

I: Presumably, the order a child comes in the family is also an important factor in determining personality. What about the first-born?

K: Well, parents are usually thrilled when the first child is born, so it's the focus for a lot of attention and parental aspiration. The tendency is for the child to identify strongly with the parents, taking the line of least resistance, and doing whatever they want. Being a conscientious student at school, if this is important to the parents, is a simple way of currying favour with them, although it doesn't necessarily lead to more affection – in fact, this is often given to the younger ones. The oldest child typically dislikes and avoids change and risk-taking – many establishment figures such as politicians turn out to be oldest children, it seems.

I: That's interesting. I imagine younger siblings are less likely to do as they're told, therefore?

K: Absolutely. Younger children are often very keen to stand out in some way from the sibling born immediately before them. In what's probably an effort to attract attention, they seem to cultivate different personality characteristics or skills. So, if the oldest child is academic, for instance, the second might go to great lengths to be creative or sporty or vice versa. Alternatively, if the first child is particularly close to one parent, say the mother, the second might identify more closely with the father. People like revolutionaries or radical thinkers have often tended to be youngest children – perhaps their strategy is to rebel against the entire rest of the family.

I: And what effect does it have on a child when a new sibling comes along?

K: Mm, an interesting question. It's obviously more of a shock to the system for the oldest child, who may do their best to dominate, in an attempt to keep things as they were before and preserve the status quo. But if the new child grabs all the attention, there is a risk that the older child will feel resentful at being pushed out and react angrily towards the parents. How great an impact a new brother or sister will have, though, is mainly related to the gap between them. Children aged four and over are far less affected by the birth of a sibling because they're better equipped to cope. Whereas a gap of less than two years between children can often cause problems, even leading to psychological problems in some cases.

I: In what ways are adult relationships affected by childhood?

K: I think emotions such as humour or anger are probably inherited from our parents, and we can do little about that, obviously. And all parents do, to some degree, offload baggage from their own childhood onto their children – again inevitably. But I firmly believe that what will determine the pattern of our future relationships is what happens to us between six months and around the age of five. At this age, children begin to adopt specific patterns of behaviour learned from the people who take care of them, and connections start to form in the brain. As a result of this, even as adults we continue to relate to people in the way we learned to as children. So, whether you are, for example, secure or trusting, or always expect the worst, will stem from this time.

I: Does this mean that, as a psychotherapist, you can do nothing to help an adult?

K: Fortunately, I believe I can. After all, the best thing you can do for your child is to understand yourself. My aim is to help people to become aware of why they think and behave in the way they do, in order that they can make any necessary changes themselves, and that's how I can really help. Unfortunately, people often keep things repressed because they're scared of confronting their feelings, so the process of uncovering can take years. But change is always possible if you can recognise that you are misinterpreting the signs – when you think everyone hates you, or whatever.

I: Kate, we'll have to stop there. Thank you very much.

Speaking, page 65, Exercise 3b

In this part of the test, I'm going to give each of you three pictures. I'd like you to talk about them on your own for about a minute, and also to answer a question briefly about your partner's pictures.
You will each have photos showing different relationships.
Teresa, it's your turn first. Here are your photos. Please let Alex see them. I'd like you to compare and contrast the photos, saying how important you think the relationships are and how they might change. Don't forget, you have about one minute for this.

Speaking, page 65, Exercises 3c and 4

Well, two of the photos show relationships between parents and children. One is a mother with a very young baby, and the second one is a father with a young child. The third photo is a young couple.
Er, in the photo of the mother with the baby there's definitely a feeling of tenderness, whereas in the picture with the father and the young son there's more excitement and fun – they're engaged in an activity that they're both enjoying. The last photo is slightly different as it shows a personal moment in a couple's relationship. They are holding hands and sitting close together, and they look very happy.
How might these relationships change? Well, obviously at some point the baby is going to become less dependent on its mother, and therefore that relationship is bound to change. And with the father and son, the child may well become interested in other things that the father's not particularly bothered about. So, if they don't have common interests, that could affect their closeness. For the couple, who knows how things might change? Anything could happen in the future for this couple!

Exam practice 2

Listening, Paper 4 Part 3

I = Interviewer, M = Megan Turner

I: My guest today is Megan Turner, who gave up her high-powered London job as a client-services manager in 2002 to set up her own business selling ethically produced clothes. Megan, welcome.

M: Hi.

I: What made you do it?

M: After years working with multinational corporations, I sat down and thought: 'Is this really what I want to do for the rest of my life?' Yes, I could afford to buy nice clothes and go on luxury holidays, but was I happy? In the company I worked for, I felt completely anonymous, just another cog in the wheel. Each morning, I'd put on my business suit and join the rush-hour crowds. And each morning, I wished I was walking in the opposite direction. I'd always wanted to set up my own company and, after meeting Minisha, my business partner, who shared, um, this dream, I decided to hand in my notice and go for it.

I: So you set up your own company, with your own money. It must have been exciting.

M: Yeah. We, er, ploughed our savings into our tiny business which we called *Ethically Me* – making an ethically produced range of clothes. We, er, hired a designer, Minisha sourced fabrics from around the world, and our first range of clothes came out the following spring. Um, I'd read a book about ethical production, and it felt good to invest in something I really believed in. But, er, to be honest, the greatest pleasure came from being my own boss, trusting my own judgement and making my own mind up about things, you know, being at the heart of the business.

I: You make it sound all very easy – like anyone could do it.

M: It's been an incredible learning curve for both of us. Of course, some days I felt like hiding under the duvet – and I've often broken out in a cold sweat when I've realised how quickly our savings were being gobbled up. But, um, we've been overwhelmed by the generosity of many helpful people. A photographer is taking the pictures for our catalogue for free, and a web company has offered to build our website for nothing, because they both believe in what we're doing.

I: So what does 'ethically produced' mean in terms of clothes?

M: Ah. I'd always been horrified by the stories of the poor working conditions, low pay and child labour used to make the clothes sold in some well-known shops. Well, our principle is one of fair and equitable treatment for all people associated with the company. That means customers as well as suppliers *and* the wider community in the places where the clothes are made – which is often in less-developed countries. We want to reinvest in the communities there to ensure those communities survive. I mean, the priority is fair and ethically sound trading policies, but, er, we also do things like giving 5% of our profits to charities, for example, and a significant proportion of our clothes are made from organic fabrics.

I: So how do your customers know what they're buying really is ethical?

M: Well, they can look at the labels. There's a system of what we call 'version numbers'. For example, if there's V1 on the label, this means that child labour hasn't been used; V5 on the label would mean environmentally friendly manufacturing practices, where we look at things like the dyes used to colour or bleach fabrics. Um, there are seven V-numbers in all. We enter into an agreement with each of our suppliers, wherever they are in the world, that they will keep to agreed health and safety standards – and they have to allow independent experts to go and check on these things. We renew these agreements each year, but only if our standards are being met.

I: So you have no regrets about starting the business?

M: There are sacrifices involved, and you need nerves of steel, but when I consider what I've achieved, I know it's been worth it. It's hard not having a regular pay cheque going into my bank account each month, but I've learnt to adapt. Instead of eating out with friends, I'll suggest a lunch at home, and the annual skiing trip has been replaced by walks in the countryside. I do get a bit lonely working from home – it's the only real drawback. But, er, hopefully, our company will take off, and one day we'll be able to employ staff and I'll have someone to chat to. However, I've vowed that they'll never end up feeling as I did. It's amazing to think that I'm in control of my own destiny – and that, whatever happens, is the most important thing a person can have.

I: Megan, thank you for joining us.

M: Thank you.

Module 5

Listening 1, page 75, Exercise 2a

When people are asked to think of names of big multinational companies, one that invariably comes up is Nike, the sports goods manufacturer. Originally, when it was started up in the USA in the 1960s, the company was called Blue Ribbon Sports … the name doesn't have quite the same impact, does it? But it wasn't until 1971 that the company changed its name to Nike, the name of the Greek goddess of victory, and their famous logo was invented by a graphic-design student in the same year. She got a fee of just $35 for designing it, and it's been used by Nike ever since.
Well, the company's grown hugely since then, and is currently the largest seller of sports clothing and equipment in the world. One of

their most important products is footwear, such as running shoes and football boots, and the factories which manufacture these tend to be located in developing countries, but Nike also has factories in Europe and North America, and including manufacturers, suppliers and retailers, they employ close to a million people worldwide.
However, it's not all been plain sailing. Nike is one of several global companies targeted by those campaigning against the negative effects that multinational companies can have on society. In Nike's case, they've been accused of violating human rights in their factories in developing countries, including, in some cases, the use of child labour, and there have also been complaints of poor working conditions in some of their factories.
Nike has responded to these accusations by putting into place a code of conduct for all of its suppliers. Among other things, this forbids the use of child labour, and lays down health and safety policies. It's displayed in each factory in the local language, so that every employee can read it, and it's also on their website. The company's also involved in various global community projects, including one in Indonesia known as 'Opportunity International', which provides loans of money to women in rural areas who want to set up small businesses of their own. There's still ongoing criticism of the company's policies, but the company claims that very few of its competitors have taken such rapid and effective measures to improve the situation for their employees and also to benefit the wider community.

Listening 1, page 75, Exercise 3a

When people are asked to think of names of big multinational companies, one that invariably comes up is Nike, the sports goods manufacturer. Originally, when it was started up in the USA in the 1960s, the company was called Blue Ribbon Sports … the name doesn't have quite the same impact, does it? But it wasn't until 1971 that the company changed its name to Nike, the name of the Greek goddess of victory, and their famous logo was invented by a graphic-design student in the same year. She got a fee of just $35 for designing it, and it's been used by Nike ever since.

Listening 1, page 75, Exercise 4a

Well, the company's grown hugely since then, and is currently the largest seller of sports clothing and equipment in the world. One of their most important products is footwear, such as running shoes and football boots, and the factories which manufacture this tend to be located in developing countries, but Nike also have factories in Europe and North America, and including manufacturers, suppliers and retailers, they employ close to a million people worldwide.
However, it's not all been plain sailing. Nike is one of several global companies targeted by those campaigning against the negative effects that multinational companies can have on society. In Nike's case, they've been accused of violating human rights in their factories in developing countries, including in some cases the use of child labour, and there have also been complaints of poor working conditions in some of their factories.
Nike has responded to these accusations by putting into place a code of conduct for all of its suppliers. Among other things, this forbids the use of child labour, and lays down health and safety policies. It's displayed in each factory in the local language, so that every employee can read it, and it's also on their website. The company's also involved in various global community projects, including one in Indonesia known as 'Opportunity International', which provides loans of money to women in rural areas who want to set up small businesses of their own. There's still ongoing criticism of the company's policies, but the company claims that very few of its competitors have taken such rapid and effective measures to improve the situation for their employees and also to benefit the wider community.

Listening 2, page 79, Exercise 2

I've chosen to talk about the history of Easter Island because it's a striking example of how human societies depend on their environment, and what happens if they destroy it.
The island's most famous features are, of course, its enormous stone statues. Over 500 years ago, the people of Easter Island constructed one of the world's most advanced societies of its time on this tiny Pacific island. But their very existence depended on the limited resources available locally, because the island is one of the most isolated anywhere on Earth – 1,000 miles away from its nearest inhabited neighbour.
The islanders lived in closely related tribes, each ruled by its own chief. The focus of social life were the stone platforms, called *ahu*, which each tribe built. Over 300 of these platforms were constructed in total, mainly around the coast. These were used for religious and ceremonial events, and were designed to put statues on. The statues were obviously sacred to the islanders, but the fact that they all face inwards away from the sea towards the fields and villages indicates that they were seen as offering protection, too, although against what we cannot be sure.
At first, the islanders had no problems finding food. Although they don't appear to have cultivated much in the way of grain crops, or kept many domesticated animals, they did grow potatoes, as well as catching fish and trapping birds. These three things, all of which were in plentiful supply, made up their staple diet. This left them ample time to carve the enormous statues which played such a big part in their lives.
Because of easy access to stone, the statues were always at least five metres high, but over the years, as inter-tribal competition increased, so did the number and size of the statues. It was this desire to outdo each other in building statues that proved to be the beginning of the end for the islanders.
The amazing thing is how these huge statues got from the quarry where they were carved to the stone platforms on which they were to sit. The only form of transport on the island, after all, was manpower. Well, the explanation is probably that the statues were transported on tree trunks, which acted as rollers. As a consequence, by the year 1600, the island was almost totally deforested.
By this date, many of the trees which once covered the island would already have been cut down as the population grew. People needed fuel for cooking and wood for the construction of houses and boats. But it was the incredible number of tree trunks needed to move over 800 massive statues that is thought to have spelled disaster.
The inability to erect more statues must have put an end to social and ceremonial life, but the absence of trees also led to soil erosion, so that plant and animal species became extinct, whilst people were unable to build boats they needed to go fishing, or to escape from the island. When the first Europeans arrived in 1722, they found the islanders inhabiting caves, on the very edge of starvation. Unfinished or toppled statues were to be seen everywhere.
The cultural demands of the islanders had proved too great for the limited resources available on the island, and when the environment was ruined, the society very quickly collapsed. The story serves as a grim warning to the modern world, as our own aspirations outstrip the available resources.

Speaking, page 81, Exercise 2b

Now, I'd like you to discuss something between yourselves, but please speak so that we can hear you. Here are some photos illustrating environmental problems in our world today. Talk to each other about the relative importance of each problem and then decide which is the most urgent one to address. You have about four minutes for this.

Speaking, page 81, Exercises 2c and 3a

S1 = Student 1, S2 = Student 2, E = Examiner

S1: OK, well, where shall we begin? Shall I start? Well, let's talk about this photo – it's about genetically modified crops. I've read that GM crops are difficult to control – they can spread and contaminate other crops – organic crops and so on.
S2: I don't like the idea at all, I must say, because it …
S1: And we don't know what long-term effect it might have on our health. The more we tamper with food, the more risks we run with our health. Having said that, I'm not that worried, really.
S2: Which problem *do* you find worrying, then?
S1: I think air pollution – all these toxic car fumes making us ill. Yet there are more and more roads being built every day.
S2: Mm. I do agree with you on that, and I don't think we're doing anywhere near enough to restrict traffic in our cities.
S1: I think people should be made to pay charges if they want to use their cars in the centre. I'm worried about global warming, too –

our summers seem to be getting hotter and hotter all the time. Look at the floods in this photo – there'll be more and more, as our climate changes.

S2: If we did more to develop alternative sources of energy, it would help, but to me the situation seems to be pretty much as it was years ago. Don't you think that we should …

S1: I don't think that the problem of waste is nearly as bad as they make out, though. We do seem to be getting better about disposing of our rubbish. A lot of it is recycled now.

S2: Yes, but wouldn't you agree we've just become a throwaway society? For example, look at …

S1: It's the issue of conservation of resources again in my opinion. It's all part of the same thing, isn't it? Just think of the way they're destroying the forests in some countries – entire species have already died out because of loss of habitat.

S2: Mm, yes, I suppose so … Anyway, we have to decide which problem we think is the most urgent to do something about. What do *you* think?

S1: I'd say air pollution is by far the most serious problem on a day-to-day basis. Definitely.

S2: Yes, and because of the long-term effect on the climate as well. Yes, I agree that this is probably worth fighting for.

E: Thank you. So which problem have you decided is the most urgent one to address?

Module 6

Listening 1, page 91, Exercises 2 and 3

M = Martin, S = Sandra

M: Oh I'm not sure I'll be doing that again! I was useless. Not to mention that I feel as if I've been run over by a bus.

S: But that's the whole point. It's not meant to be easy. You're supposed to feel as if you've worked hard. That's why we're doing it, remember? To raise our pulses and work our hearts, like the doctor said. Otherwise it wouldn't be worth our while coming! The problem with you is that you're the kind of person who buys an expensive gym subscription and then thinks you don't need to bother actually going. You've got to put a bit of time and effort into getting fit – it's not going to happen overnight.

M: I realise that, but I thought it could have been a bit more – I don't know – gentle, somehow. You didn't get much chance for a breather. It's the way you have to remember so many things at the same time – you know, which way to turn, what do to with your arms, when to change partners. I'm far too clumsy for this, if the truth be known – I kept tripping over my own feet.

S: You and me both – don't worry about it. That's probably the hardest thing – to think of everything at the same time. It's a bit like learning to drive, except we have to keep in time to the music as well. But the teacher was very patient, I thought. Anyway, I've not had such a good laugh for ages. I thought I might ask if she does any other classes, too.

M: Oh come off it! You're joking, I hope.

Listening 1, page 91, Exercise 4

M = Man, W = Woman

M: I don't deny that the idea of more organic farming is environmentally justifiable but it's still a fact that there is not yet sufficient public interest to make it a viable proposition in the market place.

W: So how come last year sales of organic food jumped by thirty per cent? Surely that's a clear indication that a growing number of consumers are waking up to the realisation that it is not only better for you, but it's tastier, too. In fact the industry is now worth £1.6 billion and Britain is the third highest consumer in the world – after the US and Germany.

M: Yes, but hang on – the fact remains that it's not 40% of our total farm produce – it's still only 4% and I would be wary of saying that 96% of what our farms produce is second best just because it's not organic. And given the fact that there's no conclusive evidence either way that it's nutritionally better for you than mass produced farm food, the issue has to be whether you personally believe it tastes better. It's a lifestyle choice and you have to be prepared to pay more for it.

W: You do, but then again you have to ask yourself why food produced by non-organic farming methods costs so much less. Organic food is more expensive because it takes longer to produce and is more labour intensive. Animals are kept in free-range conditions and fed natural diets, for a start, which must mean it's healthier, surely? And several recent medical reports have established a link between additives used in conventional farming and asthma and heart problems.

M: You might have a point there, but it's early days and it has yet to be proved ...

Listening 2, page 95, Exercise 1

Extract 1

L = Liz, C = Carl

L: What stands out in your mind about going to your first football match, Carl?

C: Er – well, there was all the pushing and shoving at the gates, and wondering what I'd do if I got separated from my dad. My dad was a bit worried about all the stuff being shouted during the game – you know, the er colourful language mainly directed towards the other side, or one of ours who missed a goal – he kept looking at people as if to say 'Hey, watch your tongue, there's a seven-year-old here' but he needn't have bothered because I wasn't really listening. I was just so mesmerised by the sheer volume of people there – like a sea of blue shirts that went on forever.

L: My first match was at Anfield, watching my brother's team, Liverpool. I wasn't that bothered about footie at the time but the atmosphere was just so overwhelming – all the chanting and singing – that I was in tears for most of the time. Tim thought I'd be fed up after ten minutes but the time went quite fast. I spent most of it just watching the supporters – all that passion!

C: I'm like that myself. I still get so worked up I always feel exhausted afterwards.

Extract 2

I = Interviewer, S = Steve

I: Your police force was the first in the country to start focusing on cold cases – re-opening investigations into crimes which had never been solved. Why was that?

S: Well, there'd just been a nasty attack, which we realised had strong similarities to one which we'd been unable to solve twenty years before. The case in 1987 had had a lot of publicity – feelings were running very high in the community because the victim was so young – but we'd had to give up on it, which was frustrating. But by using the advanced scientific techniques just becoming available we were able to match a single hair found on the scene all those years ago with one from our current investigation, and finally bring the offender to justice. Of course since then many other forces have followed our example but ours was the first really big operation in this country.

I: What was your initial reaction to being asked to work on the case?

S: Well, the thing is, I'd been rather hoping to spend my years coming up to retirement doing something relatively undemanding, like working as a community officer so I wasn't over the moon about it. But it wasn't entirely unexpected as I'd been the police officer who'd investigated the original attack and knowing a lot about the background already meant I had lots of advantages – I wasn't exactly starting from scratch.

I: And I suppose you could never really forget the case ...

Extract 3

P = Presenter, B = Becca

P: And next on the line we have Becca Thomas, who wants advice on a very thorny problem she has to deal with.

B: Hi, yes, good morning. My nine-year-old daughter has mentioned that there's a girl in her class who is picking on another child in the playground – you know, calling her names, telling other kids not to play with her and so on. My daughter obviously feels sorry for this child, although they're not special friends, but the difficult thing is that the mother of the child doing the bullying is a really good friend of mine and I lie awake all night wondering whether or not I should let her know.

P: It's a tricky one – she might feel you're judging her daughter just on the basis of what your daughter says, which she will probably not be too happy about. On the other hand, assuming what your daughter says is true, letting this bullying go on is not really an option. If you speak to the teacher without saying anything to your friend, you run the risk she'll find out you've gone behind her back – but one solution would be to encourage your daughter to tell the child being bullied to tell a teacher if there are any more problems ...

B: But if I did that, surely ...

Speaking, page 96, Exercise 3a

In this part of the test, I'm going to give each of you three pictures. I'd like you to talk about them on your own for about a minute, and also to answer a question briefly about your partner's pictures. Thomas, it's your turn first. I'd like you to compare and contrast these photos of museums – A, B and C – saying how successful these places might be in encouraging young people to develop an interest in the past. Don't forget, you have about one minute for this.

Speaking, page 96, Exercise 3b

S = Student, E = Examiner

S: Well, all the photos show children, they all appear to be totally engrossed in what they are looking at. In this one, the little boy is listening to someone talking about the, er, exhibit – it looks like a kind of gun – I think he's the guide, and he's dressed up in a uniform, so this brings it to life for the boy and makes it real. The dinosaur looks very realistic, which also brings it to life for the children in the second picture.
In this photo, the girls are actually doing something, rather than just looking. I think this kind of idea is becoming much more popular – where young people fill in worksheets or do hands-on kind of activities. It's fun, more involving. What I don't like about interactive exhibits, though, is that queues often build up, which is frustrating. I think a time limit should be set for each activity so that people don't have to wait too long.
It's not just the fact that it's interactive, though, that makes a museum interesting. It's not enough. The content of the museum is important, too – it has to be relevant to young people.
So I do think that places where young people can get involved or where the exhibits are presented in an interesting way can really inspire them to develop an interest in the past.

E: Thank you. Now, Elena, can you tell us which museum you would find most interesting?

Exam practice 3

Listening, Paper 4 Part 1

Extract One

M: I mean, let's face it, natural history filming hasn't in essence changed a lot in the past 50 years. All that's changed is the quality of the equipment; the approach remains much the same.

F: And yet more technological fixes seem to be the only way forward that film-makers are prepared to contemplate. But is that really the answer? Increasingly the technology itself becomes the source of interest rather than the subject of the film.

M: So you get people thinking: 'How did they film that?' rather than 'What's that animal doing?'

F: And if you ask me, whilst computer-generated dinosaurs may be fun for kids, how long is it before such technical temptations come closer to home? When air travel for a full film-crew begins to look both environmentally and financially questionable, will computer-generated snow leopards sneak in? Will we know? Would we care?

M: Well, I don't know about that ...

Extract Two

M: So where did the idea for ARKive come from?

F: Well, it all started in the 1980s, when people got talking at the film festival called Wildscreen, where conservationists and naturalists could come together once a year to see the latest wildlife films. What people realised was that most good wildlife footage was seen once or twice at conferences, and then consigned to a cupboard. If you were a scientist who wanted to see a film again or find some pictures, then there was no database to help you. Anyway, at that time the technology to build such a database cheaply and effectively didn't really exist, so at first all the talk came to nothing.

M: So what's changed?

F: Well, the need to preserve pictures and sound recordings of endangered species has become even more urgent. We've been fortunate enough to secure long-term investment in the project from a major wildlife charity. And the other thing, of course, is the technology itself. New developments, particularly broadband connections and digital storage systems, have turned the idea into reality. Because if visual material can be stored in digital form, and accessed easily via the Internet, it's not only preserved, but becomes freely and easily available to everyone. To my mind, that's the beauty of ARKive.

Extract Three

M: A new gas pipeline is being built locally and we've both been helping with lots of surveys of bats, small mammals and reptiles to see what wildlife will be disturbed by the construction of the pipeline. If our recommendations are followed, then certain areas – like species-rich woods or hedgerows – will be avoided by the contractors who are putting the pipeline in. Our role is to act as a consultant to those contractors rather than representing other interested parties such as conservation lobbies, but it's still worthwhile, isn't it, Debbie?

F: Oh yes – knowing that I make a difference to rare and endangered species is very satisfying. I don't really enjoy having to handle wildlife in a physical sense, but fortunately, that doesn't happen very often. The long hours can be hard as early mornings and late nights are part of the job if you're dealing with bats as well as birds. You need good animal identification skills, and the ability to write reports is certainly part of the job, as record-keeping is of the essence – but I don't mind that so much.

Module 7

Listening 1, page 107, Exercise 2

Today, we're looking at the life and work of musician Evelyn Glennie, the world-famous Scottish percussionist, who has been deaf since childhood.
When I was about 15, I went to see Evelyn Glennie in concert, and it had a great impact on me, not just because of her deafness and her undoubted virtuosity as a musician, but because of her originality. It was like nothing else I'd seen or heard. This was partly down to the kind of music she performs, which is a mixture of different traditions and cultures, and partly because of the style in which she plays. It's very unusual to be able to play so delicately one minute and then so forcefully the next. It really takes your breath away!
Evelyn is also renowned for being a great visual performer with great individuality. She wears very unusual clothes – nothing like the stereotypical musician in a suit or smart dress. Depending on what the music is, she might dress up as, say, an alien or a cat, and this, together with her unique choice of instruments, all adds to the excitement.
Evelyn plays all the normal percussion instruments – xylophone, cymbals and so on – as well as drums from all over the world, such as bongos and steel drums. But what I find particularly fascinating is when she uses everyday objects such as flowerpots and spoons, to come up with sounds which you've never heard before. Apparently, she owns about 1,500 instruments, and when she travels, she takes up to 600 different kinds with her at a time, playing about 60 in any one concert. It must cost a fortune to transport them all!
Evelyn became deaf at the age of 12, but she's never made a fuss about it and only took up percussion afterwards. The reason she always plays barefoot is so she can feel the vibrations of her instruments and the orchestra. But when she applied to music colleges after leaving school, she had 16 applications turned down – just because she was deaf. This didn't put her off, though, and eventually, at 17, she was taken on as a student by the Royal Academy of Music. Her achievements since then speak for themselves. She is a top

international musician and has transformed the role of percussive instruments within the conservative world of classical music, winning awards in the process. Not bad going for someone still under 40! On top of her hectic professional life, she's also in charge of a research centre which aims to educate people in music as a form of communication and recreation. She is associated with around 150 charities, which have the benefit of her help and support.
Recently Evelyn was voted 'Scotswoman of the Decade', and I suspect she's inspired lots of people like me to become musicians, not just because of her ability but because of the kind of person she is. She's a real role model, and if ever I think I can't do anything, I just think of her and all that she's achieved against the odds.

Listening 2, page 111, Exercise 2

Speaker 1
Until I got the bug, I was always perfectly happy lying on the sofa; pizza in one hand, TV remote in the other. Given the choice between an hour on my Playstation and a jog round the park, there's no doubt which I'd go for. But that all changed the day my friends all clubbed together and bought me a gym membership as a joke birthday present. It must've cost them a bit, so I thought I'd better go, and I sort of got sucked in. In some ways, I wish I hadn't, because it's taken over my life, really. I'm now working out five nights a week and, apart from my girlfriend, no one ever calls any more.
Speaker 2
It all started with my brother complaining about some soap opera he'd missed because I was watching a game show. I thought it'd be a quick way of finding out what had happened for him, but in no time at all, I was hooked. Three nights this week it's been well after midnight when I've torn myself away from the screen. My parents think I'm working on stuff from college, and I let them think that. I know it's wrong, but we get on really well, and I don't want them to know I've got a problem. I haven't even used my credit card yet, but one link just leads to another, and I've always loved window shopping.
Speaker 3
I'm not short of friends, but after a hard day at the office, I'm in no mood for socialising. I just collapse in front of the TV with a ready-meal on a tray. I know it's not clever, and I'm full of good intentions about watching my waistline and getting more exercise, though in fact I don't actually have many days off sick. My girlfriend's a real health freak who's always telling me I'm throwing money away on exactly the wrong kind of stuff. We had a big row about it yesterday, so I'll have to change – problem is, it's so easy ordering it all online from the local supermarket, and I just can't break the habit. I wish I could!
Speaker 4
I used to tell myself it was just something to do – you know, I'd meet up with friends, see what was new, then stop off at a burger bar or something. But then I realised I had a problem and I couldn't fool myself any longer – I'd feel quite cheated if I went home with nothing. It doesn't matter what it is – a book, a computer game, something to wear – it just gives me a thrill, even though it's temporary. Half the time, I don't need the stuff, which really infuriates my parents. They make me take things back and try and get my money back, though I never actually overspend, so that's not the issue.
Speaker 5
I sometimes become aware of people looking at me angrily, as if I'm disturbing them, particularly older people, who think it's anti-social – sort of thing my gran would say! But I just hate to feel out of touch with the office or the family while I'm away on business. But half the time, I'm just ringing for the sake of it. I know it sounds silly, like something out of a soap opera, but I always feel quite nervous if I'm out without it, and I keep checking that it's still in my pocket. It's becoming a real problem, actually, because my last bill was astronomical, and there's no way I can conceal that from my wife.

Speaking, page 113, Exercise 3a

In this part of the test, I'm going to give each of you three pictures. I'd like you to talk about them on your own for about a minute, and also to answer a question briefly about your partner's pictures.
Stella, it's your turn first. Here are your photos of important moments in people's lives. Please let Boris see them. I'd like you to compare and contrast the photos, saying what impact these events might have on these people's lives. Don't forget, you have about one minute for this.

Speaking, page 113, Exercises 3b and 4

Right, well, ... I think both having a baby and retiring from your job are crucial milestones in a person's life. So what impact could these events have? Well, with a new baby, the parents will suddenly have far less freedom – they won't be able to go out whenever they want and so on.
What about retirement? He doesn't look very happy – not everyone looks forward to retirement. You don't have a structure to your days any more, so you have to find other ways of filling your time. On the other hand, it's a great opportunity to do all those things you didn't have time to do before.
The third photo is a milestone too, but in a different way because winning the lottery is very unexpected, whereas at least a baby and retirement are things you can plan for! The couple look very happy. They can stop working and travel all over the world. But perhaps they might be worried too – maybe they think people will only like them for their money now.
But in general, I'd say all these events can have a very positive effect on people's lives.

Module 8

Vocabulary, page 122, Exercise 3b

materialistic, materialism
influential, influence
preoccupied, preoccupation
content, contentment
popular, popularity
controllable, control
spiritual, spirituality
idyllic, idyll
beneficial, benefit

Listening 1, page 123, Exercise 2

Speaker 1
Well, I know people say it's a kind of cultural revolution which will change the way we live, but I doubt it somehow. I wonder just how many people it will affect in reality. Somehow I don't think it'll impinge on me much – I can't see people sending their kids to history classes in the middle of the night. I know teenagers like staying up late, but even so! It must be the same for lots of other professions, too. I mean, who's going to want to make an appointment to have a tooth filled, for example, at three in the morning? Having said that, I frequently find myself marking exercise books at that time!
Speaker 2
Twenty-four-seven's obviously sound from the economic standpoint. It allows greater flexibility in the labour market, which will help many employers like myself. If you ask me, the concept of a fixed workplace will soon be outdated, too – already many people at our company work from home at least a couple of days a week because a lot of our work is done online. But there's a flipside to this of course – in that people may feel under pressure to be available for work at all times, regardless of their circumstances, and there's a danger that some bosses might take advantage. I can see myself being tempted when I need people to help me get that urgent deal sorted out.
Speaker 3
I've always had flexible working hours as I've had to take my turn being on emergency call. It doesn't happen that much but pets and zoo animals don't take any notice of surgery hours! Fortunately, I'm lucky enough to have a husband who can run his design business from home if needs be, and can cover with childcare. Some people are not so lucky and child-care services don't usually run outside normal working hours yet, which would cause a problem for many working parents if this does take off. I would imagine we'd extend our appointment system into the night if 24-hour working does come in but I'm not really too bothered one way or the other – there are advantages on both sides – it's six of one and half a dozen of the other.
Speaker 4
I'm a night owl – well you have to be if you work on newspapers – so flexible working suits me down to the ground. My job involves both

writing and sub-editing and I actually volunteer to do nights quite a lot as it is – work goes on round the clock. For me the main plus is how much faster I can get across the city out of normal hours – you can avoid the queues and rush-hours on busy roads on the way to the Evening Post offices – and of course more cars at night would mean not so many during the day, which would have to be an advantage for everyone. I went to an art exhibition in Los Angeles a few months ago at four in the morning – you could just walk around and have the place to yourself – it's great!

Speaker 5

I have worked anti-social hours in my time – people are always wanting lifts to and from airports at ridiculous times – but I always added extra to the sum on the meter for it because it's just not good for you – it plays havoc with your body clock. I think we have to accept we're a daytime species – asking people to work at night is like throwing them in the sea and asking them to stay there for a week – it's possible but there are risks. Particularly for people who consider themselves night owls – they feel good at night but they may well regret it in the long run. That's why now I refuse to take jobs after about 10 p.m., even if I can charge more for it.

Listening 2, page 127, Exercise 2

I = Interviewer, MJ = Martin Taylor, MJ = Mary Johnson

I: Juvenile crime is one of the biggest problems we face in Britain, and today we're going to hear about a groundbreaking experiment which is helping to address the issue. With us in the studio today we have Dr Martin Taylor, the man who helped to get the project off the ground, and Mary Johnson, whose child, Glenn, was one of the first children to benefit from it. Martin, let's start with you. What is the thinking behind *On Track*?

MT: Well, criminologists have been aware for some time that it is possible to say with 80 per cent accuracy the factors which might result in certain youngsters becoming potential future offenders without obviously being able to predict whether he or she would get into, let's say, armed robbery or whatever. It's a long-term initiative which relies on everyone pulling together to stop this happening – members of the health service, local schools, the police, social services , but especially the children at risk and their parents. If the scheme is to work with any degree of success, we have to have their cooperation or we may as well not bother – that's why the scheme has to be voluntary.

I: And how do you actually identify which kids these might be?

MT: The project targets pre-teenage youngsters from around four years old in areas like the one Mary and her son Glenn live in. That is, places with a high incidence of unemployment and poverty. Of course, this doesn't necessarily mean a child will turn to crime but it's one of the factors to be taken into account, along with where they are educated. Having said that, of far more significance in my view is how well a child relates to his parents and siblings on a day to day basis. Having seen Glenn in action at home, I felt there was a good chance he'd be involved in crime by the time he reached adulthood.

I: How did you react to being told that your son was at risk, Mary?

MJ: The thing is that we'd been getting more and more exasperated with his behaviour by the day. He was constantly rude to all of us and if we tried to discipline him, it just made matters worse – he'd just shout back and get aggressive. We just didn't know how to handle him and things were going downhill rapidly. What shocked me was seeing how many young kids were involved in petty crime; throwing stones, vandalising cars, that sort of thing. I felt they'd put pressure on Glenn to do the same. So I felt quite glad in a way because it meant that something was finally going to be done – the school certainly didn't seem to be doing much. I mean obviously it's not great to know your son's a hooligan, but I can't say it was entirely unexpected.

I: Did *On Track* initially get in touch with you, Mary?

MJ: They're not allowed to do that, although they had, apparently, sent information home with the kids at school, which of course Glenn didn't pass on. I had heard something about it through a friend but never got round to getting the number. Then one day – just as I was feeling really low – I came across a leaflet from the library which said that it was a voluntary scheme aimed at keeping young people out of trouble. It offered support to both children and parents, so we decided to go for it. We had nothing to lose, really.

I: And how were they able to help you?

MJ: One of the things they do is visit your home and encourage the family to talk things through. That didn't work with Glenn because he's dead against any kind of counselling, and wouldn't join in. But the parenting course we went on was a real eye-opener. Doing roleplays of situations we might find ourselves in with other people in the same boat as us made us realise we weren't very skilled parents at all. I used to get very wound up with his behaviour but the course taught me to keep calm and communicate rather than shout. I was encouraged to give him loads of attention when he behaved well, rather than just being negative all the time. This was a revelation to me.

I: And did he respond well to this?

MJ: Better than I could have dreamt. That's what he obviously wanted and needed, but I hadn't realised. He still has his off days but he doesn't call me names anywhere near as much and we can actually sometimes have a conversation instead of just shouting. There's a long way to go and academically he's still not making any progress. But the incredible thing for me is he will now sometimes give me a hug and show me he does actually care, which was unheard of before. In just two years, by getting involved with the project, I feel I've given him a real chance of a new life.

Speaking, page 130, Exercise 4a

Now, I'd like you to discuss something between yourselves, but please speak so that we can hear you. Here are some photos which show different aspects of police work. Discuss how challenging each aspect is, and decide which **two** photos would be most suitable for a police recruitment brochure. You have about four minutes for this.

Speaking, page 130, Exercises 5b and 6a

E = Examiner, S1 = Student 1, S2 = Student 2

E: Do you think there is more crime now than in the past, or are people just more afraid of crime?

S1: Well, I think there was actually just as much of it before – it's just that everyone is much more aware of it nowadays because of the media coverage. The newspapers send out panic signals – you know, as if muggers were waiting around every corner.

S2: Yes, I wish people would realise that crime is really quite low. What they should be worrying about more is, for example, encouraging people to drive safely and …

S1: I agree absolutely. Road accidents are still the biggest killers – people driving too fast, or perhaps not concentrating because they're talking on their mobiles. It's time the police put more effort into preventing people dying on the roads.

E: Thank you. Do you think violent films and computer games encourage people to commit crimes?

S2: Well, that's a difficult one. Um … I think the effect of these games and films is probably exaggerated to some extent, but I'd really rather they didn't show violent films on TV when children are likely to see them. What do you think?

S1: Well, yes, probably for most people there's no problem – they can separate reality from ... er ... games, but suppose someone with violent tendencies was watching a film like this or playing a game – they might get ideas.

E: Thank you.

Exam practice 4

Listening, Paper 4 Part 1

I = Interviewer, T = Tom, A = Alison

I: Next week, adventurer Tom Westfield is setting off on his latest expedition, crossing the Atlantic by an especially perilous route in an open boat. Meanwhile, back in London, his personal assistant, Alison Nunn, will be in charge of operations. They both join me in the studio today. So, Tom, turning to you first, it seems an adventurer needs a PA these days?

T: That's right, although it's a relatively new thing for me. I took two and a half years to plan this trip, organising everything from sponsorship and fundraising to testing the boat. I'd be the first to admit that my particular skill is not administration. Being self-employed, it didn't occur to me at first to take on a PA, but it was an excellent move because it's freed me completely to do what I'm best at, while Alison deals very efficiently with what I'm not good at. But I also like the comradeship of working with someone, of having someone on my side. She's really the unsung heroine of this trip.

I: So, Alison, what attracted you to the job?

A: Well, I had a good position as a graduate PA in advertising, I'd worked in insurance and accountancy, but I was never quite comfortable in the world of commerce and finance, and I was looking for something different. I'd been doing voluntary work in my free time, and I actually met Tom by chance at a charity ball. I had just competed in the London Marathon to raise money for a children's hospital, and I felt very energised. We talked about life's challenges, and Tom said he really needed someone to sort out his life, so I said, 'What about me?' He told me to think about it and get in touch again if I was serious. I took my time deciding and read his book before going to work for him. He's very special, so driven by his passion. I certainly made the right move.

I: That was the book you wrote after you'd sailed single-handedly round the world?

T: Yes. That first book was called *Reaching my Goal*, and covers the early part of my life. I've always been very active, and I've had a go at most extreme sports – you know, hang-gliding, kite-surfing, all that. But, while at university, I had a nasty horse-riding accident where I hurt my back. The doctors said, you know, 'No more physical challenges for you – you'll have to take it easy from now on.' But I just couldn't accept it. The book talks about me getting back to fitness and the goals I set myself along the way. I thought sailing round the world was the ultimate one, but actually I didn't stop after that.

I: So, what drives you, Tom?

T: One of the things that I do to raise money for my trips is talking to kids in schools and colleges about motivation. I found that preparing for those talks did lead me to question my own motives for doing these trips. But I came to the conclusion that basically we live in a fascinating world and we're barely scratching the surface of what's achievable – and that's what drives me.

I: So, Alison, tell us what you're doing to support Tom on this latest trip.

A: I'll have my hands full pretty much 24 hours a day, just running the logistics. That means things like contacting coastguards, arranging refuelling, etc. I'll basically be acting as the headquarters. The expedition from North America to the UK will take three to four weeks. This particular route hasn't been completed before – it's very dangerous. There'll be five men, including Tom, in a boat which is essentially a large inflatable dinghy and it has no shelter.

I: And how are you feeling Tom – with just a week to go?

T: Before you go on an expedition, you go through the whole range of emotions. At times, I've felt terrified at the thought of what I've taken on; at others, anxious about the details – I suppose I'm rather insecure in some ways, I like to feel in control of everything. But now we've reached this stage, I just feel rather vulnerable, as if I've done all I can and it's just down to luck with the weather. Once you start, you go into survival mode and, as time goes by, you get tired, wet and fed up and you start missing home.

Module 9

Listening 1, page 139, Exercise 2

I = interviewer, J = Jodie, G = Gary

I: In today's programme, we're looking at job interviews; how to present yourself if you want to land the job in question. With me are Jodie Bradwell, a recruitment consultant with a top London agency, and Gary Smart, a university careers officer. So, Jodie, I've read that first impressions really count in interviews. Is that the case?

J: Well, research has shown that when we meet people for the first time, we very quickly make up our minds about them. What the researchers found was that in job interviews especially, it's the impression people give as they walk through the door that sticks, no matter what they may go on to say and do. It's an alarming thought, and I'm not 100% convinced, but those were the findings. As a candidate, your best defence against this, of course, is to make sure you send out the right messages from the moment you walk in; the way you greet people, what you wear, anything that can influence the interviewer's opinion.

I: Gary, would you go along with that?

G: Well, if we're talking about graduate recruitment, I would hope that it's not quite so superficial. I mean, companies invest heavily in their recruitment procedures, and what we seem to be saying here is that all you need is a smart suit and a firm handshake and you're in, and I think that's too simplistic. In any case, if you present a false impression of yourself in an interview – wear clothes you'd never be seen dead in elsewhere, adopt a … a different way of walking and talking – you're hardly going to feel comfortable and at ease in the situation, so it could be counter-productive.

I: Jodie?

J: Gary's right, of course. It's important to be yourself in an interview, and what we're talking about here is presenting yourself in the best light – not pretending to be somebody you're not. The thing to remember about this research, though, is that it's talking about subconscious impressions. Of course, the interviewers don't mean to base everything on a first impression, but it seems that they are influenced, on a deep level, by certain aspects of a candidate's behaviour, even if that's not what they notice at the time – you know, not what they discuss when that person's just walked out the door.

I: So, let's imagine that a company has narrowed the field down to, say, six candidates – all with an equal chance on paper. How would you, Gary, advise those people to behave at interview? What about the issue of dress?

G: Well, according to another study, 70% of employee turnover results from people not fitting in with the culture of a company, rather than a lack of ability or skills. So the interviewers are looking to see whether or not someone seems to be the sort of person who'll get on in the sort of place it is. So candidates need to do their homework on this. The general rule of thumb is find out what people wear in the job – it's often smart casual in offices these days – and go up a notch. You want to look as if you've made an effort, but also that you're going to fit in with the culture.

I: And Jodie, what about body language? Can we really alter the signals we send out about ourselves?

J: Well, at interviews candidates are rather nervous and often too bound up with their own feelings to think about how the interviewers may perceive them. So, I'd say be yourself, but do think about how you're going to behave. It's important to look people in the eye, for example, and to smile. Candidates who project an image of vitality and energy come across as more capable to the interviewers, more inspiring than those who seem flat and characterless. So, stay upbeat, sit up straight and speak clearly. Think about the total message you're communicating to the people opposite you – because it's not just what you say.

I: Gary?

G: I'd go along with that. We do role-plays with students to get them thinking about how body language affects verbal communication. You know, we get a volunteer to sit on a chair with their arms folded, legs crossed, looking at the floor – all classic symptoms of interview nerves – and then ask them questions. Then we get the same person to sit in a relaxed way, looking straight at the interviewer, and ask them some more questions. What always happens is that in the second position, the person's voice sounds different, the tone lifts, and they suddenly sound sure of themselves, more energetic and enthusiastic, and these are all the qualities you need to project in an interview.

I: Have you ever tried that, Jodie?

J: I've seen it done on a video – it's really striking. But I'd like to add that, sure, it's worth knowing what the optimum body-language signals are, but don't get hung up on them. If you're projecting the right qualities, and feeling the appropriate emotions, the body language will follow naturally. But if you sense during an interview that you're flagging, um, that you're sounding less positive than you would like to, you should monitor your body language and adapt it to lift your mood and your verbal tone.

I: Moving on to what to say and what not to say, perhaps you could …

Listening 2, page 143, Exercise 2

We've all got at least one T-shirt. They're cheap, stylish and easily replaceable. But how many of us know the history of this familiar garment?

Well, it all started during the First World War. Sailors in the United States Navy noticed that their French counterparts were wearing cotton undershirts which were much cooler and more comfortable than the woollen ones they were issued with. The Americans quickly adopted the French habit, and cotton shirts shaped like a letter T soon became standard issue for all ranks of the US forces. A modern classic had arrived.

The 1930s saw great advances in the cloth industry. The mass production of cotton cloth made it cheaper and more widely available, and something resembling the modern T-shirt became a popular item of underwear for the masses in the USA. The shirts were made of the same material as they are now, although they were much looser, had sleeves down to the elbows and a V-shaped neck. The modern classic shape, of course, is with a round neck and much shorter sleeves than was originally the case.

Exposing one's underwear in public before the 1950s was unheard of, however. In the 1930s and 40s, this was felt to be appropriate only for people such as sportsmen or workmen, who needed freedom of movement to play or work. However, news pictures of T-shirted soldiers fighting in the Second World War contributed to the gradual acceptance of the garment as outerwear.

The person who did most to cement the popularity of the T-shirt, though, was probably an actor. In the early 1950s, the play, *A Streetcar Named Desire* starred the young Marlon Brando, wearing a skin-tight T-shirt which showed off every muscle. Suddenly, every young man wanted one, and not a loose floppy one either! But it was when the T-shirt got taken up by film stars in Hollywood that it became associated not so much with physical strength as with rebellion, especially once teenagers had seen James Dean wearing one in the film *Rebel Without a Cause*. By 1955, it had become fashionable for young men to wear the T-shirt without another shirt covering it, and by the end of the 1950s, the impact of the movies was such that even women were wearing them.

Oddly enough, it wasn't until the Budweiser beer company began giving away T-shirts bearing its logo in the mid-sixties that everyday brands realised the advertising potential of the garment. But the idea quickly caught on, with logos for everything from Coca-Cola to cigarettes appearing on the front of T-shirts.

In the 1970s, the T-shirt evolved yet again, this time as a vehicle for political protest. Inexpensive and stylish, it was an ideal way to comment on issues in the news. Feminists wore T-shirts with legends like *A woman needs a man like a fish needs a bicycle*, and when black activist Angela Davis was imprisoned, *Free Angela* T-shirts appeared worldwide.

So, when we look at the T-shirt today, …

Module 10

Listening 1, page 155, Exercise 2

Extract 1

L = Liz, R = Richard

L: I didn't know you were into popular comedy. In fact I thought you didn't watch much TV.

R: You're right, I'd much rather listen to music or read – and most of the stuff on at that time of night is rubbish anyway. And I'd got it into my head that Time Out was going to be like that other series my fifteen-year-old used to go on about. You know, The Family. Ground-breaking, maybe, but it left me cold. Anyway, I was determined to see what all the fuss was about – people are always quoting bits from it in the office the morning after and I felt a bit left out because I had nothing to say. I found it very funny, actually.

L: I just crack up every time I see that middle-aged couple and the hairdresser – I mean, we all know people like this, don't we? The actors have got them just right. But in a way I wish they'd have other roles as well – it runs the risk of getting a bit repetitive.

R: Isn't that part of the appeal, though? You get to know the people and how they'll react and you're just waiting for the punchlines. My only criticism is that some of the sketches dragged on a bit at times, and ...

Extract 2

I = Interview, S = Sasha

I: Many people say that they find British humour hard to get used to.

S: Well, whatever they think, the British haven't got the monopoly on irony – actually I've found that saying things like 'Well, thanks for all your help!' when they haven't helped, is a fairly universal form of humour. The thing that has really struck me here though is the way this kind of irony creeps into nearly every conversation, it's almost second nature to people. Also unique here is the way you're keen not to be 'over the top'. For example, yesterday a friend spent about ten minutes describing a really awful holiday she'd had, to which someone commented very dryly, 'So, you wouldn't go back there, then?!' I don't think I'll ever be able to do that!

I: Why do you think humour is such an important part of British life?

S: I think it's all tied up in the desire the British have not to take themselves too seriously. It's OK to be earnest about other matters but never about yourself. There's also the dread of appearing to be showing off. So if you say you've won a race or something, you should follow it up by saying something like, 'Of course, my opponents all had broken legs.' It's not the kind of humour which is going to have people rolling around – it's too subtle for that and that's the last thing they'd want anyway.

Extract 3

A = Assistant, C = Customer

A: I'm afraid when you placed the order there was no way of knowing that stocks of this particular product were going to be held up.

C: So just when were you planning to let me know? I've been waiting three weeks now for my order to arrive and what feels like another three weeks waiting in a queue at the end of a phone line until somebody decided they would speak to me. It now transpires that you have known for two weeks it would be delayed, but nobody had the courtesy to let me know. To add insult to injury, incidentally, I paid a fortune on Express delivery to get the order within two days!

A: You will not of course be charged for …

C: What I'm really astounded by is why a business like yours would risk their reputation in this way. Surely word of mouth recommendation is the best form of publicity and you're certainly not going to get this from me! Quite the contrary! If it weren't that I'd wasted so much time on this business anyway, I'd be writing to your manager as well. As for the money I've spent hanging on the line – well, it's no wonder so many people phone in to these consumer rights programmes on the TV.

Listening 2, page 159, Exercise 2

I = Interviewer, P = Paul Daniel, J = Jane Gilchrist

I: In the television documentary *Operatunity*, viewers followed the fortunes of a group of aspiring singers taking part in a talent competition organised by the English National Opera company. With us in the studio today is Paul Daniel, from the company, and Jane Gilchrist, chosen as one of the six finalists. Paul, what was the idea behind the competition?

P: Well, it wasn't meant to be like *Pop Idol*, because we weren't out to manufacture a star. But we did suspect there was a wealth of untapped talent out there, and our primary aim was to bring out the best in non-professional singers; open up avenues for them.

The winner will go forward to star in an opera on stage, but all the finalists are benefiting from coaching from some of the company's top singers and music staff. At the back of our minds was also the hope that televising the whole process might also increase understanding of opera in general.

I: ... and there must be a lot of preconceptions about opera out there. Jane, do you think you fit the stereotypical image of an opera singer?

J: Far from it. The thing is, people expect opera singers to lead exotic lives, whereas mine is anything but – you know, I'm just a supermarket cashier. I did grow up surrounded by music – Mum used to sing in a choir, and Dad would play guitar – but we were enthusiasts rather than serious musicians. It was actually my school which put me in touch with an excellent amateur operatic society. It's thanks to them that I got some training, and I've been singing in their productions ever since.

I: And were you never tempted to make it your career?

J: Well, I was offered a place with a professional opera company at the age of 21, on the condition that I did three years at a music college first. Anyway, I was persuaded, probably quite rightly, that I wouldn't be able to afford the fees, so I got married instead and went on to have four children. I mean, I don't regret having a family, of course, but, you know, all this time I've harboured a nagging feeling that maybe I blew my big chance of a career in music.

I: So how did you feel when this competition came along so many years later?

J: Well, a good friend of mine turned up at my door one day and shoved some papers at me and told me to fill them in. I told myself I mustn't blow this second chance. Mind you, I didn't think I stood much of a chance. You had to make a video of yourself which I found a bit daunting, but in the end I managed it OK.

I: Presumably, Paul, you and the other judges had a clear idea of what you were looking for at that stage?

P: Absolutely. Out of 3,000, about 100 were invited to an audition. Just by studying the video, we could eliminate people whose voices weren't strong enough. Obviously it's important to be able to act as well as sing, because each role is different, but even that's irrelevant if the person concerned hasn't got what it takes in terms of determination, energy and stamina. You've got to be able to take the knocks as well as having the musical range. Unfortunately, some of the people who were outstanding on the video were less impressive 'live'.

I: And for the 20 who passed the audition, there was an intensive weekend's coaching with the other finalists. How did that go, Paul?

P: Well, when you're working with professionals, you're dealing with people who have been trained over many years and who are used to the cut and thrust of rehearsals – we all expect positive criticism from each other. For the *Operatunity* finalists, things were very different. They were, of course, much less used to the speed and intensity that professionals deal with on a daily basis, and we had to match their enthusiasm with very careful feedback. They needed longer to adapt what they had prepared, and often needed help and support at a more basic level. But, coming from different walks of life, they were full of experiences and were the kind of characters that us professionals don't get to work with – that was a real bonus.

I: And throughout, the TV cameras have been recording everything. Did that affect you, Jane?

J: It was a tense time over the weekend – we obviously wanted to do our best – and yet the TV people were constantly popping in and out. Oddly enough, the cameras never bothered me, really – it's amazing how you seem to unconsciously block them out. Mind you, I did feel a bit defenceless at times, if I did or said something silly, I'd start wondering whether they were going to show that bit on TV. But I needn't have worried. I mean, it wasn't as if we went into this not knowing about the documentary. No, it could have been a lot worse!

I: Well, this weekend the winner will be chosen, and we wish Jane all the best and hope ...

Exam practice 5

Listening, Paper 4 Part 4

Speaker 1

Directing your own production, whether it's a live event or on screen is the dream of people working in the arts. So I jumped at the chance of directing this new six-part romantic-comedy. It'll be broadcast at primetime in the autumn schedules. The writing's full of wit and humour, so I'm checking the story; making sure there's something there beneath the laughs – because something so reliant on brilliant dialogue needs visual appeal too. My advice to young hopefuls is to try to get on with everybody because it's a personality-driven industry. I entered the industry via a film-studies degree, working as lowly script supervisor on a number of low-budget feature films before getting my present post.

Speaker 2

This work's been my life; I've done everything from pop videos to classical ballet. At the moment, I'm directing a major new work with my own company, which is really exciting. Once the show opens, I'll be ensuring that the choreographer's original creative vision is maintained, no matter how many different times we perform it, and I'll be working with the performers as they develop their roles. I'd warn aspiring directors that they might have to be prepared to work for next to nothing in order to gain the necessary experience, however. So listen to what older people have to tell you. Even if their opinions go against your own instincts, they do know what they're on about.

Speaker 3

I'm hyperactive with a low boredom threshold – that's what attracted me to directing. There are three of us in the unit, making light entertainment programmes, everything from pop-music quizzes to soap operas. My words of wisdom for young media graduates today would be: broadcasting is a terribly competitive industry, so don't let all the rejections you get at the beginning get you down. I began as a researcher on national radio which, as it turned out, was fortunate. I was in a fairly small department, so soon worked my way up to assistant producer on a live comedy show. When the resident director left for a job in TV, I lost no time in jumping into his shoes.

Speaker 4

I've no formal training and perhaps that's why I've no aspirations to become 'part of the industry', as it were. I'm more interested in experimentation, and I'd advise young people to ignore all the hype they hear on media-studies courses. Always work intuitively, that's my advice; you should welcome uncertainty and challenge – that's how you learn. I won a best new director award at the Edinburgh Festival last year for an investigative piece I did on the pressures of being a musician on tour with an orchestra. But I'm still learning. I'm going to be more ruthless about how much material I shoot in future – if there's too many possibilities at the editing stage, it gets too complicated.

Speaker 5

My ambition's always been to make feature films, though I actually did a theatre studies course in Sydney and worked as a dancer to pay my way. Then I came over to Europe, where I was lucky enough to get my present job as artistic director of a small touring company dedicated to producing and promoting works by contemporary Australian playwrights, so it was perfect. Although it's all good experience, I'd say to newcomers that you should decide what you want to do and stick with it – don't be sidetracked, because I'd still like to get into films. I like the idea that there's a finished product – the stage is very here-and-now, great for the audience, but what if your work's never seen by the right people?